Communications
in Computer and Information Science   2170

Rationale

The CCIS series is devoted to the publication of proceedings of computer science conferences. Its aim is to efficiently disseminate original research results in informatics in printed and electronic form. While the focus is on publication of peer-reviewed full papers presenting mature work, inclusion of reviewed short papers reporting on work in progress is welcome, too. Besides globally relevant meetings with internationally representative program committees guaranteeing a strict peer-reviewing and paper selection process, conferences run by societies or of high regional or national relevance are also considered for publication.

Topics

The topical scope of CCIS spans the entire spectrum of informatics ranging from foundational topics in the theory of computing to information and communications science and technology and a broad variety of interdisciplinary application fields.

Information for Volume Editors and Authors

Publication in CCIS is free of charge. No royalties are paid, however, we offer registered conference participants temporary free access to the online version of the conference proceedings on SpringerLink (http://link.springer.com) by means of an http referrer from the conference website and/or a number of complimentary printed copies, as specified in the official acceptance email of the event.

CCIS proceedings can be published in time for distribution at conferences or as post-proceedings, and delivered in the form of printed books and/or electronically as USBs and/or e-content licenses for accessing proceedings at SpringerLink. Furthermore, CCIS proceedings are included in the CCIS electronic book series hosted in the SpringerLink digital library at http://link.springer.com/bookseries/7899. Conferences publishing in CCIS are allowed to use Online Conference Service (OCS) for managing the whole proceedings lifecycle (from submission and reviewing to preparing for publication) free of charge.

Publication process

The language of publication is exclusively English. Authors publishing in CCIS have to sign the Springer CCIS copyright transfer form, however, they are free to use their material published in CCIS for substantially changed, more elaborate subsequent publications elsewhere. For the preparation of the camera-ready papers/files, authors have to strictly adhere to the Springer CCIS Authors' Instructions and are strongly encouraged to use the CCIS LaTeX style files or templates.

Abstracting/Indexing

CCIS is abstracted/indexed in DBLP, Google Scholar, EI-Compendex, Mathematical Reviews, SCImago, Scopus. CCIS volumes are also submitted for the inclusion in ISI Proceedings.

How to start

To start the evaluation of your proposal for inclusion in the CCIS series, please send an e-mail to ccis@springer.com.

Seiki Saito · Satoshi Tanaka · Liang Li ·
Satoshi Takatori · Yuichi Tamura
Editors

Methods and Applications for Modeling and Simulation of Complex Systems

23rd Asia Simulation Conference, AsiaSim 2024
Kobe, Japan, September 17–20, 2024
Proceedings

Springer

Editors
Seiki Saito
Department of Informatics and Electronics,
Faculty of Engineering
Yamagata University
Yamagata, Japan

Satoshi Tanaka
College of Information Science
and Engineering
Ritsumeikan University
Kyoto, Kyoto, Japan

Liang Li
College of Information Science
and Engineering
Ritsumeikan University
Osaka, Japan

Satoshi Takatori
Research Organization of Open Innovation
and Collaboration
Ritsumeikan University
Osaka, Japan

Yuichi Tamura
Faculty of Intelligence and Informatics
Konan University
Hyogo, Japan

ISSN 1865-0929   ISSN 1865-0937 (electronic)
Communications in Computer and Information Science
ISBN 978-981-97-7224-7   ISBN 978-981-97-7225-4 (eBook)
https://doi.org/10.1007/978-981-97-7225-4

This Springer imprint is published by the registered company Springer Nature Singapore Pte Ltd.
The registered company address is: 152 Beach Road, #21-01/04 Gateway East, Singapore 189721, Singapore

Preface

This volume contains the papers from the 23rd Asia Simulation Conference (AsiaSim 2024), an annual event organized by the Asian Simulation Federation (ASIASIM), which includes CSF (China Simulation Federation), JSST (Japan Society for Simulation Technology), KSS (Korea Society for Simulation), SSAGSG (Society for Simulation and Gaming of Singapore), and MSS (Malaysian Simulation Society). Since its inception in the 2000s, the conference has been hosted annually in different Asian countries. AsiaSim serves as a forum for scientists, academicians, and professionals worldwide. Its purpose is to promote modeling and simulation in industry, research, and development by providing a platform for regional and national simulation societies in Asia.

This year, AsiaSim was held in Kobe, Japan, in conjunction with the 43rd JSST Annual International Conference on Simulation Technology. The conference featured research on a wide range of topics, including modeling languages, simulation methods and software applications, simulation-based design and analysis, deep learning, digital twins, transportation, manufacturing, visualization, virtual reality, and more. Each submission underwent a rigorous review process by multiple reviewers, followed by a final review by the program committee. Out of over a hundred submissions, 28 papers were accepted and orally presented in online and in-person sessions. The accepted papers are included in this CCIS volume.

We extend our gratitude to the keynote speakers for their valuable insights. We also thank the external reviewers for their time, effort, and prompt responses. Additionally, we are grateful to the Program Committee and Organizing Committee members for their dedicated contributions to the success of the conference. Finally, we thank the conference participants for their significant contributions.

September 2024

Seiki Saito
Satoshi Tanaka
Liang Li

Organization

General Chair

Takahiro Kenmotsu   Doshisha University, Japan
Susumu Fujiwara   Kyoto Institute of Technology, Japan

Conference Chair

Yuichi Tamura   Konan University, Japan

International Organizing Chair

Satoshi Tanaka   Ritsumeikan University, Japan

International Organizing Co-chair

Liang Li   Ritsumeikan University, Japan

Program Chair

Seiki Saito   Yamagata University, Japan

Best Paper Award Committee Chair

Axel Lehmann   Universität der Bundeswehr München, Germany

Publication Chair

Hiroaki Ohtani   National Institute for Fusion Science, Japan

Local Committee Chair

Nobuaki Ohno   University of Hyogo, Japan

Program Committee Members

Wentong Cai	Nanyang Technological University, Singapore
Byeong-Yun Chang	Ajou University, South Korea
Yoshihisa Fujita	Nihon University, Japan
Takashi Hara	National Institute of Technology, Kagoshima College, Japan
Kyoko Hasegawa	Tokai University, Japan
Taku Itoh	Nihon University, Japan
Masami Iwase	Tokyo Denki University, Japan
Hideki Kawaguchi	Muroran Institute of Technology, Japan
Yun Bae Kim	Sungkyunkwan University, South Korea
Yuanjun Laili	Beihang University, China
Jongsik Lee	Inha University, South Korea
Tomoko Mizuguchi	Kyoto Institute of Technology, Japan
Zaharuddin Mohamed	Universiti Teknologi Malaysia, Malaysia
Hiroaki Nakamura	National Institute for Fusion Science, Japan
Ryusuke Numata	University of Hyogo, Japan
Jinsoo Park	Yong In University, South Korea
Yahaya Md Sam	Universiti Teknologi Malaysia, Malaysia
Ryosuke Seki	National Institute for Fusion Science, Japan
Xiao Song	Beihang University, China
Hiroto Tadano	University of Tsukuba, Japan
Chako Takahashi	Yamagata University, Japan
Satoshi Takatori	Ritsumeikan University, Japan
Yuki Takemura	National Institute for Fusion Science, Japan
Hiroshi Tamura	Chuo University, Japan
Gary Tan	National University of Singapore, Singapore
Yongmeng Teo	Nanyang Technological University, Singapore
Satoshi Togo	University of Tsukuba, Japan
Miyuki Yajima	National Institute for Fusion Science, Japan
Yuki Uchida	National Institute of Technology, Nagaoka College, Japan

Ryosuke Ueda   Tohoku University, Japan
Norhaliza Abdul Wahab   Universiti Teknologi Malaysia, Malaysia
Shugo Yasuda   University of Hyogo, Japan
Masayuki Yokoyama   National Institute for Fusion Science, Japan
Masafumi Yoshida   National Institute of Technology, Ube College,
                                    Japan
Lin Zhang   Beihang University, China

Contents

Interdisciplinary Simulation and Machine Learning

Networks and Complex Systems

Modeling, Simulaiton, and Visualization of Digital Twin

Methods for Simulation and Modeling

Simulation of Overall Performance of Jack up Platforms Under Multiple Working Conditions

Jie Gao[✉], Yujia Shang, Lu Ding, and Xiawei Feng

Instrumentation Technology and Economy Institute, Beijing, China
3081196409@qq.com

Abstract. Jack up platforms are important offshore oil drilling equipment. Due to the harsh working environment, their safety must be ensured during design and manufacturing. Based on the truss leg structure of CJ 50 jack up platform, a finite element model of a three-legged jack up platform was established using CAE simulation software Abauqs. The overall performance of the jack up platform under jacking conditions, pre-load conditions, and survival conditions was simulated. The results show that under the design sea conditions of the jack up platform, the CJ 50 type jack up platform model established can meet the safety performance requirements of jacking conditions, pre-load conditions, and survival conditions. Among them, the stress caused by pre-load conditions on the jack up platform is the largest, followed by jacking conditions, and finally pre-load conditions. The maximum stress occurs at the contact position between the legs and the lower guide plate in the jacking house area. The research results can provide a basis for the design and safe operation of jack up platforms.

Keywords: CJ 50 jack up platform · truss leg · CAE simulation · Multiple operating conditions · Safety performance

1 Introduction

With the increase of population, the demand for oil and gas resources in China is growing day by day, leading to a gradual increase in the intensity of marine oil and gas resource extraction. Compared with other types of oil and gas drilling platforms, jack up platforms require less steel, have a relatively low cost, and are mainly used in shallow sea areas. They can withstand extreme loads under various sea conditions. Given the many advantages of jack up platforms, they have been widely used in the world's offshore waters. In order to improve the reliability of the platform, reduce damage, and improve operational safety, many scholars [1, 2] have conducted extensive research on the structure and performance of self elevating platforms in the past half century of development.

Hendriyawan et al. [3] used the finite element method to calculate the effect of old footprints on the vertical bearing capacity of spudcans. Tolooiyan Ali et al. [4] established a conical spudcan for use along the Tunisian coast, and used a finite element method based on arbitrary Lagrangian Euler to estimate the penetration of the spudcan.

S. Saito et al. (Eds.): AsiaSim 2024, CCIS 2170, pp. 3–13, 2024.
https://doi.org/10.1007/978-981-97-7225-4_1

Gao Jie et al. [5] established a simulation model of a jack up platform using finite element software, and calculated the critical sliding distance and critical RPD of the pile shoes of the jack up platform in different directions considering bending moment. Yi MyungSu et al. [6] conducted parameter sensitivity analysis on different environmental parameters of jack up platforms. Through sensitivity analysis of environmental characteristics, the significance and sensitivity of the influence of various environmental parameters on leg strength were elucidated. Do VietDung et al. [7] studied how to use particle swarm optimization based on fuzzy controllers (FC) to adapt the network controlled jacking system to the effects of internal and external disturbances with time delays. We have developed a Fuzzy Particle Swarm Optimization (FPSO) controller and a fuzzy controller, demonstrating the advantages of the proposed method. The applicability of the proposed method has been demonstrated through simulation and experimental results using Matlab software and embedded systems. Metrevoli Yu et al. [8] combined finite element methods with data-driven methods and proposed a deep learning based simulation method for jack up platforms. The above studies have all considered the impact of specific working conditions on the overall or local performance of jack up platforms, and have not analyzed the safety performance of jack up platforms under operating or extreme working conditions.

This article takes the CJ50 three leg jack up platform as the object, establishes an overall simulation model of the jack up platform, and obtains the safety performance requirements of the jack up platform under jacking conditions, pre-load conditions, and survival conditions according to the design specifications of different sea conditions. Among them, pre-load conditions cause the greatest stress on the jack up platform, followed by survival conditions, and finally jacking conditions. The maximum stress occurs at the contact position between the truss legs and the lower guide plate in the jacking house area. The research results can provide a basis for the design and safe operation of the jack up platform.

2   Overall Simulation Analysis

2.1   Overall Simulation Model

The three leg jack up platform includes the main hull, jacking house area, and truss leg model. The bow of the ship has a single leg, and the tail has two symmetrically distributed legs. Each jacking system in the jacking house area contains three sets of gear racks, and the entire platform has 54 gears that match the gear racks on the leg chord. Due to the large volume of the actual jack up platform, its safety performance cannot be verified through experiments. In order to calculate the reliability of the platform, based on the structural characteristics and design sea conditions of the platform, finite element simulation software Abaqus was used to establish finite element simulation models of the jack up platform under three working conditions: jacking conditions, pre-load conditions, and survival conditions, as shown in Fig. 1. The three legs adopt a beam element model, while the main hull adopts a shell element model and provides cross-sectional properties to simulate the actual size of the platform structure. The design sea conditions of the CJ 50 jack up platform are shown in Table 1.

Table 1. Sea conditions for designing jack up platforms.

parameter	CJ 50-X120-F		
working condition	Jacking Condition	Pre-load Condition	ABS survival
Total length of legs	165.1 m	165.1 m	165.1 m
Water depth	121.9 m	121.9 m	106.7 m
Air gap	15 m	2 m	17 m
Wave height	4 m	4 m	18.3 m
Wave period	8 s	8 s	15.6 s
Wave theory	Airy	Airy	Airy
Wave angle	180°, 210°, 240°	180°, 210°, 240°	180°
Wind speed	12 m/s	12 m/s	51.4 m/s
Current velocity	0.6 m/s	0.6 m/s	0.77 m/s
Jacking weight	16030 t	31320 t	16030 t

The bow leg is defined as Leg1, and the starboard and port legs are defined as Leg2 and Leg3, respectively; Chords and slant support corresponding to each leg are defined as Chord1, Chord2, Chord3, and Brace1, Brace2, and Brace3 as shown in Fig. 2 (Fig. 3).

2.2   Connections Setting Up

In order to analyze the overall response with a detailed leg model, the legs are connected to the ship through Gap elements [9], and the gears and racks are connected through linear springs, as shown in Figs. 4 and 5. The connection between legs and the hull is:

LG = Lower Guide
FIX = Fixed System
PIN = Jacking Gear
UG = Upper Guide (see Fig. 4).

The initial gap between the leg and the guide plate is δ_0. And only withstand axial pressure after contact [9].

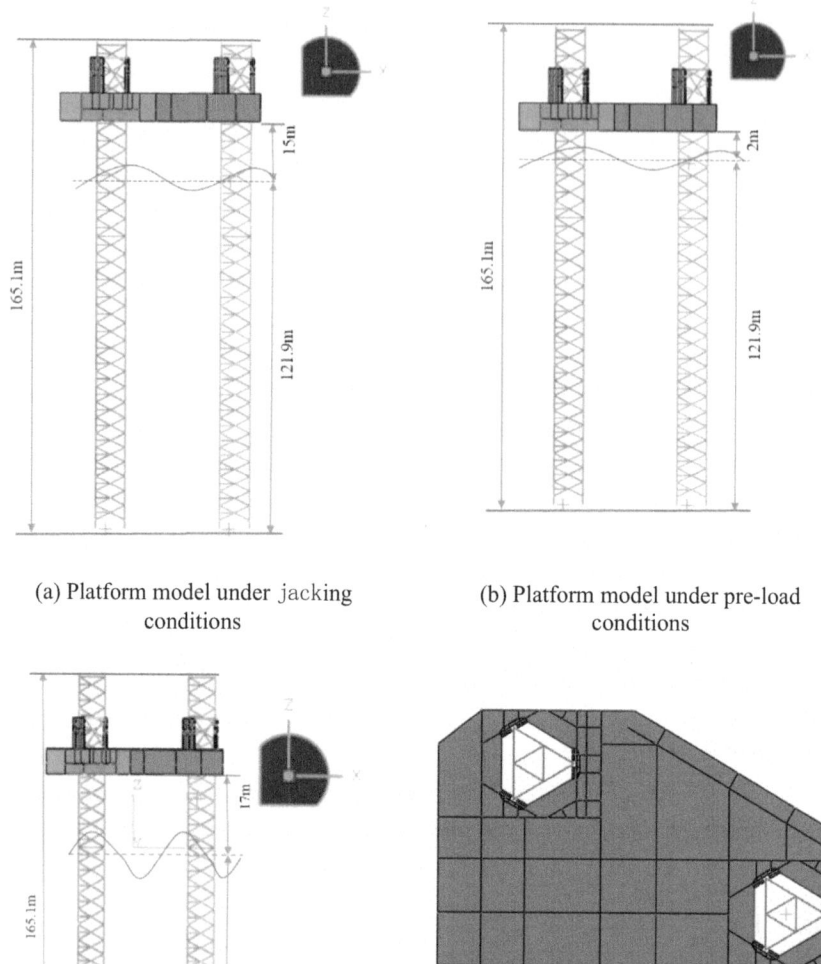

(a) Platform model under jacking
conditions

(b) Platform model under pre-load
conditions

(c) Platform model under survival
conditions

(d) Top view of the platform

Fig. 1. Platform model under different operating conditions.

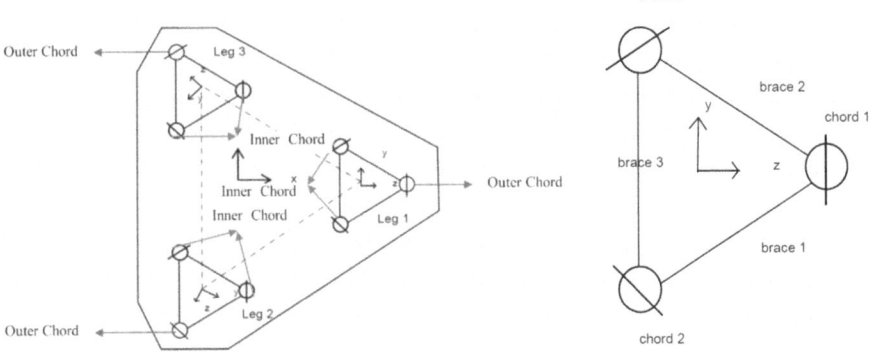

Fig. 2. Schematic diagram of platform and leg model.

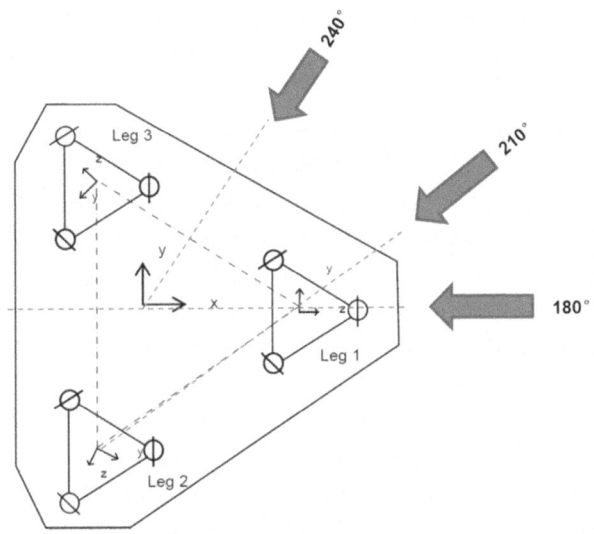

Fig. 3. Load schematic diagram.

2.3 Overall Finite Element Simulation Results

Modal and stress analysis was conducted on the model under jacking conditions, pre-load conditions, and survival conditions based on three design conditions of a jack up platform.

2.3.1 Modal Analysis

Modal analysis is a method for studying the dynamic characteristics of structures, generally applied in the field of engineering vibration. Among them, modal refers to the inherent vibration characteristics of mechanical structures. Based on subspace modal analysis method, Abaqus is used to perform modal analysis on the platform and obtain

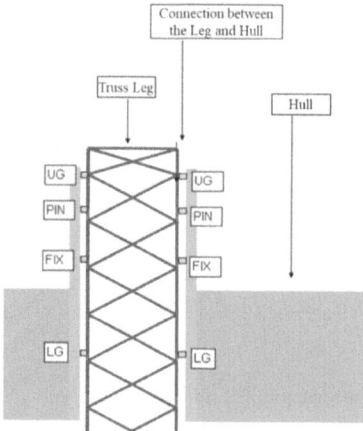

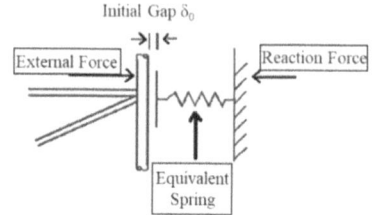

Fig. 4. Detailed model of leg to hull connection [10]

Fig. 5. Contact model between legs and guide plates [9]

the lower order frequency of the platform. The first frequency of jacking Condition is 0.04253 Hz, pre-load Condition is 0.04550 Hz, survival Condition (350 ft bottom pin) is 0.08250 Hz, and Survival Condition (300 ft bottom spring) is 0.16221 Hz. The corresponding cycles are 23.5128 s, 21.9780 s, 12.1212 s, and 6.165 s. The simulation results and first-order modes are shown in Table 2 and Fig. 6.

Table 2. The natural vibration period of a jack up platform in three different states

Condition	Natural vibration period
Jacking Condition	23.5128 s
Pre-load Condition	21.9780 s
Survival Condition	12.1212 s

2.3.2 Jacking Condition Result Analysis

Jacking condition refers to the normal jackinging state of a jack up platform when it is unloaded. The sea conditions and environmental loads are shown in Table 1. Under the jacking condition, the depth of the leg entering the water is 121.9 m, the air gap is 15 m, the wind speed is 12 m/s, and the ocean current speed is 0.6 m/s. Airy wave pattern is used, and the wave angle, wind direction, and ocean current direction are all 180°, 210°, and 240° directions. The load capacity is 16030 t. The maximum stress distribution of the structure obtained through finite element simulation is shown in Fig. 7.

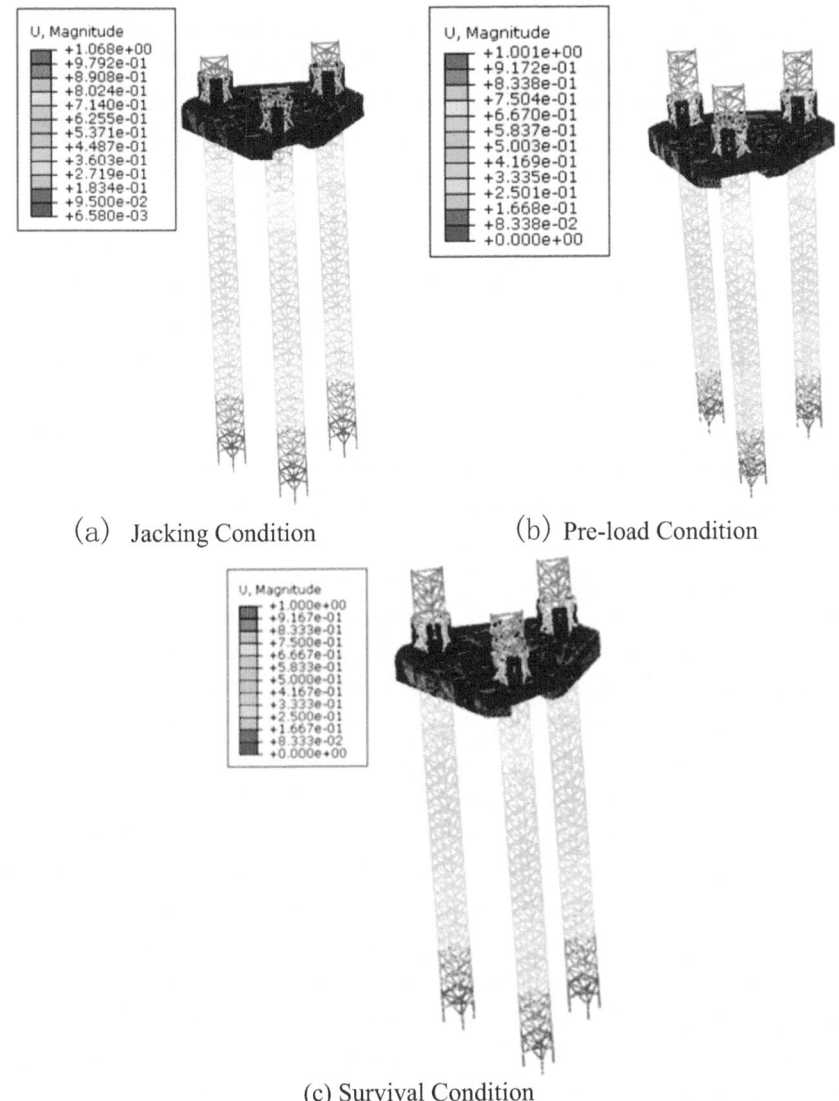

(a) Jacking Condition   (b) Pre-load Condition

(c) Survival Condition

Fig. 6. First order modal of platform structure under jacking conditions, pre-load conditions, and survival conditions.

Through finite element simulation, it was found that under the jacking state, the maximum stress occurred at the contact position between the leg and the lower guide plate in the jacking house area under the load in three directions. Among them, the maximum stress of the platform structure at 180° wind direction is generated near the jacking house area of Leg2 and Leg3, with a maximum stress of 336.7 Mpa; When the wind direction is 210°, the maximum stress of the platform structure is located near the jacking house area of Leg2, with a maximum stress of 402.9 Mpa; When the wind

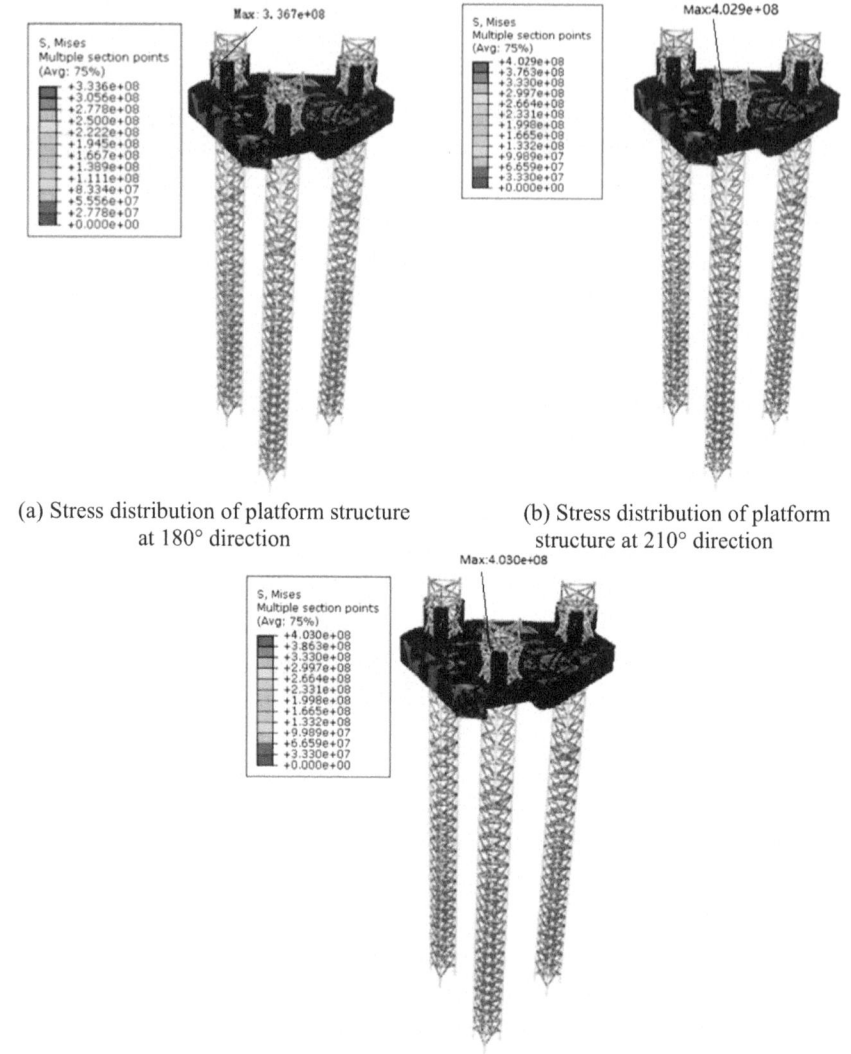

(a) Stress distribution of platform structure at 180° direction

(b) Stress distribution of platform structure at 210° direction

(c) Stress distribution of platform structure at 240° direction

Fig. 7. Stress Cloud under jacking condition

direction is 240°, the maximum stress of the platform structure is located near the jacking house area of Leg2, with a maximum stress of 403 Mpa.

2.3.3 Pre-load Condition Result Analysis

In the pre-load condition, it refers to a state in which a jack up platform is in a pressurized state when inserting piles. The sea conditions and environmental loads are shown in Table 1. Under the Pre-load condition, the depth of the legs entering the water is 121.9 m, the air gap is 2 m, the wind speed is 12 m/s, and the ocean current speed is 0.6 m/s.

The Airy wave pattern is used, and the wave angle, wind direction, and ocean current direction are all 180°, 210°, and 240° directions, with a load capacity of 31320 t. The maximum stress distribution of the structure obtained through finite element simulation is shown in Fig. 8.

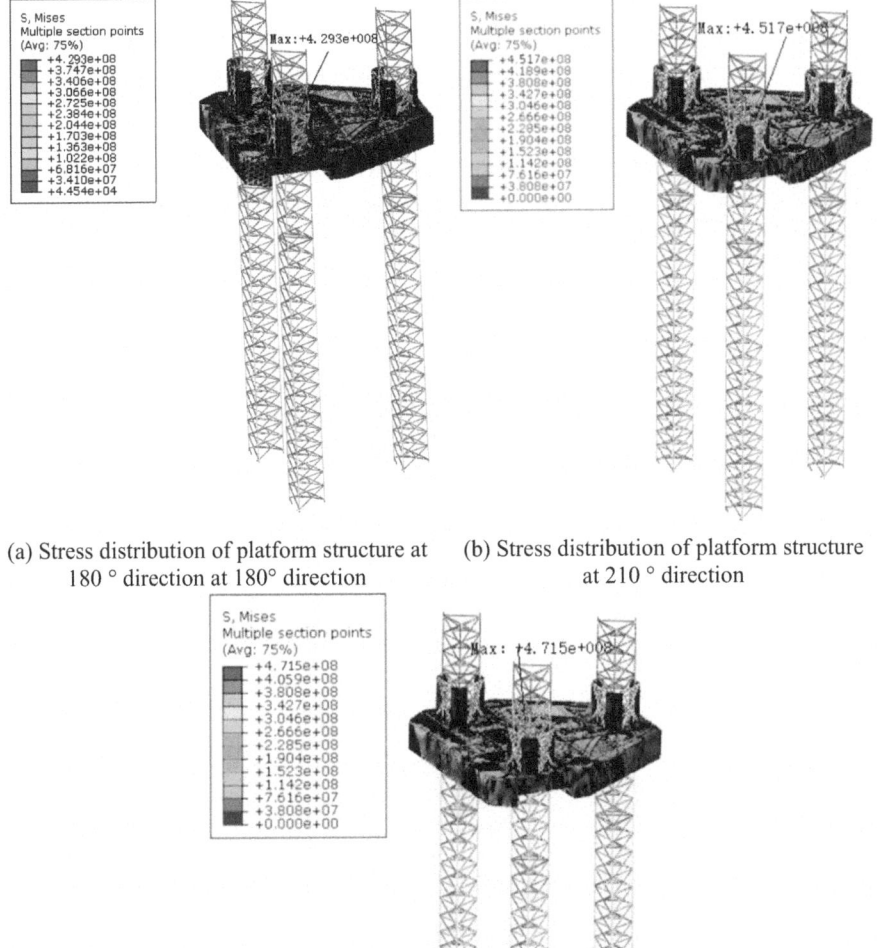

(a) Stress distribution of platform structure at 180 ° direction at 180° direction

(b) Stress distribution of platform structure at 210 ° direction

(c) Stress distribution of platform structure at 240 ° direction

Fig. 8. Stress Cloud under pre-load condition

In the pre-loaded state, through finite element simulation results, the maximum stress occurs on the rear two legs at a wind direction of 180°, with a maximum stress of 429.3 Mpa; At a wind direction of 210°, the maximum stress of Leg2 is 451.7 Mpa; at a wind direction of 240°, the maximum stress on the leg is 471.5 Mpa.

2.3.4   Survival Condition Result Analysis

In survival condition, it refers to a state in which a jack up platform is in a pressurized state when inserting. The sea conditions and environmental loads are shown in Table 1. Under pre-load condition, the depth of the legs entering the water is 121.9 m, the air gap is 17 m, the wind speed is 51.4 m/s, and the ocean current speed is 0.77 m/s. Airy wave pattern is used, and the wave angle, wind direction, and ocean current direction are 180°. The load capacity is 16030 t. The maximum stress distribution of the structure obtained through finite element simulation is shown in Fig. 9.

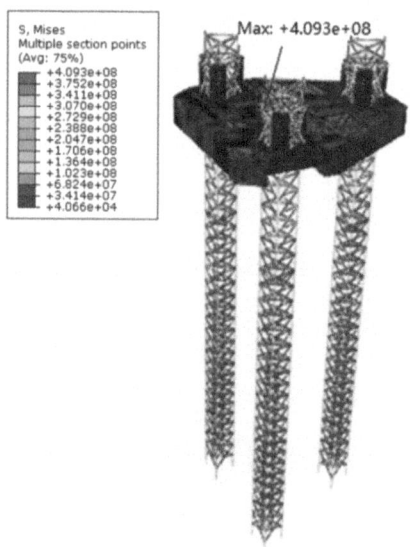

Fig. 9. Stress distribution of platform structure under 180° load direction in survival condition

Due to the symmetry of the structure, in the survival condition state, through finite element simulation, the maximum stress occurs near the jacking house area of the two tail legs at the wind direction of 180°. The maximum stress at the contact position between the legs and the lower guide plate is 409.344 MPa and 409.478 MPa, respectively.

3   Conclusions

A finite element simulation model for the jacking condition, pre-load condition, and survival condition of a full-size jack up platform was established based on the CJ 50 type jack up platform. The overall safety performance of the jack up platform under different design conditions was analyzed, and the conclusions are as follows:

(1) Through modal analysis, it was found that the first frequency of jacking condition is 0.04253 Hz, pre-load condition is 0.04550 Hz, and survival condition corresponds to periods of 23.5128 s, 21.9780 s, and 12.1212 s.

(2) Under the three working conditions of jacking condition, pre-load condition, and survival condition jack up platform on, the maximum stress occurs at the contact position between the leg and the lower guide plate in the jacking house area under different load directions. Among them, the stress generated by pre-load condition is the highest, and the stress generated by jacking condition is the lowest.

(3) Due to external loads, the jack up platform, especially the leg position, experiences uneven stress concentration, which can easily cause local damage. Therefore, during the operation process, it is necessary to promptly check the degree of damage to the jacking system and legs to ensure the safe operation of the platform.

The research conclusion can guide the CJ 50 drilling platform. The on-site operation of the CJ 50 platform is not closely related to geological conditions and has universal applicability.

Acknowledgement. This work was supported by the National Key R&D Program of China (Grant No. 2022YFF0610400). The support is gratefully acknowledged.

Disclosure of Interests. The authors declare that they have no known competing financial interests or personal relationships that could have appeared to influence the work reported in this paper.

References

1. Welaya, Y.M.A., Elhewy, A., Hegazy, M.: Investigation of jack up leg extension for deep water operations. Int. J. Naval Archit. Ocean Eng. **7**(2), 288–300 (2015)
2. Mirzadeh, J., Kimiaei, M., Cassidy, M.J.: Performance of an example jack up platform under directional random ocean waves. Appl. Ocean Res. **54**, 87–100 (2016)
3. Hendriyawan, V.A, Denny, T., et al.: Effects of prior footprints on the bearing capacity of spudcan foundations: a case study. In: IOP Conference Series: Earth and Environmental Science, vol.1249, no. 1 (2023)
4. Tolooiyan, A., Gavin, K., Dyson, A.P.: Estimation of spudcan penetration in variable sand deposits with the Arbitrary Lagrangian Eulerian finite element method, p. 281 (2023)
5. Gao, J., et al.: Spudcan slip prediction of jack up rig near the footprint: a coupled RPD-moment based solution, vol. 18, no. 5, pp. 695–706 (2023)
6. Yi, M.S., Park, J.: Global structural behavior and leg strength for jack up rigs with varying environmental parameters, vol. 11, no. 2, pp. 405–405 (2023)
7. Do, V.D., et al.: Jacking and energy consumption control over network for jack up rig: simulation and experiment, vol. 29, no. 3, pp. 89–98 (2022)
8. Yu, Y.M.: Deep learning based simulation of jack up rig. In: IOP Conference Series: Earth and Environmental Science, vol. 872, no. 1 (2021)
9. Gao, J., Duan, M.L., Li, J.B.: Influence of spudcan lateral slip on key components of jack up platform. Paper Presented at the 31st International Ocean and Polar Engineering Conference, Rhodes, Greece (2021). Author, F., Author, S.: Title of a proceedings paper. In: Editor, F., Editor, S. (eds.) CONFERENCE 2016. LNCS, vol. 9999, pp .1–13. Springer, Heidelberg (2016)
10. SNAME5-5A. Guidelines for site specific assessment of mobile jack up units (2007)

DFMM: An Object Tracking Approach Based on Deep Feature Modification

Kai Chen[1], Yujie Huang[1(✉)], Xiaodong Zhao[1], Ziyuan Wang[1], and Pengfei Wang[2]

[1] Nanjing University of Aeronautics and Astronautics, Nanjing 210016, China
yj_huang@nuaa.edu.cn
[2] Nanjing Research Institute of Electronic Engineering, Nanjing 210007, China

Abstract. In complex tracking environments, existing trackers primarily encounter issues of redundant deep convolutional features and a shortage of positive samples in the target tracking process. To address these challenges, an attention mechanism model DFMM, the Deep Feature Modification Model, is proposed based on the fusion of spatial and channel domains. This model comprises three consecutive sub-modules: spatial self-attention, channel attention, and spatial attention. Building upon this, a deep convolutional network adaptable to various visual algorithms is constructed. Additionally, strategies for feature extraction and enhancement based on feature modification are designed to mitigate problems such as redundant feature negative feedback and a lack of positive samples. Experimental results demonstrate that integrating the feature modification module in mainstream ResNet target classification tasks significantly reduce Top-1 and Top-5 error rates without incurring additional computational overhead or necessitating network structure adjustments, achieving lightweight integration. Furthermore, incorporating the feature modification module in multiple related tracking algorithms enhances tracking performance and addresses discriminator overfitting issues.

Keywords: Object Tracking · Attention Mechanism · Feature modification

1 Introduction

With the advancement and demand in intelligent transportation and robotics, target tracking has gradually become one of the fundamental research problems in the field of computer vision. The development of deep learning has provided significant impetus for improving the performance of various target tracking models. In real-world complex environments, the process of tracking moving targets faces numerous challenges, such as the continuous variation of background noise, occasional occlusions of the targets, as well as variations in scene illumination and target deformation [1, 2]. Addressing these challenges requires multidimensional research into target tracking methodologies to develop robust and high-performance trackers.

Visual object trackers rely solely on the target features in the first frame. Some methods primarily utilize the deep features of convolutional neural networks (CNN)

S. Saito et al. (Eds.): AsiaSim 2024, CCIS 2170, pp. 14–29, 2024.
https://doi.org/10.1007/978-981-97-7225-4_2

to train trackers in an end-to-end manner [4–7], leveraging offline training of CNN models on large-scale datasets [8, 9] to acquire robust recognition capabilities. Another category of methods is based on correlation filters (CF), which predominantly employ circulant matrices as the parameters for correlation filtering and perform online learning to generate dense sampling. Correlation filter-based trackers [10–13] rely on powerful representations with a large number of parameters and frequent online updates.

However, most trackers roughly merge various features together, which introduces a significant number of redundant features at each level of the network, masking hierarchical features. These redundant features not only increase the online training time of the discriminator but also elevate the learning difficulty of the discriminator. Additionally, prior research [3, 13] has confirmed that a large number of trainable parameters increase the risk of severe overfitting. Moreover, due to the large parameter size and sparse features of CNN, deeper networks such as ResNet [14] have shown more effective performance for tracking tasks.

2 Related Works

In recent years, there has been extensive development in target tracking models. MDNet [16] further integrated offline multi-domain training and online updating of classifiers to identify specific targets. Following the end-to-end paradigm, some works have employed siamese matching structures to further learn similarity metrics, considering deep correlation filters as part of the network. DaSiamRPN [6] proposed a framework based on SiamRPN to learn features capable of recognizing distractors and explicitly suppress them during the inference of online tracking processes. SiamVGG [17], building upon SiamFC [4], replaced the base network AlexNet [19] with VGG [18] to enhance tracking performance. These methods typically initialize templates based on the first frame or employ simple moving average strategies to update templates.

Within the realm of computer vision, the correlation filter algorithm has garnered widespread attention in visual tracking owing to its high computational efficiency in the Fourier domain. KCF [10] employed ridge regression and multi-channel features to address correlation filter parameters. DeepSRDCF [22] integrated the deep convolutional features of CNNs into SRDCF [21], which overcame boundary effects by employing negative Gaussian penalty weights on filter parameters. C-COT [12] further transformed feature maps of different resolutions into continuous spatial domains to achieve higher accuracy. Subsequently, ECO [13] improved the performance and efficiency of the C-COT tracker. Based on ECO, CFWCR [23] normalized each individual feature extracted from different layers to attain more reliable results.

3 Proposed Method

Effective target features significantly enhance the performance of trackers. Due to the increasing depth of feature networks from shallow to deep layers, there inevitably exists considerable redundancy within deep features, while genuinely effective information is relatively sparse. Simultaneously, as images serve as inputs to convolutional networks, a substantial amount of background information within convolutional network features interferes with the effective feature information [24, 25]. We introduce DFMM, an effective approach that departs from existing deep networks as feature extractors and instead utilizes an attention mechanism that integrates channel and spatial domains based on ResNet [14] to construct a weight activation mechanism. During the feature selection process, useful features for target discrimination are selected through weight parameters in both domains, thereby constructing a plug-and-play, adaptable target classification network suitable for various existing tracking models.

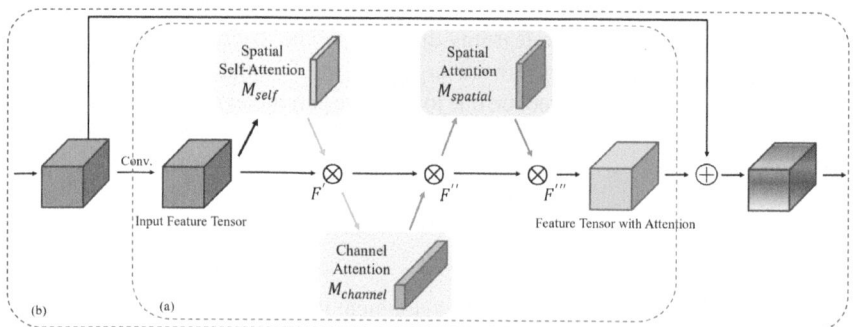

Fig. 1. The overall structure of the attention mechanism module integrating channel and spatial domain.

3.1 The Model of Feature Modification Attention Mechanism

We emphasize constructing attention mechanisms from the perspective of attention domains, primarily including spatial-wise and channel-wise domains. The spatial-wise attention mechanism transforms spatial information from the original image into another space, utilizing network learning to preserve it; while the channel-wise attention mechanism assigns a weight to each signal on every channel, representing the relevance of each channel to key information. By combining these two domains, an improved attention mechanism model can be designed to emphasize useful features and disregard irrelevant adverse features.

We propose DFMM. It includes three consecutive sub-modules, allowing each branch structure to effectively learn key information meaningful for target discrimination in both channel and spatial domains, as illustrated in Fig. 1(a). Spatial self-attention can be regarded as the self-excitatory response of feature tensors. Its purpose is to preliminarily filter out abnormally large values in a certain channel of the feature tensor. Meanwhile, spatial attention operates on the tensor after channel-wise attention has performed element selection along the channel dimension, allowing the use of max-pooling and average-pooling to perform fine-grained feature selection based on spatial positions. Thus, the feature modification module adopts the sequence of preliminary feature selection– coarse-grained feature selection– fine-grained feature selection to construct the feature modification module.

Given an intermediate feature map of a convolutional network, each convolutional block in the feature modification module can adaptively adjust, ultimately obtaining modified convolutional features that are positively significant for target classification discrimination, i.e., emphasizing useful information while suppressing or even negating the impact of less useful information. Given an intermediate feature map $F \in \mathbb{R}^{C \times H \times W}$ as input, the feature modification module sequentially derives a two-dimensional spatial self-attention map $M_{self} \in \mathbb{R}^{1 \times H \times W}$, a one-dimensional channel attention map $M_{channel} \in \mathbb{R}^{C \times 1 \times 1}$, and a two-dimensional spatial attention map $M_{spatial} \in \mathbb{R}^{1 \times H \times W}$. The process of obtaining the overall attentional convolutional feature output can be summarized as follows:

$$\begin{cases} F\prime = M_{self}(F) \otimes F \\ F'' = M_{channel}(F') \otimes F' \\ F''' = M_{spatial}(F'') \otimes F'' \end{cases} \tag{1}$$

where $\otimes$ denotes element-wise multiplication, F', F'', F''' represent the results after passing through three consecutive sub-modules, with F''' being the final modified feature. During the multiplication process, attention weights are propagated (copied) accordingly: channel attention weights are propagated along the spatial dimension.

Spatial Self-attention Module. The intermediate features of convolutional networks are three-dimensional tensors, with the channel axis representing the "content" of the image and the spatial axes representing the "position" of the image. Channel attention can be understood as an initial coarse-grained selection of image content, while spatial attention mechanisms can be understood as fine-grained selection of two-dimensional features in the image. If a certain channel of convolutional features contains strong noise information, relying solely on pooling operations can easily lead the network to learn the noise information. Therefore, we propose constructing a spatial self-attention module before the channel attention module to perform preliminary fine-grained selection of two-dimensional spatial information for each channel of convolutional features.

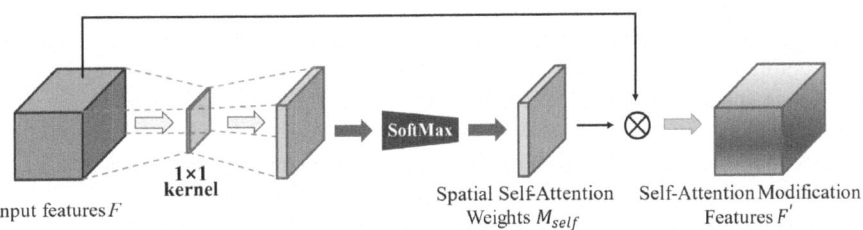

Fig. 2. Structure diagram of spatial self-attention module.

The structure of the spatial self-attention module is illustrated in Fig. 2. The target features undergo 1×1 convolutional operations to reduce dimensionality, resulting in a feature tensor with a channel count of 1. Subsequently, an activation function sigmoid is applied to transform the feature values to the [0, 1] range. Following this, spatial self-attention weights M_{self} are obtained using a SoftMax function. These attention weights are then multiplied element-wise with the input features to obtain the self-attention modified features, which can be understood as the self-excitatory response output of the input features. The calculation process of M_{self} is described in formula (2):

$$M_{self}(F) = SoftMax(\int^{1@1\times1} (F)) \tag{2}$$

where, $\int^{1@1\times1}$ represents the 2D convolution operation performed on F, where the size of the convolution kernel is 1×1, and the number of convolution kernels is 1.

Channel Attention Module. The channel attention map is generated based on the relationships between feature channels. Each channel of the feature map can be defined as a feature detector [26]. Utilizing channel attention allows for the emphasis on content that positively contributes to image discrimination while disregarding content that has a negative impact. To obtain attention weight parameters in the channel domain, the spatial dimensions of the features need to be compressed. To summarize spatial information, average pooling has typically been employed. Meanwhile, max pooling can gather important clues about unique object features to infer more refined channel attention. Empirical research indicates that utilizing these two functions can significantly enhance the network's representational capacity compared to using them individually. This demonstrates the effectiveness of the feature modification module design.

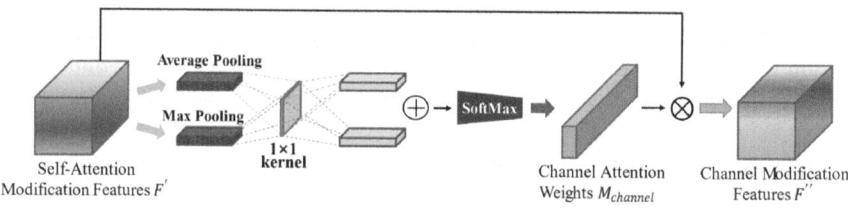

Fig. 3. Overall flowchart of a single channel attention module.

As shown in Fig. 3, firstly, the input self-attention modified features $F\prime$ undergo max-pooling and average-pooling operations to aggregate spatial domain information, resulting in spatial feature descriptors: $F_{avg}^{channel}$ and $F_{max}^{channel}$, referred to as the average-pooling and the max-pooling features, respectively. Subsequently, a shared convolutional network and SoftMax function are utilized to generate channel attention map $M_{channel} \in \mathbb{R}^{C\times1\times1}$ from $F_{avg}^{channel}$ and $F_{max}^{channel}$. The shared convolutional network consists of C 1×1 convolutional kernels. After applying this shared convolutional network to the feature descriptors $F_{avg}^{channel}$ and $F_{max}^{channel}$, element summation is performed on the descriptors, followed by the SoftMax function to obtain the final channel attention map $M_{channel}$. The calculation of channel attention is described in formula (3):

$$M_{channel}(F) = SoftMax\left(\int^{C@1\times1} (AvgPool(F)) + \int^{C@1\times1} (MaxPool(F))\right) \tag{3}$$

where, $\int^{C@1\times1}$ represents the 2D convolution operation performed, where the size of the convolution kernel is 1×1, and the number of convolution kernels is 1.

Spatial Attention Module. As shown in Fig. 4, the spatial attention feature map is primarily generated through the spatial relationships between each channel feature element. Spatial attention focuses on "where" the informative parts are, serving as a complement to channel attention. Average pooling and max pooling operations are first applied along the channel axis, and their results are concatenated along the channel dimension to form descriptor features that are effective in both maximum response and mean response aspects. [28] has been proposed that concatenating operations along the channel axis is effective in highlighting informative regions. Upon cascading the descriptor features, a convolutional layer is applied to generate the spatial attention map $M_{spatial} \in \mathbb{R}^{1 \times H \times W}$, which encodes values to emphasize or suppress information.

The spatial attention module aggregates channel information from the feature map using two pooling operations, resulting in two two-dimensional maps: $F_{avg}^{spatial} \in \mathbb{R}^{1 \times H \times W}$ and $F_{max}^{spatial} \in \mathbb{R}^{1 \times H \times W}$. Here, $F_{avg}^{spatial}$ represents the average merged features within the channels, while $F_{max}^{spatial}$ represents the maximum merged features. Subsequently, a standard convolutional layer is applied to concatenate them and perform convolution, thereby generating the corresponding two-dimensional spatial attention map, as described in formula (4):

$$M_{spatial}(F) = SoftMax\left(\int^{1@1\times 1} ([AvgPool(F); MaxPool(F)])\right) \qquad (4)$$

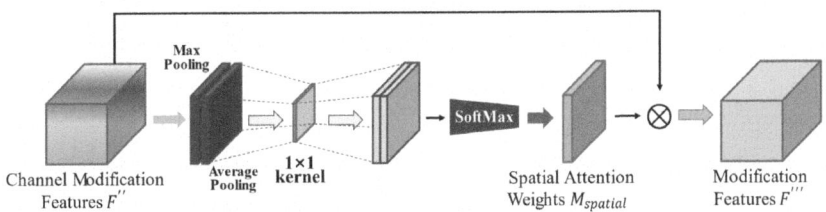

Fig. 4. Overall flowchart of a single spatial attention module.

Depth Residual Module Optimization. Due to the feature modification module's ability to maintain input-output sizes while having a minimal parameter count, it can seamlessly integrate into existing convolutional networks. We integrate the proposed feature modification module into the residual modules of excellent deep network ResNet to construct a target classification network with attention mechanisms, as illustrated in Fig. 1(b). On one hand, this module can improve the accuracy of the target classification net-work. On the other hand, based on the channel attention weight parameters and spatial attention weight parameters within the module, it can select deep convolutional features that are positively correlated with target discrimination for target tracking.

3.2 Deep Convolutional Features Extraction Based on Feature Modification

Despite the exceptional performance of deep convolutional networks, deep features are computationally expensive compared to handcrafted features, which may hinder their real-time applicability. As the depth of convolutional networks increases, they inevitably contain a considerable amount of redundant information, with relatively fewer genuinely useful features. Yet, using all these features solely for training discriminators can be wasteful of resources and may introduce features that negatively impact discrimination within the convolutional layers.

We address these challenges by selectively extracting effective features from each level of the convolutional network based on the attention mechanism integrated with the channel and spatial domains, as discussed in Sect. 3.1 (ResNet-50 + feature modification). To ensure that different levels of features contribute effectively to target tracking, we analyze the specific performance contributions of different levels of feature information. Additionally, a feature enhancement strategy is employed to increase the number of positive samples during the target tracking process.

Convolution Feature Selection in the Same Level. In the construction of deep convolutional networks, as the network layers deepen, problems such as gradient vanishing, gradient explosion, and network degradation arise due to the presence of redundant layers. To address this, ResNet [14] ensures that the input and output remain unchanged after passing through redundant layers using a residual block structure.

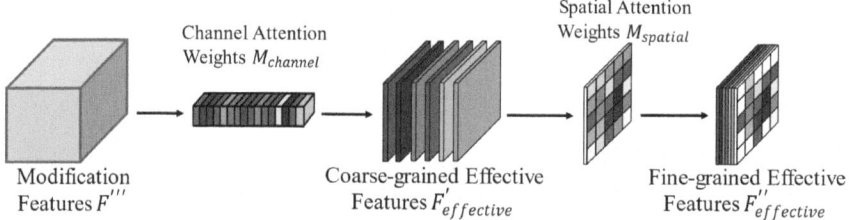

Fig. 5. The extraction process of effective features in the feature tensor.

However, in practical networks, not only do redundant layers, but also feature tensors in non-redundant layers contain redundant information in both the channel and spatial domains. Existing methods directly use a complete feature tensor to train correlation filters, resulting in both resource wastage and negative impacts on discriminator learning. Therefore, a more intelligent network structure is required to autonomously learn useful channel and spatial distribution information within feature tensors at the same level, while also autonomously suppressing negative channels and redundant spatial information within feature tensors.

We select effective feature information for target discrimination from both the channel and spatial axes of the feature tensor through the channel attention weight $M_{channel}$ and spatial attention weight $M_{spatial}$ in the feature modification module. As illustrated in Fig. 5, firstly, channels with channel attention weight $M_{channel}$ higher than threshold $T_{channel}$ in the ResNet residual block's feature tensor F'' are selected, resulting in $F_{effective}'$. Subsequently, elements in $F_{effective}'$ corresponding to spatial attention weight $M_{spatial}$ lower than threshold $T_{spatial}$ are set to 0, ensuring that the finally selected effective features $F_{effective}''$ contain fewer redundant information in channels and less noise and background information in spatial dimensions. The definitions of $T_{channel}$ and $T_{spatial}$

are shown in formula (5) and (6) respectively.

$$T_{channel} = \frac{1}{length\left(M_{N@1\times1}^{channel}\right)} = 1/N \tag{5}$$

$$T_{spatial} = \frac{1}{length\left(M_{1@h\times w}^{spatial}\right)} = \frac{1}{w \times h} \tag{6}$$

The Selection of Convolutional Layer Positions within Convolutional Networks. We introduce a hierarchical functionality into the tracking framework. For deep features, they mainly contain high-level semantic information of the target, but as the network deepens, the resolution of the convolutional features decreases, leading to decreased accuracy. For shallow networks, they mainly represent the texture and color information of the target, but as soon as the target undergoes deformation, shallow features undergo significant changes, exhibiting poor robustness.

Therefore, the core improvement is to treat the two types of features differently, utilizing deep features to enhance the robustness of the tracker while using shallow features to improve its accuracy. Since the feature differences between adjacent layers are similar, selecting adjacent layers introduces more redundancy and interference. Choosing the better performance level from each stage depicted as the feature extraction layer, not only enhances the specificity of the features but also readily adapts to abrupt changes in scale.

For ResNet50 with feature modification, the convolutional features selected based on feature modification demonstrate stable performance across occlusion, illumination variations, and simple scenes. Moreover, they also exhibit differences among convolutional features at different levels. Similar to ResNet50, selecting the better performance level from each stage depicted as the feature extraction layer, not only enhances the specificity of the features but also readily adapts to the scale variations of the target. However, unlike ResNet50, the features selected after passing through the feature modification network are fewer in number but more effective in performance. They can simultaneously improve the performance of the tracker in terms of accuracy and real-time capability.

Enhanced Visual Features of the Target. In generic object tracking tasks, only the first frame contains the actual target, while the remaining training samples are generated by the tracker during the tracking process. This leads to a scarcity of real target samples during training, and the generation of samples entirely depends on the cyclic matrix. Consequently, such samples only contain information about horizontal and vertical movements, lacking information about target rotation, deformation, and occlusion. We propose the following enhancement methods:

- Mirror: Horizontal flipping.
- Rotation: Fixed rotation at 12 angles from $-60°$ to $60°$.
- Translation: Horizontal and vertical translation by n pixels before feature extraction, equivalent to shifting the feature map by n/s pixels, where s is the stride of the convolutional kernels in the convolutional network.
- Blur: Gaussian blur filtering to simulate common motion models and scale changes in tracking scenes.

- Dropout: Channel-wise feature dropout, where 20% of the feature channels are randomly set to 0, while the remaining channels are amplified to maintain sample energy.

4 Experiments

Experiment on Classification Effect of Feature Modification Module. For the proposed feature modification module, classification experiments are conducted on ImageNet-1K to rigorously evaluate it. Additionally, comparisons are made with existing network models such as SE-ResNet, based on channel attention, and CBAM, based on both channel and spatial domains, across various experimental metrics. This experiment evaluated feature modification module, SE-ResNet and CBAM based on the number of parameters, GFLOPs, Top-1 error rate and Top-5 error rate.

Table 1. Classification results of different network structures on ImageNet-1K.

Network structures	Parameter quantity	GFLOPs	Top-1 error (%)	Top-5 error (%)
ResNet18	11.69M	1.814	29.60	10.55
ResNet18+SE	11.78M	1.814	29.41	10.22
ResNet18+CBAM	11.78M	1.815	29.27	10.09
ResNet18+DFMM	11.76M	1.814	**29.12**	**9.95**
ResNet34	21.08M	3.664	26.69	8.60
ResNet34+SE	21.96M	3.664	26.13	8.35
ResNet34+CBAM	21.96M	3.665	25.99	8.24
ResNet34+DFMM	21.96M	3.664	**25.45**	**8.11**
ResNet50	25.56M	3.858	24.56	7.50
ResNet50+SE	28.09M	3.860	23.14	6.70
ResNet50+CBAM	28.09M	3.864	22.66	6.31
ResNet50+DFMM	28.09M	3.862	**21.01**	**6.01**
ResNet101	44.55M	7.570	23.38	6.88
ResNet101+SE	49.33M	7.575	23.35	6.19
ResNet101+CBAM	49.33M	7.581	21.51	5.69
ResNet101+DFMM	49.33M	3.579	**20.69**	**5.07**

Table 1 summarizes the experimental results. The ResNet network with the feature modification module outperforms all other methods significantly, indicating the module's ability to generalize well across various models on large-scale datasets. Furthermore, our model improved accuracy over SE-ResNet and CBAM, demonstrating the effectiveness of the fusion attention method, which can generate richer descriptive features effectively. The results also indicate that, in terms of parameters and computations, the overall overhead of the DFMM is minimal, enabling lightweight plug-and-play integration into various convolutional networks.

In the qualitative analysis, we applied the class activation heatmap Grad-CAM [31] to different networks using images from the ImageNet set. The class activation heatmap uses gradients to compute the importance of spatial locations in convolutional layers, which can clearly show the regions that have a positive impact on target discrimination. As shown in Fig. 7, the class activation heatmap masks generated by the proposed network better cover the regions of the target objects. In other words, networks integrated with the feature modification module can effectively utilize the information in the regions of target objects and gather features from them, while the scores for the target categories also increase accordingly.

The spatial self-attention weight parameters are generated through 1×1 convolutions. This convolutional network continuously learns the spatial distribution of the target itself, reducing the influence of background noise. This is a capability not possessed by previous models. Meanwhile, we replace all fully connected layers and 3×3 convolutions in CBAM with 1×1 convolutions. This operation ensures that the model's parameter count remains unchanged after adding spatial self-attention, while also accelerating computation speed and improving model prediction accuracy. Additionally, this module is relatively independent and can be seamlessly integrated into most existing convolutional networks (such as AlexNet, VGG, ResNet, DenseNet) for immediate use.

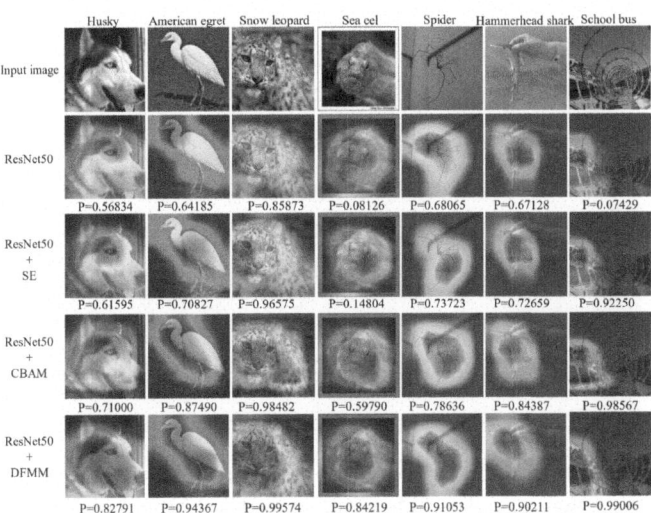

Fig. 7. Different levels of convolutional feature performance of ECO trackers in three scenarios.

Effective Convolutional Feature Tracking Verification Experiment. For the effective convolution features constructed, extracted, and enhanced as proposed, they are applied to existing excellent tracking algorithms based on correlation filters in different datasets for feature optimization. Comparative analysis is conducted to validate the effectiveness of extracting target convolution features based on the fusion of spatial and channel domains attention mechanisms. This paper conducts experiments on the most widely used OTB100 [32] and VOT [33–36] series datasets.

The existing excellent algorithms based on correlation filters, namely ECO [13], C-COT [12], DeepSRDCF [22], Staple [29], and DSST [37], are enhanced by incorporating modified convolution features or replacing their own deep features with corrected convolution features to obtain optimized algorithms, denoted as ECO-M, C-COT-M, DeepSRDCF-M, Staple-M, and DSST-M, respectively.

In the OTB dataset, the average center position error of all tracking results within a video sequence is computed to summarize the overall performance of trackers. OTB100 employs a threshold of 20 pixels for scoring. Additionally, tracking results are evaluated against ground truth bounding boxes using the success rate metric, which measures the overlap ratio. The success rate curve illustrates the ratio of successful frames as the threshold varies from 0 to 1. Evaluating trackers based on success rate values at specific thresholds may not be representative. Therefore, considering the Area Under the Curve (AUC) of each success plot is essential for ranking tracking algorithms.

In the VOT dataset, three main evaluation metrics are primarily used: Accuracy (A), Robustness (R), and Expected Average Overlap (EAO). Accuracy measures the overlap between the predicted bounding box by the tracker and the ground truth bounding box. The tracking accuracy at time t is defined as the overlap between the predicted bounding box A_t^P and the manually annotated ground truth bounding box A_t^G.

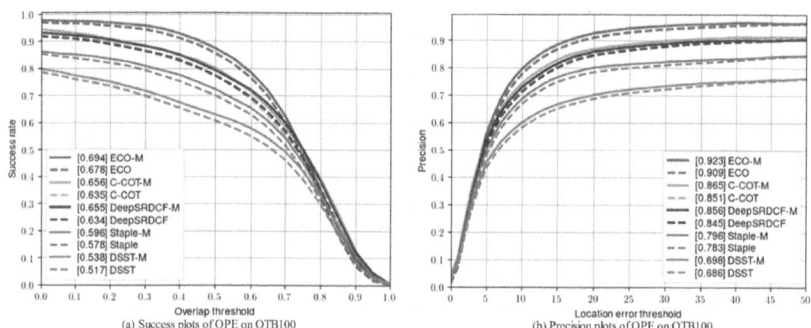

(a) Success plots of OPE on OTB100   (b) Precision plots of OPE on OTB100

Fig. 8. Success rate plot and precision rate plot on the OTB100 benchmark.

Robustness quantifies the number of times the tracker loses the target during tracking, measured by the failure rate. The failure rate indicates the number of times the tracker deviates from the target and requires reinitialization once the overlap falls below a threshold, defining it as losing the target. It can be expected that if the tracker fails at a specific frame, it is likely to fail again upon reinitialization in the next frame. To mitigate this direct correlation, the tracker is reinitialized after a period of failure.

The EAO aims to capture both good A and R in a tracker, but directly combining the two scores using a weighted sum may be unfair. Therefore, the ideal EAO is defined as the average value obtained by summing from 1 to a desired maximum value of N frames in a video sequence, i.e., the expected average coverage, as shown in formula (8):

$$\Phi_{N_s} = \frac{1}{N_s} \sum_{i=1:N_s} \Phi_i \qquad (8)$$

Figure 8 illustrates the success and precision rates of existing excellent modification filter algorithms and their corresponding optimized algorithms with modified features on the OTB dataset of 100 video sequences. It is evident that with respect to target localization success rates, the modified features maintain a gain of approximately 2% compared to the contrast algorithms; similarly, they also demonstrate a stable gain of about 1% in target localization precision rates.

Table 2. Performance comparison on the VOT2016 and VOT2020 video sequence benchmark.

Trackers	VOT2016			VOT2020		
	A	R	EAO	A	R	EAO
ECO-M	0.5295	16.5817	0.3293	0.5080	13.9009	0.3294
ECO	0.4847	15.0437	0.3089	0.4978	13.5112	0.3077
C-COT-M	0.5470	17.9393	0.3251	0.5378	14.4228	0.3292
C-COT	0.5403	23.8950	0.2941	0.5057	15.2288	0.3039
DeepSRDCF-M	0.5423	22.7120	0.2771	0.4804	20.9212	0.2541
DeepSRDCF	0.5220	20.3462	0.2756	0.5016	23.9644	0.2282
Staple-M	0.5702	19.2720	0.3206	0.4985	14.6296	0.2890
Staple	0.5333	20.9812	0.2933	0.5405	19.8836	0.2733
DSST-M	0.53698	26.0329	0.2323	0.4721	30.1225	0.1780
DSST	0.5245	44.8138	0.1805	0.4005	63.0723	0.0976

The VOT challenge involves competition among short-term visual tracking algorithms consisting of 60 sequences. Table 2 displays the A, R as well as EAO on the VOT2016 and VOT2020 dataset. The proposed modified features generally yield gains in EAO scores over the base correlation filter algorithms. It is noteworthy that the EAO scores of C-COT-M and Staple-M with corrected features surpass those of the ECO algorithm.

Meanwhile, Fig. 9 shows the accuracy and robustness plot (a), EAO plot (b), success plot rate (c), EAO curve plot (d) of the modified tracking algorithm and its base one on the VOT2016 benchmark. It can be seen that modifying the features can not only improve the accuracy of target tracking, but also improve the robustness. In particular, it is worth noting that the accuracy of the Staple algorithm is greatly improved after adding the modified features, since the Staple itself does not have deep features, which also confirms the effectiveness of deep features in object feature representation.

Figure 10 presents partial visualization examples of the methods with modified features compared to the base algorithms on three example sequences of OTB100 and

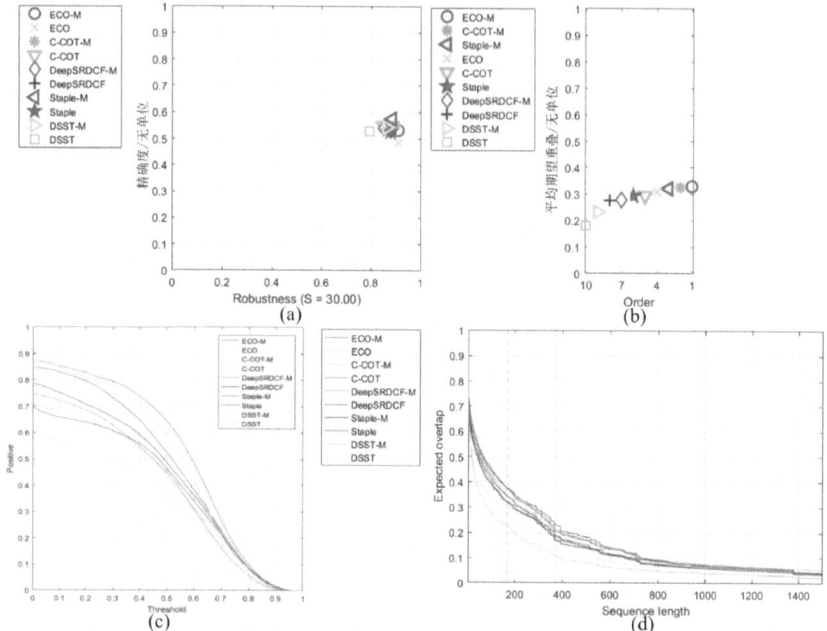

Fig. 9. Accuracy and robustness (a), EAO (b), success rate (c), EAO curve (d) on the VOT2016 benchmark.

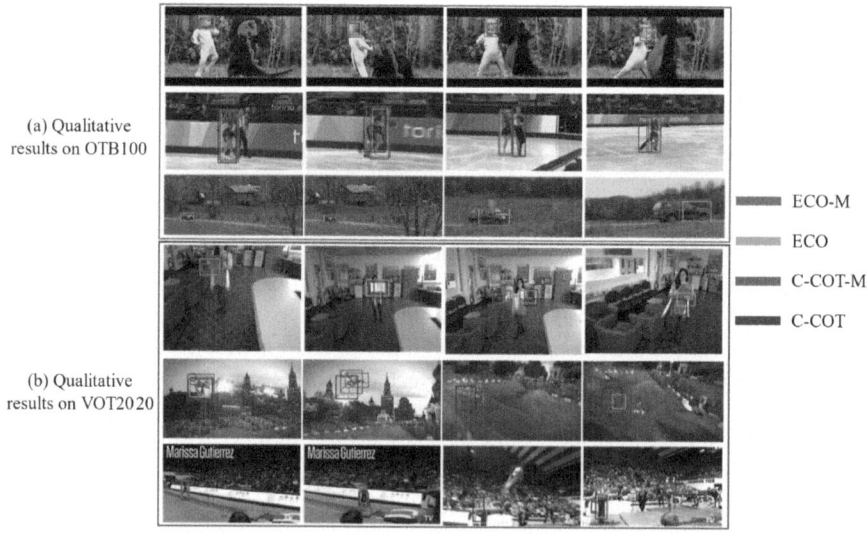

Fig. 10. Qualitative results of algorithms on three video sequences of OTB100 and VOT2020.

VOT2020, respectively. In cases of scale variation (top row), deformation (middle row), and out-of-plane rotation (bottom row), the base algorithms tend to overfit heavily to specific target regions, resulting in poor target estimation. Conversely, trackers with corrected features are better equipped to address the overfitting issue of the discriminator, thereby enabling more accurate prediction of target appearance.

5 Conclusion

In order to solve the problems of deep convolutional feature redundancy and lack of positive samples in tracking process, an attentional mechanism feature correction model based on the fusion of spatial domain and channel domain is proposed. After feature optimization, a variety of basic algorithms achieve better tracking performance, including accuracy, success rate, robustness, etc., and effectively solve the overfitting situation of the discriminator in different scenarios. At the same time, the overall overhead of the feature correction module is small in terms of parameters and calculations, and it can achieve lightweight plug and play in various convolutional networks.

Acknowledgement. This work was supported by the National Natural Science Foundation of China (52202417); China Postdoctoral Science Foundation (2022TQ0155,2022M721605); Open Project Program of State Key Laboratory of Virtual Reality Technology and Systems, Beihang University (VRLAB2023A02); Young Elite Scientists Sponsorship Program by CAST (2023QNRC001); Young Elite Scientists Sponsorship Program by JSTJ (JSTJ-2023-XH032). Authors thank reviews for their valuable comments.

References

1. Zhang, Y., et al.: Bytetrack: multi-object tracking by associating every detection box. In: Avidan, S., Brostow, G., Cissé, M., Farinella, G.M., Hassner, T. (eds.) ECCV 2022. LNCS, vol. 13682, pp. 1–21. Springer, Cham (2022). https://doi.org/10.1007/978-3-031-20047-2_1
2. Wang, Q., Zheng, Y., Pan, P., Xu, Y.: Multiple object tracking with correlation learning. In: Computer Vision and Pattern Recognition, pp. 3876–3886. IEEE (2021)
3. Stadler, D., Beyerer, J.: Improving multiple pedestrian tracking by track management and occlusion handling. In: Computer Vision and Pattern Recognition, pp. 10958–10967. IEEE (2021)
4. Bertinetto, L., Valmadre, J., Henriques, J.F., Vedaldi, A., Torr, P.H.: Fully-convolutional Siamese networks for object tracking. In: Hua, G., Jégou, H. (eds.) ECCV 2016. LNCS, vol. 9914, pp. 850–865. Springer, Cham (2016). https://doi.org/10.1007/978-3-319-48881-3_56
5. Li, B., Yan, J., Wu, W., Zhu, Z., Hu, X.: High performance visual tracking with Siamese region proposal network. In: Computer Vision and Pattern Recognition, pp. 8971–8980. IEEE (2018)
6. Zhu, Z., Wang, Q., Li, B., Wu, W., Yan, J., Hu, W.: Distractor-aware Siamese networks for visual object tracking. In: Ferrari, V., Hebert, M., Sminchisescu, C., Weiss, Y. (eds.) ECCV 2018. LNCS, vol. 11213, pp. 103–119. Springer, Cham (2018). https://doi.org/10.1007/978-3-030-01240-3_7
7. Wang, Q., Gao, J., Xing, J., Zhang, M., Hu, W.: DCFNet: discriminant correlation filters network for visual tracking. arXiv preprint arXiv:1704.04057 (2017)
8. Real, E., Shlens, J., Mazzocchi, S., Pan, X., Vanhoucke, V.: Youtube-boundingboxes: a large high-precision human-annotated data set for object detection in video. In: Computer Vision and Pattern Recognition, pp. 5296–5305. IEEE (2017)

9. Russakovsky, O., et al.: ImageNet large scale visual recognition challenge. Int. J. Comput. Vision **115**, 211–252 (2015)
10. Lei, Z., Yanjie, W., Honghai, S., Zhijun, Y., Shuwen, H.: Robust visual correlation tracking. Math. Probl. Eng. (2015)
11. Luque-Baena, R.M., Ortiz-de-Lazcano-Lobato, J.M., López-Rubio, E., Domínguez, E., Palomo, E.J.: A competitive neural network for multiple object tracking in video sequence analysis. Neural process. Lett. **37**, 47–67 (2013)
12. Danelljan, M., Robinson, A., Shahbaz Khan, F., Felsberg, M.: Beyond correlation filters: Learning continuous convolution operators for visual tracking. In: Leibe, B., Matas, J., Sebe, N., Welling, M. (eds.) ECCV 2016. LNCS, vol. 9909, pp. 472–488. Springer, Cham (2016). https://doi.org/10.1007/978-3-319-46454-1_29
13. Danelljan, M., Bhat, G., Felsberg, M.: ECO: efficient convolution operators for tracking. In: Computer Vision and Pattern Recognition, pp. 6638–6646. IEEE (2017)
14. He, K., Zhang, X., Ren, S. and Sun, J.: Deep residual learning for image recognition. In: Computer Vision and Pattern Recognition, pp. 770–778. IEEE (2016)
15. Li, H., Li, Y., Porikli, F.: DeepTrack: learning discriminative feature representations online for robust visual tracking. IEEE Trans. Image Process. **25**(4), 1834–1848 (2015)
16. Nam, H., Han, B.: Learning multi-domain convolutional neural networks for visual tracking. In: Computer Vision and Pattern Recognition, pp. 4293–4302. IEEE (2016)
17. Li, Y., Zhang, X., Chen, D.: SiamVGG: visual tracking using deeper Siamese networks. arXiv preprint arXiv:1902.02804 (2019)
18. Simonyan, K., Zisserman, A.: Very deep convolutional networks for large-scale image recognition. arXiv preprint arXiv:1409.1556 (2014)
19. Krizhevsky, A., Sutskever, I., Hinton, G.E.: ImageNet classification with deep convolutional neural networks. Commun. ACM **60**(6), 84–90 (2017)
20. Bolme, D.S., Beveridge, J.R., Lui, Y.M.: Visual object tracking using adaptive correlation filters. In: Computer Vision and Pattern Recognition, pp. 2544–2550. IEEE (2010)
21. Danelljan, M., Hager, G., Shahbaz Khan, F., Felsberg, M.: Learning spatially regularized correlation filters for visual tracking. In: European Conference on Computer Vision, pp. 4310–4318. Springer, Cham (2015)
22. Danelljan, M., Hager, G., Shahbaz Khan, F., Felsberg, M.: Convolutional features for correlation filter based visual tracking. In: Proceedings of the IEEE International Conference on Computer Vision Workshops, pp. 58–66 (2015)
23. He, Z., Fan, Y., Zhuang, J., Dong, Y., Bai, H.: Correlation filters with weighted convolution responses. In: Proceedings of the IEEE International Conference on Computer Vision Workshops, pp. 1992–2000 (2017)
24. Yu, E., Li, Z., Han, S., Wang, H.: RelationTrack: relation-aware multiple object tracking with decoupled representation. IEEE Trans. Multimed. (2022)
25. Zeng, F., Dong, B., Zhang, Y., Wang, T., Zhang, X., Wei, Y.: MOTR: end-to-end multiple-object tracking with transformer. In: Avidan, S., Brostow, G., Cissé, M., Farinella, G.M., Hassner, T. (eds.) ECCV 2022. LNCS, vol. 13687, pp. 659–675. Springer, Cham (2022). https://doi.org/10.1007/978-3-031-19812-0_38
26. Zeiler, M.D., Fergus, R.: Visualizing and understanding convolutional networks. In: Fleet, D., Pajdla, T., Schiele, B., Tuytelaars, T. (eds.) ECCV 2014. LNCS, vol. 8689, pp. 818–833. Springer, Cham (2014). https://doi.org/10.1007/978-3-319-10590-1_53
27. Zhou, B., Khosla, A., Lapedriza, A., Torralba, A.: Learning deep features for discriminative localization. In: Computer Vision and Pattern Recognition, pp. 2921–2929. IEEE (2016)
28. Li, H., Lin, Z.: Accelerated proximal gradient methods for nonconvex programming. In: Advances in Neural Information Processing Systems, pp. 379–387 (2015)

29. Bertinetto, L., Valmadre, J., Golodetz, S., Miksik, O., Torr, P.H.: Staple: complementary learners for real-time tracking. In: Computer Vision and Pattern Recognition, pp. 1401–1409. IEEE (2016)
30. Woo, S., Park, J., Lee, J.Y., Kweon, I.S.: CBAM: convolutional block attention module. In: Ferrari, V., Hebert, M., Sminchisescu, C., Weiss, Y. (eds.) ECCV 2018. LNCS, vol. 11211, pp. 3–19. Springer, Cham (2018). https://doi.org/10.1007/978-3-030-01234-2_1
31. Selvaraju, R.R., Cogswell, M., Das, A., Vedantam, R., Parikh, D., Batra, D.: Grad-CAM: visual explanations from deep networks via gradient-based localization. In: Computer Vision, pp. 618–626. IEEE (2017)
32. Wu, Y., Lim, J., Yang, M.H.: Online object tracking: a benchmark. In: Computer Vision and Pattern Recognition, pp. 2411–2418. IEEE (2013)
33. Kristan, M., et al.: The visual object tracking VOT2016 challenge results. In: Hua, G., Jégou, H. (eds.) ECCV 2016. LNCS, vol. 9914, pp. 777–823. Springer, Cham (2016). https://doi. org/10.1007/978-3-319-48881-3_54
34. Mishra, D., Matas, J.: The visual object tracking VOT2017 challenge results. In: International Conference on Computer Vision (ICCV), pp. 22–29. IEEE (2017)
35. Kristan, M., Leonardis, A., Matas, J., Felsberg, M., Fernandez, G.: The sixth visual object tracking VOT2018 challenge results. In: Leal-Taixé, L., Roth, S. (eds.) ECCV 2018. LNCS, vol. 11129, pp. 3–53. Springer, Cham (2018). https://doi.org/10.1007/978-3-030-11009-3_1
36. Kristan, M., Leonardis, A., He, L.: The eighth visual object tracking VOT2020 challenge results. In: Bartoli, A., Fusiello, A. (eds.) ECCV 2020. LNCS, vol. 12539, pp. 547–601. Springer, Cham (2020). https://doi.org/10.1007/978-3-030-68238-5_39
37. Chen, Y., Wang, J., Xia, R., Zhang, Q., Cao, Z., Yang, K.: The visual object tracking algorithm research based on adaptive combination kernel. J. Ambient Intell. Hum. Comput. **10**, 4855–4867 (2019)

Semi-physical Modeling and Simulation for Industrial Product Design Based on X Language

Kunyu Xie[1,2], Lin Zhang[1,2](✉), Zhen Chen[1,2], and Pengfei Gu[1,2]

[1] Beihang University, Beijing 100854, China
zhanglin@buaa.edu.cn

[2] Key Laboratory of Intelligent Manufacturing Systems Technology, Beijing 100854, China

Abstract. With the increasing complexity of research and development in industrial product (IP), systems engineering methods have been widely adopted, particularly the model-based system engineering (MBSE) method. Semi-physical simulation techniques can significantly expedite the transition from an IP model to the final prototype, thereby ensuring a higher success rate in IP design. However, the mainstream MBSE modeling language, SysML, lacks support for simulating semi-physical scenarios. Therefore, it becomes necessary for the system engineer to integrate semi-physical simulation tools. To address this issue, this paper proposed a semi-physical modeling and simulation method for IP based on the X language, which is a system modeling and physical modeling integrated modeling language. The semi-physical simulation of the X language semi-physical model is derived by the XDEVS-simulator. Compared to existing SysML-based methods that necessitate the use of different software tools at various stages, the X language-based method enables system engineers to employ a unified language and software throughout modeling, digital simulation and semi-physical simulation phases in IP design. This effectively reduces complexity and enhances efficiency in IP design. Finally, a case study involving mechanical arm design is used to validate the effectiveness of the proposed method.

Keywords: The X language · Industrial product · XDEVS · Semi-physical simulation · MBSE

1 Introduction

Industrial product (IP) plays a pivotal role in industrial manufacturing, such as crane, communication devices, and machine tools. Therefore, the research and development of an efficient IP is very important for manufacturing productivity. The overall IP design process encompasses requirement gathering, functional analysis, architectural decomposition, behavioral design, and requirement verification, as show in Fig. 1. In the initial

L. Zhang—This work was supported by the National Key R&D Program of China (2023YFB3308200).

phase of IP design, it is imperative to gather requirements based on the perspectives of IP stakeholders. Subsequently, system engineers analyze and design IP functions to fulfill these requirements. By considering the collaboration and interaction among these functions, an system architecture can be derived. This system architecture of the IP typically encompasses multiple levels ranging from systems down to subsystems and ultimately indivisible components and parts. Finally, during the concluding stage of modeling, the detailed behavior of these components and parts will be designed.

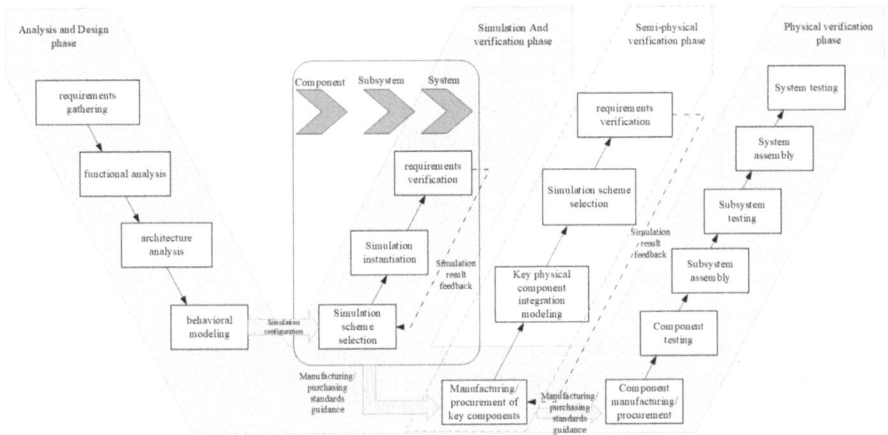

Fig. 1. The whole process of IP design verification

After modeling, the IP model needs to undergo digital simulation verification in order to ensure the rationality of the design scheme. This step is crucial for ensuring that the manufactured IP prototype system meets all design requirements. However, even with the use of digital simulation, there are still instances where key components procured or manufactured based on verified IP models fail to meet design requirements when assembled into prototype IP. The reason is that the IP model in the design process tends to oversimplify parameters or mechanisms, which can potentially result in misleading selection of IP indicators. In order to avoid this problem, it is imperative to conduct pre-assembly testing on the key components of the IP. Semi-physical simulation serves as the primary method for conducting these tests. By employing semi-physical simulation, system engineers can seamlessly integrate critical components into digital models to facilitate pre-assembly testing, thereby ensuring smooth execution of subsequent IP prototype system testing and enhancing the probability of successful design.

Due to the significance of semi-physical simulation based on Model-Based System Engineering (MBSE) in the IP development process, extensive research has been conducted based on the mainstream MBSE modeling language SysML. However, due to the limitations of SysML in supporting semi-physical simulation, the semi-physical design of IP based on SysML has been divided into two steps: system-level modeling using SysML and semi-physical software-based simulations. These two steps are integrated through model transformation or code generation.This method involving multiple softwares makes the work of system engineers quite inconvenient. To solve this problem,

Zhang et al. proposed X language [1, 2], which supports MBSE and can integrate system level modeling and physical level modeling of IP. Combined with the X language modeling and simulation software XLab and MBSE methodology X-SEM [3], X language can provide system-level modeling and digital simulation integration for IP design. However, the previous IP design methods based on X language are limited to digital simulation.

In this paper, we present a method for semi-physical design of IP based on the X language, aiming to achieve IP semi-physical design using a unified modeling language and enhance the efficiency of research and development of IP through MBSE. Specifically, we first present the semi-physical IP modeling method based on the X language, and then present the semi-physical simulation method based on the X language simulation engine XDEVS-simulator [4]. The following contents of this paper are as follows: In Sect. 2, we introduce the related research on the semi-physical design of IP based on MBSE. In Sect. 3, the semi-physical modeling and simulation method based on X language is introduced. In Sect. 4, a mechanical arm design case is employed to validate the proposed method. Finally, a conclusion and future work are given in Sect. 5.

2 Related Work

In the related work section, we will give some related research of IP design based on MBSE and semi-physical simulation, and give a brief introduction to the X language.

2.1 Semi-physical Design Method of Industrial Product Based on MBSE

With the increasing demand for IP, many scholars and engineers have carried out related research on the design method of IP based on MBSE. Most of these methods focus on the digital simulation, but some scholars have also studied the semi-physical simulation verification of IP. Gao et al. designed a architecture of the satellite communication simulation platform by using Object-Oriented System Engineering Method (OOSEM) and SysML language, and tested the effectiveness of the structure by using the method of semi-physical simulation in the verification process [5, 6]. Hu et al. built a UAV semi-physical simulation system architecture based on VxWorks platform [7]. By adding semi-physical simulation to the UAV design stage, it effectively provided convenience for UAV test and evaluation in the laboratory stage, and solved the problem of high-test cost in the UAV design process. Earle et al. designed a real-time simulation engine RT-CADMIUM based on the Discrete EVent System specification (DEVS) [8], which can support real-time communication with hardware devices to support semi-physical simulation verification.

At present, in the process of IP design based on MBSE, there is relatively little research on the semi-physical simulation verification process. The main reason is that SysML does not support simulation. If system engineers need to carry out simulation or further semi-physical simulation after modeling with SysML, they need to apply MAT-LAB, Modelica or other simulation software/languages to simulate the model through model integration or model transformation [2]. Undoubtedly, the utilization of diverse software tools at different stages adds complexity and intricacy to the system engineering.

2.2 X Language and Its Software Family

To solve the problem that SysML does not support simulation, Zhang et al. proposed X language [1, 2], which can support the integration of system modeling and physical simulation, so that system engineers can quickly integrate modeling and simulation processes of IP models with the X language. As shown in the Fig. 2, the language structure of X language includes two parts: text and diagram. The diagram part includes requirements diagram (X-RQD), use case diagram (X-UCD), definition diagram (X-DED), connection diagram (X-COD), activity diagram (X-ACD), state-machine diagram (X-STMD), equation diagram (X-EQD) and model structure diagram (X-MSD). These diagrams modeling forms can make engineers easier to communicate with colleagues in the early stage of product design. The text part of the X language including eight types of restricted class: the continuous class (X-CTC), the discrete class (X-DSC), the agent class (X-AGC) and couple class (X-COC) are used to model continuous, discrete, intelligent behavior and multi-domain hybrid models respectively. At the same time, the two forms of X language can be converted to each other. As shown in the Fig. 2, each class of the X language have multiple parts. For example, the couple class includes a definition part and a connection part. Each of these parts can be mapped to a diagram: a definition part mapped to a definition diagram and a connection part mapped to a connection diagram.

XLab is a BS-based software that supports X language modeling and simulation, the architecture of the model is show in Fig. 3. XLab can support the modeling of the X language diagram format and the text format, and the text format model can be compiled by XLab compiler into a simulation model executed by C++. For the compilation of hybrid models, the models built by the discrete, continuous and couple class are transformed into their corresponding XDEVS models and simulated by the XDEVS-simulator [9].

3   Semi-physical Design Method for Industrial Product Based on X Language

3.1   Semi-physical Modeling for the Industrial Product Based on X Language

For the IP modeling process, the current grammars of X language can provide support. As shown in Fig. 4, in the requirement and function design phase, the X language provides the X-RQD and the X-UCD to model the requirements and functions of IP respectively. Further in the architecture decomposition stage, the X language provides the X-COC for modeling system hierarchy and the X-MSD for organizing IP model files. The components of the IP system can be modeled in the definition part of the X-COC, and the connection between these components can be modeled in the connection part of the X-COC. The X-MSD can be used to configure the IP system modeling structure and chosing the final simulation configuration with pruning [9]. Through the process of pruning, system engineers can obtain simulation configurations of IP at various scales, enabling them to acquire the simulation configuration of specific components or subsystems. In terms of specific behavior modeling, X language provides multiple continuous, discrete and intelligent behavior description grammars such as X-CTC, X-DSC and X-AGC.

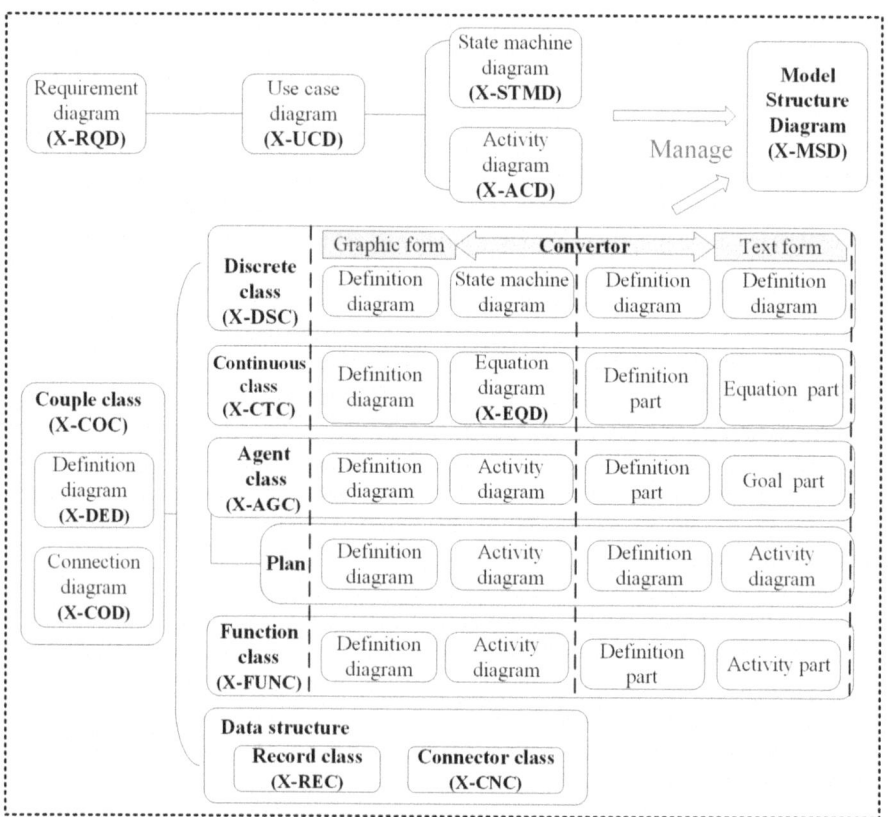

Fig. 2. The architecture of the X language

After the modeling stage, a preliminary digital model of the IP can be obtained, which is organized by X-MSD into different views (requirements, functions, architecture, and behavior) and levels (system, subsystem, and component). Additionally, the configuration for IP system simulation can be derived using pruning grammar. Subsequently, the pruned simulation model is compiled and simulated by XLab. Finally, the simulation results are obtained to verify the design requirements of the IP.

Based on the above modeling and digital simulation process, the system engineer can initially determine the key indicators and parameters of the core physical components in the IP (such at the motor's rated torque). Consequently, engineers can procure or manufacture core components based on these identified parameters. Additionally, the physical components should be customized with corresponding data acquisition module, data processing module, and communication module based on the interface of the previous X language model in order to establish effective communication with the digital model.

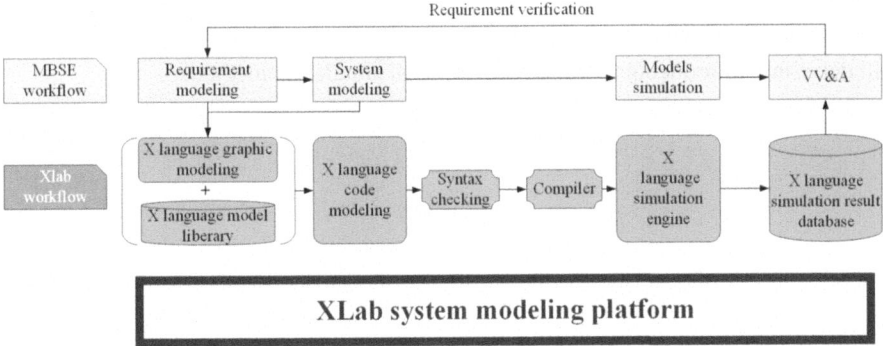

Fig. 3. The XLab software platform

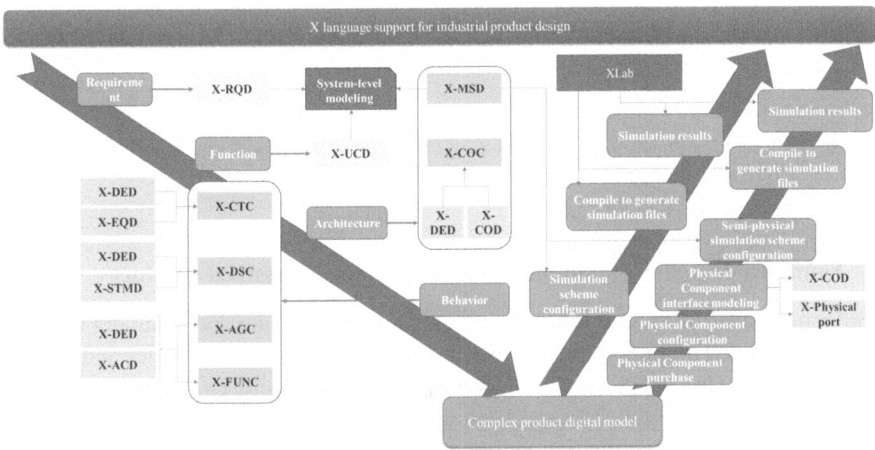

Fig. 4. The X language support for IP design

After constructing the necessary modules, the physical components can be further integrated into the digital model of X language. For this integration, the X language provides modeling grammars for establishing communication with physical devices through the physical port. By utilizing physical ports, the X language facilitates the implementation of base classes for defining the interfaces of physical devices. The Fig. 5 shows the interfaces model of a physical *motor* defined with the X language, wherein the two physical port variables enable the digital model to acquire motor speed and torque information from the physical side. Specifically, the two ports will be seamlessly integrated into the digital model through socket standardization during the semi-physical simulation. At the same time, the ports corresponding to the physical device can serve as interfaces for integrating the digital model of IP, as shown in Fig. 6. The connection in Fig. 6 indicates that the *Motor* is directly connected to the *ControlBus* of the mechanical arm. Subsequently, the physical *Motor* model can serve as a specialization of the digital Motor model in the X-MSD of the IP, and can replaced the digital Motor in the semi-physical simulation configuration generation process by pruning. At this stage, the modeling work of the semi-physical IP model concludes, while the subsequent section will delve into introducing specific semi-physical simulation methods.

```
class Motor
port:
    physical output real rotate_speed("socket");//the rotate speed of the motor
    physical output real torque("socket");// the torque of the motor
end;
```

Fig. 5. Digital interface mapping of the motor model

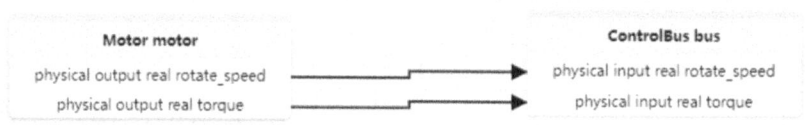

Fig. 6. Integrated connection based on X-CND between motor model and control bus model

3.2 Semi-physical Simulation for the Semi-physical Industrial Product

In order to simulate the above semi-physical model, the compiler and simulator of XLab should be modified to support the data exchange between the physical device and the digital model of IP. The architecture of the semi-physical simulation engine based on the X language is shown in Fig. 7. The real-time simulation of the XDEVS-simulator is based on the DEVS real-time simulation algorithm [8]. When driving the XDEVS-simulator for real-time simulation, the *asynchronous simulator* can monitor input from the physical device as an interrupt event. At the same time, it can send control information to the physical device through discrete events. The XLab compiler subsequently compiles both the physical ports and the corresponding model which defined these ports into TCP-servers and the *asynchronous simulator*, represented in light yellow in Fig. 7. The modification on the XLab compiler will not be further elaborated within this context. Based on the updated compiler and simulation engine, XLab can realize the semi-physical simulation verification of IP models built with X language.

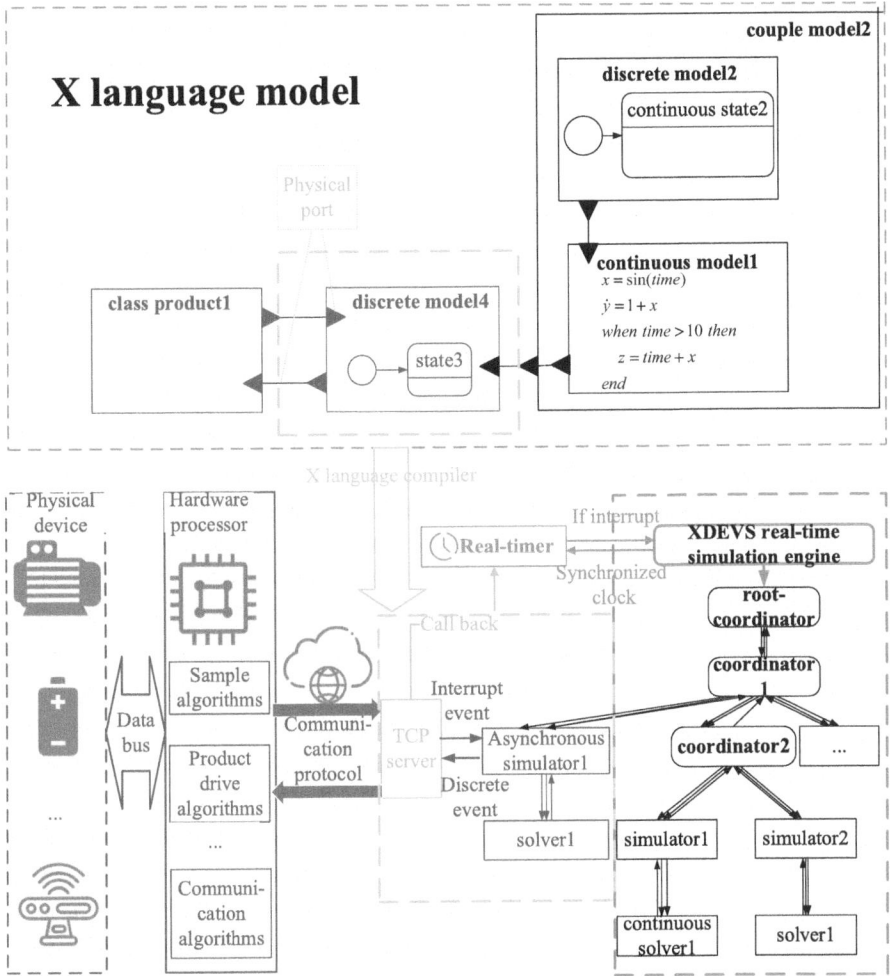

Fig. 7. The connection between the real-time simulation engine and the physical device

4 Experiment

In this section we will use a mechanical arm design case to validate the proposed method. The mechanical arm utilized in this study is a simplified version of an industrial-grade robotic arm, specifically designed to assist machine tools. It should be emphasized that real mechanical arm design is far more complex than the situation presented in this paper. The example in this paper are mainly to further illustrate the proposed method.

Initially, we conducted a preliminary analysis of the application requirements to the mechanical arm. The ultimate requirement of the mechanical arm is "provide the transportation function between the machine tools and the conveyor belt", and this requirement can be devided into the following aspects:

(1) Simplicity: reduce the design complexity of the mechanical arm as much as possible while meeting the design requirements;

(2) The construction of the mechanical arm is required to be less than 3 degrees of freedom.

(3) Fast response: the mechanical arm needs to have high speed motion capabilities;

(4) The time for the mechanical arm to complete the maximum Angle (180° per joint) rotation need to less than 10 s;

(5) High accuracy: the mechanical arm needs to ensure high accuracy during the movement;

(6) The rotation angle error of each joint of the mechanical arm should be less than 1%.

Based on these requirements, we designed this mechanical arm model with the method proposed above. The requirement model, functional model, architecture model, and behavioral models of the mechanical arm is shown in Fig. 8. In order to meet the requirements of simplicity in requirement (1) and (2), we choose a two-axis mechanical arm structure. It should be emphasized because the requirements to be verified are mainly about accuracy and speed properties, and the designed target mechanical arm is lightweight, we did not further consider the mutual dynamic relationship between the mechanical arm and the motor. Instead, we assumed that the motor could achieve comparable speed and accuracy indicators when driving the mechanical arm. However,

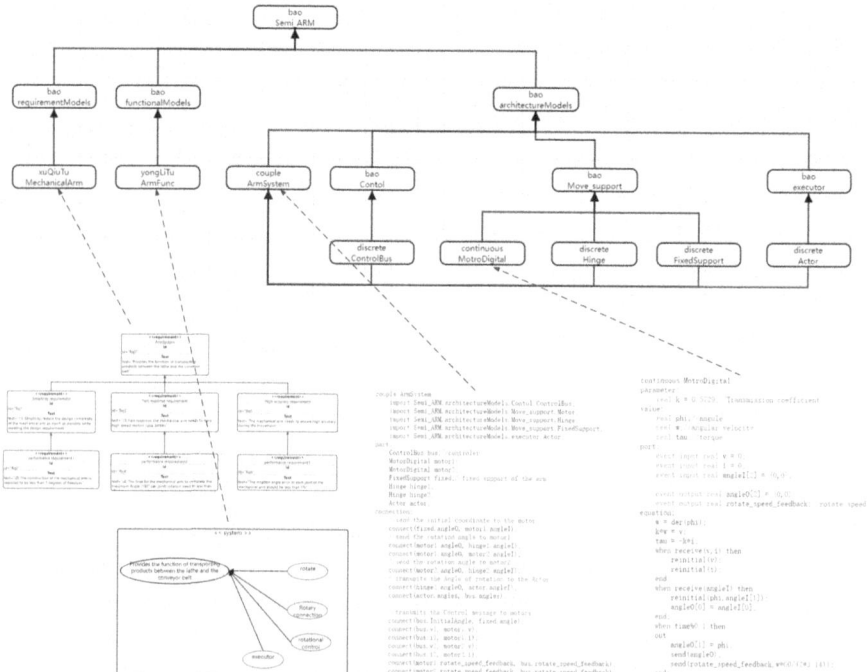

Fig. 8. The X-MSD and the requirements, function, architecture model of the mechanical arm system

for higher precision designs, it is imperative to account for their dynamic relationship. The designed mechanical arm model is then simulated and tested by XLab, because the focus of this paper is to introduce the semi-physical simulation, the design process of the model is not introduced in detail.

After completing the modeling phase, according to the method proposed in this paper, it is necessary to further confirm whether the performance of key components is suitable through semi-physical simulation. For the mechanical arm, the key core component is the motor that performs the joint movement. We will mainly carry out tests on the motor that supports the rotation of the base joint in the two-axis mechanical arm. Through the simulation in the previous stage, we determined the parameters of the motor, and carried out the corresponding procurement. The parameters and shape of the motor are shown in Fig. 9, which is a brushless DC electric motor (BLDC). In order to drive the motor, we also purchased the corresponding driver board and STM32 microprocessor. The STM32 microprocessor will load the program driven by the motor and the data acquisition program in advance. The driver program includes PWM drive signal control for the BLDC, speed PID control program for the BLDC, speed sampling program and WIFI communication program. The signal transmission relationship between the STM32 microprocessor and the X language simulation model is shown in Fig. 10. During the semi-physical simulation, the motor will work at a speed of 400 revolutions per minute (RPM), and the STM32 microprocessor will run the data acquisition program to sample the speed and angle of the motor at a sampling frequency of 10 times per second. The simulation results are shown in the Fig. 11. It can be seen that the motor takes less than 10 s to complete the rotation of 180° Angle (maximum rotation Angle), which meets the requirement (4). The rotation Angle error is 0.0086 less than 1%, which meets the requirement (6). In summary, the selected motors are able to meet the basic requirements of the design. Through the verification of the semi-physical simulation, it is shown that the motor design choice of the mechanical arm can meet the design requirements. The drive motor can be selected according to this parameter standard in the subsequent manufacturing and assembly of the mechanical arm.

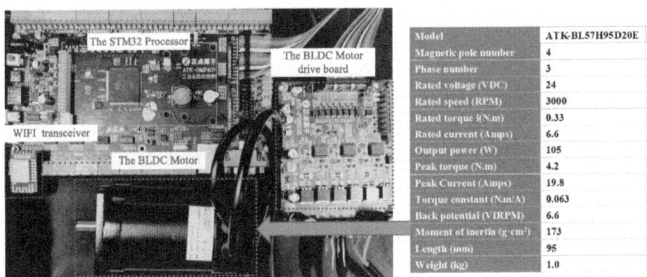

Model	ATK-BL57H95D20E
Magnetic pole number	4
Phase number	3
Rated voltage (VDC)	24
Rated speed (RPM)	3000
Rated torque ((N.m)	0.33
Rated current (Amps)	6.6
Output power (W)	105
Peak torque (N.m)	4.2
Peak Current (Amps)	19.8
Torque constant (N.m/A)	0.063
Back potential (V /RPM)	6.6
Moment of inertia (g·cm²)	173
Length (mm)	95
Weight (kg)	1.0

Fig. 9. Purchased motor, drive board and STM32 microprocessor

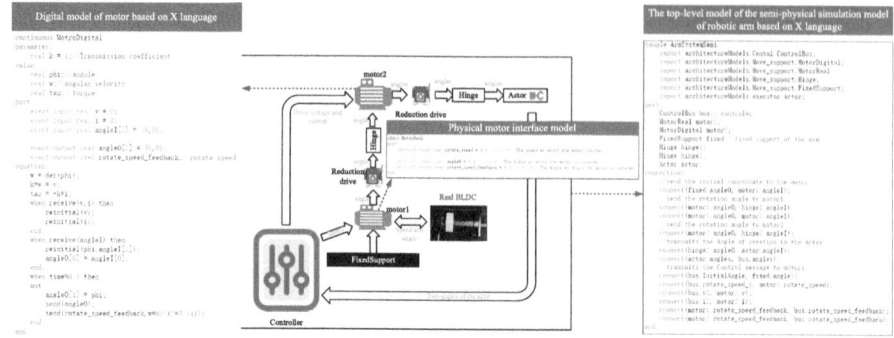

Fig. 10. The top-level connection relationship of the semi-physical model of the mechanical arm

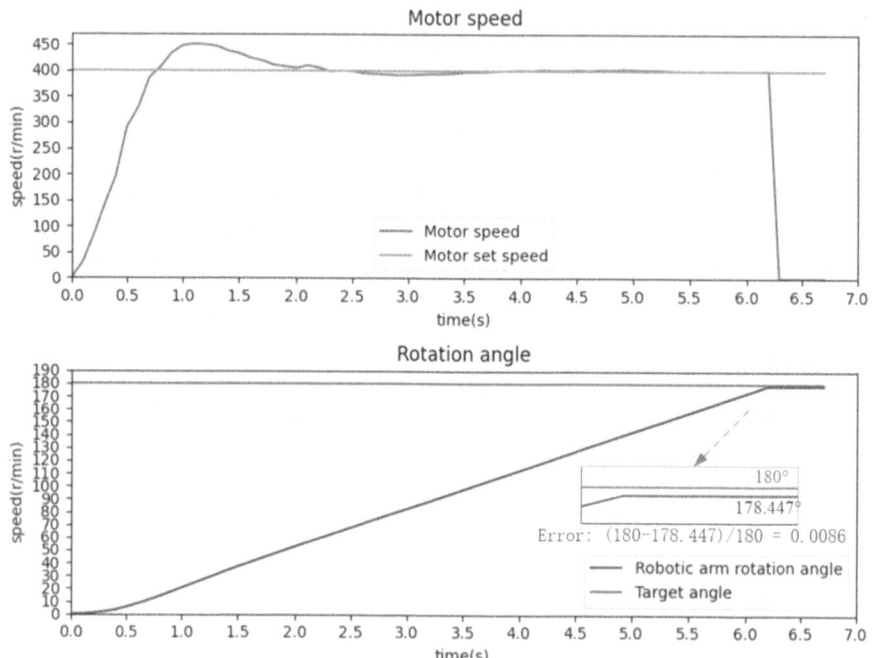

Fig. 11. Simulation results of the motor

5   Conclusion and Future Work

With the increasing complexity of IP, MBSE-based method has gained widespread adoption in IP design. Semi-physical simulation techniques can significantly expedite the transition from an IP model to the final prototype, thereby ensuring a higher success rate in IP design. However, the mainstream MBSE modeling language, SysML, lacks support for simulating semi-physical scenarios. To address this issue, this paper proposes an IP design method based on X language that comprises two parts: IP design modeling using X language and semi-physical simulation utilizing XDEVS-simulator. Compared to existing SysML-based methods that necessitate the use of different software tools at various stages, the X language-based approach enables system engineers to employ a unified language and software throughout requirements collection, functional analysis, architecture decomposition, behavior modeling, as well as subsequent digital simulation and semi-physical simulation phases in IP design. This effectively reduces complexity and enhances efficiency in IP design. Finally, a mechanical arm design case is employed to validate the proposed method.

In our future research, we will place greater emphasis on the integration of IP design, manufacturing, and operation using X language. Specifically, we aim to advance digital twin modeling and maintenance technology of IP based on X language to facilitate the comprehensive application of X language in IP design, manufacturing, assembly, operation, and maintenance processes.

References

1. Zhang, L., Ye, F., Laili, Y., et al.: X language: an integrated intelligent modeling and simulation language for complex products. In: Proceedings of the 2021 Annual Modeling and Simulation Conference, ANNSIM 2021 (2021). https://doi.org/10.23919/ANNSIM52504.2021.9552057
2. Zhang, L., Ye, F., Xie, K., et al.: An integrated intelligent modeling and simulation language for model-based systems engineering. J. Ind. Inf. Integr. **28**, 100347 (2022). https://doi.org/10.1016/J.JII.2022.100347
3. Gu, P., Chen, Z., Zhang, L., et al.: X-SEM: A modeling and simulation-based system engineering methodology. J. Manuf. Syst. **74**, 198–221 (2024). https://doi.org/10.1016/J.JMSY.2024.01.013
4. Xie, K., Zhang, L., Laili, Y., Wang, X.: XDEVS: a hybrid system modeling framework (2022). https://doi.org/10.1142/S1793962322430012
5. Gao, S., Liu, Y., Fan, L., et al.: Model-based semi-physical simulation platform architecting for satellite communication system. In: 2018 13th System of Systems Engineering, SoSE 2018, pp. 379–386 (2018). https://doi.org/10.1109/SYSOSE.2018.8428761
6. Gao, S., Cao, W., Fan, L., Liu, J.: MBSE for satellite communication system architecting. IEEE Access **7**, 164051–164067 (2019). https://doi.org/10.1109/ACCESS.2019.2952889
7. Hu, W., Cong, W., Chen, X., et al.: Design and implementation of UAV semi-physical simulation system based on VxWorks. In: Jia, Y., Zhang, W., Fu, Y., Wang, J. (eds.) CISC 2023. LNEE, vol. 1089, pp. 687–698. Springer, Singapore (2023). https://doi.org/10.1007/978-981-99-6847-3_60
8. Earle, B., Bjornson, K., Ruiz-Martin, C., Wainer, G.: Development of a real-time DEVS kernel: RT-cadmium. In: Proceedings of the 2020 Spring Simulation Conference, SpringSim 2020 (2020). https://doi.org/10.22360/SPRINGSIM.2020.CPS.002
9. Xie, K., Zhang, L., Li, X., et al.: SES-X: a MBSE methodology based on SES/MB and X language. Information **14**, 23 (2022). https://doi.org/10.3390/INFO14010023

RGFormer: Residual Gated Transformer for Image Captioning

Zehui Jin, Kai Chen[✉], Guoyu Fang, and Dunbing Tang

Nanjing University of Aeronautics and Astronautics, Nanjing 210016, China
chen_kai@nuaa.edu.cn

Abstract. Image captioning is a cross-modal task that combines computer vision and natural language processing. The model is required to generate an appropriate caption for the given image. To address this challenge, we proposed a Residual Gated Transformer, RGFormer, as an enhancement of Transformer architecture. The model based on CLIP and RGFormer, CRM, is then proposed for image captioning. CRM utilizes the encoder of CLIP to extract image features as a prefix to the caption. The prefix is projected into language space using RGFormer, a lightweight mapping network, and then fed into GPT-2 to generate captions. CLIP was trained on an extensive dataset comprising image-text pairs, which contains rich visual and semantic information and is exceptionally well-suited for vision-language tasks. The core idea of CRM is to reduce the disparity between visual and textual representations by using RGFormer to accomplish the cross-modal task. CRM could generate meaningful captions for diverse and large-scale datasets in a short training time without additional annotations or pre-training. Quantitative evaluation experiments show that CRM achieves results comparable to some advanced models on the COCO Caption dataset more efficiently.

Keywords: Image Caption · CLIP · Transformer

1 Introduction

Image captioning is a task that utilizes natural language to depict visual images. This task involves not only detecting and recognizing the main objects in the image but also expressing their relationships in natural language. For example, in the first image of Fig. 1, image captioning model first needs to detect two objects: 'dog' and 'frisbee', and then understand the relationship between them as 'dog holding frisbee in its mouth'. Image captioning is a cross-disciplinary and multi-modal cross-research direction that can be applied in many scenarios, such as automatic driving, multi-modal retrieval, intelligent surveillance, etc., and has broad application potential in the information age.

Numerous mature methods for image captioning have been proposed [1–7], predominantly adhering to the paradigm of image encoding followed by text decoding. More specifically, these methods typically extract visual features through an image encoder and then generate corresponding captions through a text decoder. Their essence lies in reduce the disparity between image and text.

© The Author(s), under exclusive license to Springer Nature Singapore Pte Ltd. 2024
S. Saito et al. (Eds.): AsiaSim 2024, CCIS 2170, pp. 42–57, 2024.
https://doi.org/10.1007/978-981-97-7225-4_4

However, due to the disparity between visual and semantic features, the training of these models typically demands substantial computational resources. They require larger-scale datasets, more complex network architectures, and a greater number of trainable parameters. Sometimes, annotations such as bounding boxes and object categories are also necessary to achieve better performance. These drawbacks constrain their applicability in practical scenarios. Therefore, researching lightweight models to address the image captioning task is of great significance.

A brown and white dog holding a frisbee in its mouth

A couple of people ridinghorses through the woods

Three young girls sitting at a table eating pizza

A giraffe bends down to eat some grass

Fig. 1. The captions describing the respective images generated by CRM

To solve the problem, we adopt a powerful vision-language pre-training model to construct a more lightweight model called CRM. Specifically, we utilize CLIP proposed by OpenAI [8], whose core idea is to map images and text into the same feature space. We utilize the image encoder of CLIP to extract features rich in both visual and semantic information from the given images. This approach does not require extensive training data or additional supervision labels. By leveraging the strong correlation between visual and semantic information, we reduce the training costs and data requirements of the model.

Inspired by Li et al. [9], we adopt a method to generate image captions using GPT-2 [10] with prefix conditioning. The specific steps are illustrated in Fig. 2. We first input the image into the encoder, which is a Vision Transformer [11] network, to obtain CLIP embeddings. These embeddings are then concatenated with a set of learnable constants, and passed through a mapping network to generate caption prefixes. We employ Transformer as the mapping network due to its powerful global representation capability. Building upon this, we propose an enhanced structure called RGFormer. Each

RGFormer block is mainly composed of a Residual Multi-Head Attention module and a feed forward layer with Gated Linear Unit. Finally, the generated caption prefixes are concatenated with embeddings obtained from encoding real tokens and input into GPT-2 to generate image captions. During training, we freeze the encoder and the language model, only fine-tuning the mapping network, thereby achieving a more lightweight model.

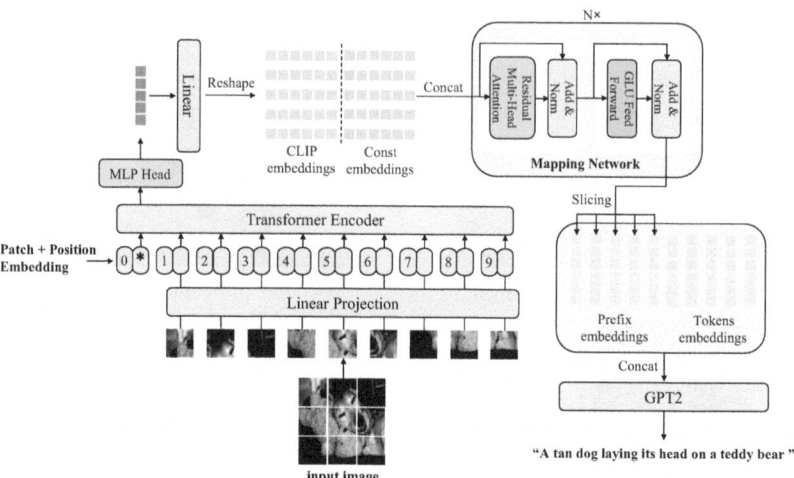

Fig. 2. The overall structure of CRM. The image encoder and GPT-2 are frozen. The mapping network RGFormer is learnable

We trained CRM on a Nvidia GTX 3090Ti GPU using the COCO Caption dataset, which took a total of 4 h. Experimental results show that our CRM exhibits good generalization ability, as shown in Fig. 1. Through quantitative evaluations, CRM achieved performance metrics comparable to some advance methods with less training time. To summarize, this paper's contributions are as follows:

- A lightweight image captioning model called CRM based on CLIP and RGFormer has been proposed, which can generate meaningful captions for diverse and large-scale datasets in a short training time.
- RGFormer, an improved Transformer based on Residual Multi-Head Attention module and GLU Feed Forward layer, which shows better performance compared to the original Transformer.

2 Related Works

2.1 Image Encoder

Image captioning models need to encode input images into a sequence of feature vectors, which are then input into a decoder to produce word sequences. Early research employed pre-trained CNN networks from classification tasks (such as VGG16 [12] and ResNet-101 [13]) as encoders to get grid-level image features. However, Anderson et al. [1] pointed out that this approach rarely considers how to determine the salient image regions. In order to generate captions more consistent with human visual habits, they first proposed using Faster R-CNN [14] to detect prominent regions in pictures and extract region-level features. This method has emerged as a standard approach for extracting image features in later studies. Although region-level features extracted by Faster R-CNN perform well in image captioning tasks, this approach requires additional detection labels for training, and extracting region-level features is time-consuming, resulting in longer training and inference time.

Rather than utilizing Faster R-CNN, which has slower processing speed and requires additional labels, we employ CLIP for feature extraction. CLIP [8] is a multimodal model based on text and images, used for jointly representing visual and semantic features. It is trained using a dataset of more than 400 million pairs of images and text in an unsupervised contrastive learning style. CLIP maps images and text descriptions into the same feature space via two encoders, thereby generating features rich in both visual and semantic information. Numerous studies have applied CLIP in diverse visual tasks, including image classification [15–17], object detection [18–20], and more.

2.2 Text Decoder

Existing image captioning models can be divided into two categories based on various text decoders: CNN-LSTM based models [1, 6, 7, 21, 22] and CNN-Transformer based models [3, 5, 23, 24]. Both types of models utilize pre-trained CNN or Faster R-CNN as encoders to extract features at grid or region level from images. The former employs LSTM [25] as the decoder, which can capture long-term dependencies and reduce gradient vanishing and exploding problems during backpropagation. However, it suffers from shortcomings in training efficiency and expressive power. The latter employs Transformer [26] as the decoder, which shows strong global representation capabilities. In addition, Transformer can be computed in parallel, unlike LSTM, which requires sequential computation over time steps. This feature enables Transformer to achieve higher efficiency.

In recent years, one of the most influential works on Transformer is BERT [27], which introduced the fundamental paradigm for subsequent language models: pretraining on large-scale datasets followed by fine-tuning on smaller datasets for various downstream tasks. In this paper, we employ GPT-2 pretraining language model as the text decoder and train a lightweight Transformer mapping network instead of fine-tuning GPT-2. Researches on establishing a shared space through vision-language pretraining are most closely associated with our work [28–30]. Zhou et al. [29] utilized an object detector to generate prefixes, which were then inputted into BERT to generate text sequences. In contrast, we utilize CLIP, which does not require an object detector or additional labels, to naturally obtain text prefixes, which are then inputted into GPT-2 to generate image captions.

3   Method

We start with the description of the image captioning problem. Given a dataset of image-text pairs $\{x^i, y^i\}_{i=1}^N$, our objective is to produce a caption that best describes the image. This caption can be represented as a sequence of tokens $\{y^i = y_1^i, \ldots y_l^i\}$, where l is the length of the longest caption in the dataset, and we pad each token sequence to length l. Our training objective is as follows:

$$\max_\theta \sum_{i=1}^N \log p_\theta\left(y_1^i, \ldots, y_l^i | x^i\right) \tag{1}$$

Here, θ represents the model training parameters. We treat the embeddings generated by CLIP for the image as the prefix of the caption, which encapsulates visual and semantic information. Then, we use an autoregressive language model to generate each word sequentially. Our objective can be described as:

$$\max_\theta \sum_{i=1}^N \sum_{j=1}^l \log p_\theta\left(y_j^i | x^i, y_1^i, \ldots, y_{j-1}^i\right) \tag{2}$$

3.1   Overview

The overall structure of our model is shown in Fig. 2. We employ a pre-trained CLIP image encoder [8], which is primarily a Vision Transformer Net [11], to extract visual information from the image x^i and obtain CLIP embeddings. Next, we concatenate the CLIP embeddings with a group of learnable constant embeddings and then pass them through a lightweight Mapping Network F to obtain the prefix embeddings:

$$\left(p_1^i, \ldots, p_k^i\right) = F\left\{Concat\left[CLIP\left(x^i\right), (c_1, \ldots c_k)\right]\right\} \tag{3}$$

Here, c represents a set of constant embeddings, which have the same length as the prefix embeddings. We use GPT-2 as the language mode. First, we tokenize the text using the GPT-2 tokenizer to obtain a sequence of tokens $\{y_1^i, \ldots, y_l^i\}$, and then we encode the tokens into token embeddings:

$$\left(t_1^i, \ldots, t_l^i\right) = gpt\left(y_1^i, \ldots, y_l^i\right) \tag{4}$$

The dimension of each token embedding t_n^i is the same as that of prefix embedding p_n^i. We concatenate p^i and t^i of image x^i:

$$z^i = p_1^i, \ldots, p_k^i, t_1^i, \ldots, t_l^i \tag{5}$$

During model training, we input the prefix-token pair embeddings $\{z^i\}_{i=1}^N$ into the language model. Given the prefix, our objective is to use an autoregressive method to predict next token. To improve training efficiency, we keep the parameters of CLIP and GPT-2 frozen and use a simple and effective cross-entropy loss to train the Mapping Network F.

$$\mathcal{L}_X = -\sum_{i=1}^N \sum_{j=1}^l \log p_\theta\left(t_j^i |, p_1^i, \ldots, p_k^i, t_1^i, \ldots, t_{j-1}^i\right) \tag{6}$$

We will next go into the detail about the specific structure of the CLIP image encoder and explain how CLIP embeddings are obtained.

3.2 Image Encoder Architecture

The CLIP Image Encoder [8] based on Vision Transformer primarily includes three modules, illustrated in Fig. 3(a).

Linear Projection of Flattened Patches: The input image is segmented into a set of patches. Taking ViT-B/32 as an example, the preprocessed 224 × 224 image is divided into 49 individual 32 × 32 patches. Each patch is then converted into a vector by a linear layer. After adding positional encoding to each vector, we input them into the Transformer Encoder.

Transformer Encoder: This module consists of multiple stacked Transformer Blocks, which are shown in Fig. 3(b). Each block includes a Multi-Head Attention Layer and a MLP block. The Multi-Head Attention Layer enables the model to assess the significance of different patches to learn relationships between them and capture dependencies across the image. The MLP block enhances the expressive power of the representation by applying a non-linear activation to the result of the attention mechanism.

Linear: The linear layer reshapes the output of the MLP head into a set of one-dimensional vectors, where each vector's dimensionality matches that of the GPT-2 word embedding vectors.

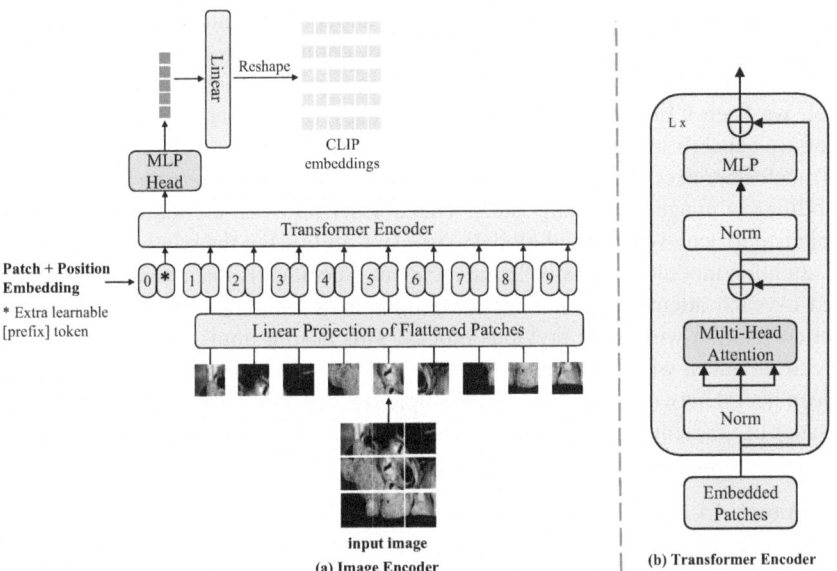

Fig. 3. The Image Encoder and Transformer Encoder

The Encoder forward propagation path is as follows:

$$z_0 = \left[x_{prefix}; x_p^1 E; x_p^2 E; \ldots; x_p^N E \right] + E_{pos} \tag{7}$$

$$z'_\ell = MSA\big(LN\big(z'_{\ell-1}\big)\big) + z'_{\ell-1} \tag{8}$$

$$z_l = MLP\big(LN\big(z'_\ell\big)\big) + z'_\ell \tag{9}$$

$$y = LN(z_L^0) \tag{10}$$

Here, x_p^n represents image patch, E represents the weight matrix of the Linear Projection layer, E_{pos} represents positional encoding, z_l represents the output of the Transformer Encoder and y represents the output of the MLP Head. Finally, we map y through a linear layer to obtain the CLIP embeddings we need:

$$(cp_1^i, \ldots, cp_k^i) = Reshape((LN(y))) \tag{11}$$

After obtaining the CLIP embeddings, we concatenate them with constant embeddings and input them into the Mapping Network F, which is shown in Eq. (3).

3.3 Mapping Network Architecture

Mapping network is the most important part of our work. In this section, CLIP embeddings extracted from image, which contain visual and semantic information, are projected into the GPT-2 space. Mapping Network is the key component for realizing cross-modal capabilities, as it bridges the vision and language spaces. To construct a lightweight model, we keep CLIP and GPT-2 frozen and use Transformer with powerful global expressive capabilities as the mapping network.

Transformer excels at handling text sequence tasks and has demonstrated powerful capabilities in image processing lately. Transformer relies on the multi-head attention mechanism to achieve powerful global expressive ability, but the relationships between each attention module are relatively independent. Specifically, the attention scores generated by each attention module are only influenced by the previous block, without direct correlation with more previous blocks. This design limits the information flow between modules, which may lead to the loss of important information. Furthermore, the deep neural networks composed of multiple stacked Transformer blocks suffer from the issues of gradient vanishing and exploding, which can lead to unstable training and convergence difficulties.

RGFormer: To address the two issues mentioned above, we propose an enhanced scheme for the Multi-Head Attention mechanism, called Residual Multi-Head Attention (RMHA). Additionally, we introduce Gated Linear Units (GLU) into the feed forward layer. Based on these two approaches, we propose RGFormer, an improved Transformer network, as shown in Fig. 4.

Residual Multi-head Attention. Multi-Head Attention distributes attention into various subspaces. In each subspace, a global attention operation is performed, allowing the model to concentrate on various facets of information. Finally, the outputs of each

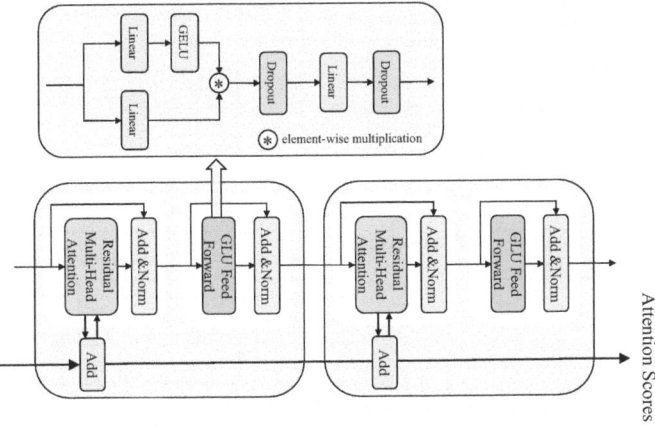

Fig. 4. The structure of our model RGFormer. The main modules are RMHA module and GLU Feed Forward layer

subspace are merged. The specific calculation process is as follows:

$$\begin{cases} Q_i = QW_i^Q, K_i = KW_i^K, V_i = VW_i^V \\ head_i = Attention(Q_i, K_i, V_i) \\ MultiHead(Q, K, V) = Concat(head_1, \ldots, head_n)W^O \end{cases} \quad (12)$$

Here, Q and K are matrices of dimension d_k. V is a matrix of dimension d_v. W_i^Q, W_i^K, W_i^V are projection matrices that project Q, K, and V into the "attention space" of each head. W^O is a weight matrix utilized to linearly transform the outputs from all attention heads. The attention function is typically implemented using scaled dot-product:

$$Attention(Q, K, V) = Softmax(\frac{QK^T}{\sqrt{d_k}})V \quad (13)$$

We propose an improved attention module called Residual Multi-Head Attention (RMHA), aiming to achieve closer collaboration between attention modules. The main idea of this module is to create a "direct" path to propagate the original attention scores, as shown in Fig. 4. Specifically, each Residual Multi-Head Attention module obtains the original attention scores of all attention heads from the previous layer's attention module and adds these "residual scores" to the attention score matrix of the current layer. The final attention scores are then calculated by *Softmax* normalization of the sum of these two scores, as Fig. 5 shows. The specific calculation process is as follows:

$$ResidualAttention(Q, K, V, Prev) = Softmax(\frac{QK^T}{\sqrt{d_k}} + Prev)V \quad (14)$$

$$Prev_N = \frac{QK^T}{\sqrt{d_k}} + Prev_{N-1} \quad (15)$$

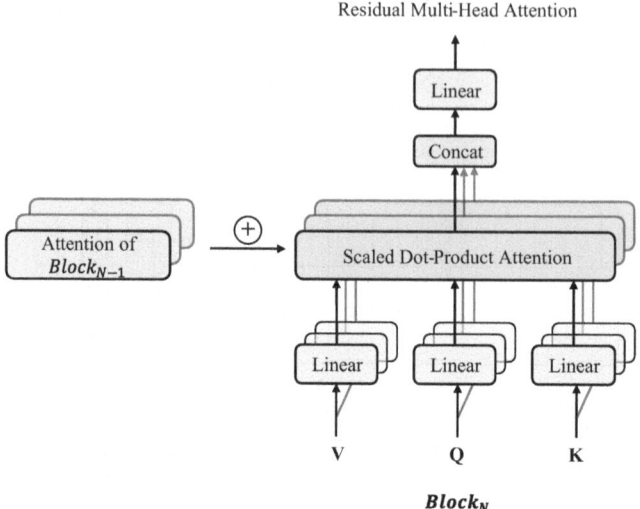

Fig. 5. Residual Multi-Head Attention (RMHA)

Here, *Prev* represents the attention scores from the previous layer's attention module. After completing the residual attention in current layer, we update *Prev* to the final attention scores and pass it to the next. This approach enables each attention module to utilize all attention scores globally, successfully enhanced information interaction between different layers and improved model performance.

GLU Feed Forward Block. Feed Forward Block includes a hidden layer, we replace it with Gated Linear Units (GLU), as shown in Fig. 4. The core idea of GLU is to introduce a gating mechanism to enhance the network's expressive power. Specifically, the forward propagation of GLU is as follows.

$$GLU(x, W_1, W_2, b_1, b_2) = \sigma(xW_1 + b_1) \otimes (xW_2 + b_2) \qquad (16)$$

Here, σ represents the activation function. In our study, we use the GELU activation. $\otimes$ denotes element-wise multiplication. GLU consists of two linear layers, which are parallel to each other, and only one of them has an activation function. During the forward propagation, the input vectors enter the two branches separately, obtaining linear and non-linear results, which are then element-wise multiplied. This approach provides a linear path for gradients while preserving non-linear functionality, which alleviates the problems of gradient vanishing and exploding in the network.

3.4　Inference

During the model's inference process, we utilize the image encoder to extract embeddings from input images and then obtain the visual prefixes through mapping network F. When decoding with GPT-2, we use prefix embeddings as conditions to predict tokens one by one. The language model uses *SoftMax* normalization to determine the probabilities of every vocabulary word for every token. Then we employ beam search to select candidates at each time step based on the probabilities, as shown in Fig. 6, aiming to enhance both diversity and accuracy of generated captions while controlling the search space size.

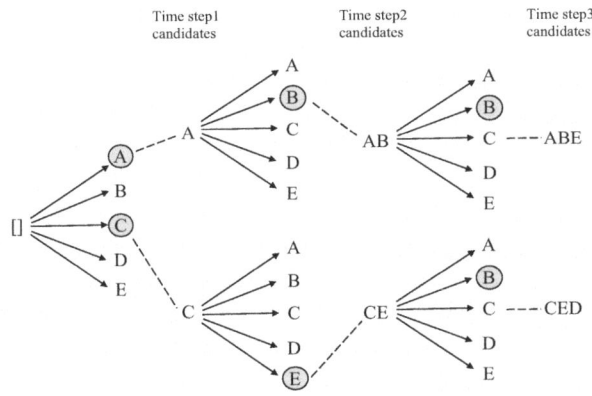

Fig. 6. Beam search algorithm

4 Results

Dataset. We use the COCO Caption dataset [32, 33] and divide it according to the method of Karpathy [34]. The training set contains 120,000 images, with each image corresponding to five captions. The validation set and the test set each consist of 5000 images.

Baselines. We compare CRM with the outstanding work of Anderson et al. [1] (BUTD) and the Vision-Language Pre-training model (VLP) proposed by Zhou et al. [29]. These two models first use a pre-trained object detection network to detect key objects and extract region-level features from the image. Then the obtained visual features are input into a text decoder to generate text sequences. BUTD uses a Long Short-Term Memory (LSTM) network for decoding, while VLP uses Transformer. The training method of VLP is similar to BERT [27], both utilizing millions of image-text pairs for extensive pre-training.

Evaluation Metrics. We validate our model's performance on the COCO Caption dataset using common metrics in image captioning tasks, including SPICE [21], BLEU [35], METEOR [36], and CIDEr [37]. Additionally, we compute the trainable parameters and training time for each model to confirm the lightweight effect of our proposed model. In complex visual-language cross-modal tasks like image captioning, model training often requires a long time, and fine-tuning models on different datasets can be challenging. By designing lightweight models, we can reduce the demand for computational resources while maintaining good performance, thereby lowering the training cost. Moreover, lightweight models are also easier to transfer to different datasets or scenarios, which could accelerate the application and deployment of the model.

Quantitative Evaluation. We provide the quantitative evaluation results of BUTD, VLP, and our model on the COCO Caption dataset as shown in Table 1. From the results, we can consider that our model's performance is very close to VLP and BUTD while requiring significantly less training time. Additionally, our model also boasts a smaller number of trainable parameters. Note that pre-training steps are not included in the VLP training time listed here. For instance, Zhou et al. [29] spent 1200 GPU hours pre-training VLP on the Conceptual Captions dataset.

Through quantitative analysis, we conclude that our model can achieve results comparable to advanced models with much less training time. This indicates that our model can be trained quickly and generalized to various datasets or application scenarios.

Table 1. Quantitative evaluation the COCO Caption dataset

Model	BLEU-4	METEOR	CIDEr	SPICE	Params (M)	Training Time
BUTDn [1]	36.2	27.0	113.5	20.3	52	960 h (M40)
VLP [29]	36.5	28.4	117.7	21.3	115	48 h (V100)
Ours	33.61	27.85	114.60	21.67	46	4 h (GTX3090Ti)

Qualitative Evaluation. Figure 7 presents the unfiltered captions generated by our model for randomly selected images from the COCO Caption test set. It's evident that these captions accurately depict the content of the images. It is noteworthy that our model is capable of recognizing some uncommon objects. For example, our model recognizes the "double decker bus" in Fig. 7, even though the Ground Truth does not mention it. This is attributed to the utilization of the CLIP image encoder pre-trained on a large and diverse dataset. Additionally, our model exhibits a degree of associative capability. Based on the identification of the object "toilet", it can associate with "in a bathroom", thereby enriching the semantic information contained in generated captions. However, we can also observe shortcomings in the model's performance. At times, it fails to recognize certain secondary objects in the images, such as "glasses" or "a cap". In contrast, VLP outperforms our model in this aspect. VLP employs the Faster R-CNN object detection network to extract region-level features from images, making it more adept at capturing such secondary objects.

In Fig. 1, we demonstrate the captions generated by our model for images collected from the internet. It is evident that our model exhibits outstanding generalization capability. To further evaluate the model's capacity to generalize across novel scenarios, we collected additional images with more complex content scenes and utilized our model to generate captions, as depicted in Fig. 8. It can be observed that when faced with images featuring a greater number of objects and more intricate content, our model is still capable of generating accurate and meaningful captions.

Prefix Length: Li et al. [9] pointed out that increasing the prefix length can enhance the model's performance in downstream tasks. After reaching a specific value, the effect of the model tends to stabilize, and this saturation value varies depending on the task. For the image captioning task, we explore the impact of different prefix lengths on model

Ground Truth	Man in glasses and a cap sitting on the toilet.	A big cat laying down in a chair on a porch.	A small dog is curled up on top of the shoes.	Two red busses are traversing next to a tall brick building.	Three cows walking along the street in a town.
BUTD	A smiling man sitting on a toilet.	A cat siting in a blue chair.	A dog sleeping with some shoes.	Red buses driving down a street.	Two cows walking along the street.
VLP	A man in a hat sitting on a toilet.	A cat sitting on top of a chair.	A dog sleeping in the shoes.	Two red busses are driving on a street.	A few cows walking down the street in a town.
CRM	A man sitting on a toilet in a bathroom.	A cat is siting outside on a chair.	A dog laying on top of the shoes.	A red double decker bus driving down a street.	Black and white photo-graph of cows walking down a street.

Fig. 7. Unfiltered results of three images randomly selected from the COCO test set (Karpathy et al. [34] split).

performance. We trained CRM with different prefix lengths for 4 epochs each on the COCO Caption dataset and evaluated them on the training and test sets using BLEU [35] and CIDEr [37] scores.

As shown in Fig. 9, small prefix lengths cannot express sufficient information, resulting in poorer results. When the prefix length is increased to 40, the model's performance on the train and test sets stabilizes, reaching a saturation point. Further increasing the prefix length does not significantly improve performance but wastes computational resources. Due to the substantial memory overhead incurred by Transformer when calculating attention weights, we set the maximum prefix length to 80 in our experiments.

Ablation Experiment: We perform ablation experiment to confirm the effectiveness of RGFormer, as presented in Table 2. We refer to the network using original Transformer as CTM. From the results of quantitative experiments, we can see that both RMHA and GLU contribute to enhancing CTM performance, with RMHA demonstrating a more pronounced effect.

BUTD: A group of people are riding skateboards

VLP: A group of kids riding skateboards on a wooden ramp

CRM: A group of young skateboarders pose for a picture

BUTD: There are some penguins and a man in a hat

VLP: A group of penguins are standing in the snow

CRM: A man standing in the snow with penguins around

BUTD: A street with many stores and a bicycle

VLP: A person walking down a street at night

CRM: A person walking down a narrow alley at night

BUTD: A man in a black hat swinging a bat

VLP: A baseball player swinging a bat at a ball

CRM: A baseball player swinging a bat during a game

Fig. 8. Captions generated for images with complex scenes

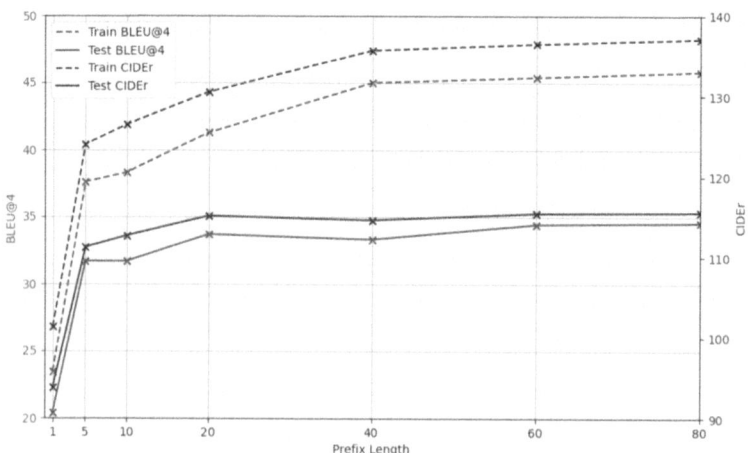

Fig. 9. The impact of different prefix lengths on model performance.

Table2. Ablation experiment of RGFormer

Model	BLEU-4	METEOR	CIDEr	SPICE
CTM	33.53	27.45	113.08	21.05
CTM + RMHA	33.59	27.73	114.27	21.56
CTM + GLU	33.50	27.62	113.49	21.23
CTM + RGFormer (CRM)	33.61	27.85	114.60	21.67

5 Conclusion

Overall, we propose a lightweight and efficient image captioning model named CRM,which is built upon the combination of CLIP and RGFormer. This model does not require additional supervision or pre-training and achieves comparable results to some advanced methods (such as BUTD and VLP) in a shorter training time. We propose RGFormer, based on the Residual Multi-Head Attention module and the Gated Linear Unit, as an improved Transformer architecture. RMHA enhances the accuracy of global attention by connecting individual attention blocks. GLU alleviates the problems of gradient vanishing and explosion through linear and non-linear combinations. Benefiting from a strong pre-trained model, CRM demonstrates excellent generalization performance not only on the COCO Caption dataset but also on other datasets and random images. This characteristic facilitates convenient deployment across various application scenarios.

Acknowledgement. This work was supported by the National Natural Science Foundation of China (52202417); China Postdoctoral Science Foundation (2022TQ0155, 2022M721605); Open Project Program of State Key Laboratory of Virtual Reality Technology and Systems, Beihang University (VRLAB2023A02); Young Elite Scientists Sponsorship Program by CAST (2023QNRC001); Young Elite Scientists Sponsorship Program by JSTJ (JSTJ-2023-XH032). Authors thank reviews for their valuable comments.

References

1. Anderson, P., et al.: Bottom-up and top-down attention for image captioning and visual question answering. In: Proceedings of the IEEE Conference on Computer Vision and Pattern Recognition, pp. 6077–6086 (2018)
2. Chen, X., Zitnick, C.L.: Learning a recurrent visual representation for image caption generation. arXiv preprint arXiv:1411.5654 (2014)
3. Li, X., et al.: OSCAR: object-semantics aligned pre-training for vision-language tasks. In: Vedaldi, A., Bischof, H., Brox, T., Frahm, JM. (eds.) ECCV 2020. LNCS, vol. 12375, pp. 121–137. Springer, Cham (2020). https://doi.org/10.1007/978-3-030-58577-8_8
4. Stefanini, M., Cornia, M., Baraldi, L., Cascianelli, S., Fiameni, G., Cucchiara, R.: From show to tell: a survey on deep learning-based image captioning. IEEE Trans. Pattern Anal. Mach. Intell. **45**(1), 539–559 (2022)

5. Tan, H. and Bansal, M.: LXMERT: learning cross-modality encoder representations from transformers. arXiv preprint arXiv:1908.07490 (2019)
6. Xu, K., et al.: Show, attend and tell: neural image caption generation with visual attention. In: International Conference on Machine Learning, pp. 2048–2057. PMLR (2015)
7. Yao, T., Pan, Y., Li, Y., Mei, T.: Exploring visual relationship for image captioning. In: Proceedings of the European Conference on Computer Vision (ECCV), pp. 684–699 (2018)
8. Radford, A., et al.: Learning transferable visual models from natural language supervision. In: International Conference on Machine Learning, pp. 8748–8763. PMLR (2021)
9. Li, X.L., Liang, P.: Prefix-tuning: optimizing continuous prompts for generation. arXiv preprint arXiv:2101.00190 (2021)
10. Radford, A., Wu, J., Child, R., Luan, D., Amodei, D., Sutskever, I.: Language models are unsupervised multitask learners. OpenAI blog **1**(8), 9 (2019)
11. Dosovitskiy, A., et al.: An image is worth 16×16 words: transformers for image recognition at scale. arXiv preprint arXiv:2010.11929 (2020)
12. Simonyan, K., Zisserman, A.: Very deep convolutional networks for large-scale image recognition. arXiv preprint arXiv:1409.1556 (2014)
13. He, K., Zhang, X., Ren, S., Sun, J.: Deep residual learning for image recognition. In: Proceedings of the IEEE Conference on Computer Vision and Pattern Recognition, pp. 770–778 (2016)
14. Ren, S., He, K., Girshick, R., Sun, J.: Faster R-CNN: towards real-time object detection with region proposal networks. In: Advances in Neural Information Processing Systems, vol. 28 (2015)
15. Conde, M.V., Turgutlu, K.: CLIP-art: contrastive pre-training for fine-grained art classification. In: Proceedings of the IEEE/CVF Conference on Computer Vision and Pattern Recognition, pp. 3956–3960 (2021)
16. Zhang, R., et al.: Tip-adapter: training-free adaption of clip for few-shot classification. In: Avidan, S., Brostow, G., Cissé, M., Farinella, G.M., Hassner, T. (eds.) ECCV 2022. LNCS, vol. 13695, pp. 493–510. Springer, Cham (2022). https://doi.org/10.1007/978-3-031-19833-5_29
17. Liu, H., et al.: CMA-clip: cross-modality attention clip for image-text classification. arXiv preprint arXiv:2112.03562 (2021)
18. Vidit, V., Engilberge, M., Salzmann, M.: Clip the gap: A single domain generalization approach for object detection. In: Proceedings of the IEEE/CVF Conference on Computer Vision and Pattern Recognition, pp. 3219–3229 (2023)
19. Teng, Z., Duan, Y., Liu, Y., Zhang, B., Fan, J.: Global to local: clip-LSTM-based object detection from remote sensing images. IEEE Trans. Geosci. Remote Sens. **60**, 1–13 (2021)
20. Gu, X., Lin, T.Y., Kuo, W., Cui, Y.: Open-vocabulary object detection via vision and language knowledge distillation. arXiv preprint arXiv:2104.13921 (2021)
21. Anderson, P., Fernando, B., Johnson, M., Gould, S.: SPICE: semantic propositional image caption evaluation. In: Leibe, B., Matas, J., Sebe, N., Welling, M. (eds.) ECCV 2016, Part V. LNCS, vol. 9909, pp. 382–398. Springer, Cham (2016). https://doi.org/10.1007/978-3-319-46454-1_24
22. Wang, W., Chen, Z., Hu, H.: Hierarchical attention network for image captioning. In: Proceedings of the AAAI Conference on Artificial Intelligence, pp. 8957–8964 (2019)
23. Ji, J., et al.: Improving image captioning by leveraging intra-and inter-layer global representation in transformer network. In: Proceedings of the AAAI Conference on Artificial Intelligence, pp. 1655–1663 (2021)
24. Luo, Y., et al.: Dual-level collaborative transformer for image captioning. In: Proceedings of the AAAI Conference on Artificial Intelligence, pp. 2286–2293 (2021)
25. Hochreiter, S., Schmidhuber, J.: Long short-term memory. Neural Comput. **9**(8), 1735–1780 (1997)

26. Vaswani, A., et al.: Attention is all you need. In: Advances in Neural Information Processing Systems, vol. 30. (2017)
27. Devlin, J., Chang, M. W., Lee, K., Toutanova, K.: BERT: pre-training of deep bidirectional transformers for language understanding. arXiv preprint arXiv:1810.04805 (2018)
28. Zhang, P., et al.: VinVL: revisiting visual representations in vision-language models. In: Proceedings of the IEEE/CVF Conference on Computer Vision and Pattern Recognition, pp. 5579–5588 (2021)
29. Zhou, L., Palangi, H., Zhang, L., Hu, H., Corso, J., Gao, J.: Unified vision-language pre-training for image captioning and VQA. In: Proceedings of the AAAI Conference on Artificial Intelligence, pp. 13041–13049 (2020)
30. Lu, J., Batra, D., Parikh, D., Lee, S.: VilBERT: Pretraining task-agnostic visiolinguistic representations for vision-and-language tasks. In: Advances in Neural Information Processing Systems, vol. 32 (2019)
31. Dauphin, Y.N., Fan, A., Auli, M., Grangier, D.: Language modeling with gated convolutional networks. In International Conference on Machine Learning, pp. 933–941. PMLR (2017)
32. Chen, X., et al.: Microsoft COCO captions: data collection and evaluation server. arXiv preprint arXiv:1504.00325 (2015)
33. Lin, T. Y., et al.: Microsoft COCO: common objects in context. In: Fleet, D., Pajdla, T., Schiele, B., Tuytelaars, T. (eds.) ECCV 2014, Part V. LNCS, vol. 8693, pp. 740–755. Springer, Cham (2014). https://doi.org/10.1007/978-3-319-10602-1_48
34. Karpathy, A., Fei-Fei, L.: Deep visual-semantic alignments for generating image descriptions. In: Proceedings of the IEEE Conference on Computer Vision and Pattern Recognition, pp. 3128–3137 (2015)
35. Papineni, K., Roukos, S., Ward, T., Zhu, W.J.: BLEU: a method for automatic evaluation of machine translation. In: Proceedings of the 40th Annual Meeting of the Association for Computational Linguistics, pp. 311–318 (2002)
36. Denkowski, M., Lavie, A.: Meteor universal: language specific translation evaluation for any target language. In: Proceedings of the Ninth Workshop on Statistical Machine Translation, pp. 376–380 (2014)
37. Vedantam, R., Lawrence Zitnick, C., Parikh, D.: CIDEr: consensus-based image description evaluation. In: Proceedings of the IEEE Conference on Computer Vision and Pattern Recognition, pp. 4566–4575 (2015)

Knowledge Graph Construction Method of Bridge Design Codes Based on Ontology and Specification Parsing

Zheng Zhang[1], Qingsong Ai[1], Junwei Yan[1], Jun Yang[1(✉)], Wei Meng[1], Quan Liu[1], and Zude Zhou[2]

[1] School of Information Engineering, Wuhan University of Technology, Wuhan, China
junyang_ie@whut.edu.cn
[2] School of Mechanical and Electronic Engineering, Wuhan University of Technology, Wuhan, China

Abstract. Bridge design codes are important references for bridge engineering, as well as the basis for evaluating the quality of bridge design and construction. However, the specification files have complex semantic structures which make semantic parsing more difficult and limit the further development of intelligent engineering construction. To reduce the difficulty of parsing specification documents in engineering construction, this article proposes a method for constructing a knowledge graph of bridge design codes to achieve knowledge structuring. It has established ontology based on bridge design codes and achieved specification paring combining ontology and natural language processing (NLP) technology. In this way, the difficulty of interpreting design codes in bridge engineering construction can be reduced and engineering efficiency can be improved.

Keywords: bridge design code · knowledge graph · ontology · NLP

1 Introduction

There are numerous national codes and industry norms in the construction industry, which play an important role in evaluating the quality of construction projects. However, the existing research on interpretation of specifications is mostly aimed at the fire safety requirements of civil residential buildings. The research on traffic facilities and building design codes is not deep enough. Structured knowledge representation methods such as knowledge graph can help computer systems store and understand codes, reducing the workload of interpreting design codes. Therefore, this article studies the construction method of a bridge design code knowledge graph.

Knowledge graph originates from the semantic web and has been widely applied in the fields of artificial intelligence and Natural Language Processing (NLP) since it was proposed by Google in 2012 [1]. It can represent structured knowledge and can better showcase things and their relations. Knowledge graphs are usually divided into general knowledge graphs and domain knowledge graphs [2]. Although the application of general

S. Saito et al. (Eds.): AsiaSim 2024, CCIS 2170, pp. 58–69, 2024.
https://doi.org/10.1007/978-981-97-7225-4_5

knowledge graphs has been limited with the rise of large language models, domain knowledge graphs still have practical significance due to their stronger interpretability and reliability [3]. Ontologies provide a way of formalizing human knowledge to enable machine interpretability [4]. It has promoted the emergence of the semantic web and played an important role in knowledge engineering and NLP.

By constructing a bridge design code knowledge graph based on ontology and text parsing, it is possible to parse specification texts with complex semantics. The study extracts and structures concepts and constraint information involved in bridge engineering, integrate bridge design code information, and can enhance the intelligence level of bridge engineering. Because the steel-concrete composite structures can achieve step-by-step production and assembly, they are suitable objects for intelligent construction in the transportation industry. This article will construct an ontology based on the Chinese national standard "Code for design of steel and concrete composite bridges" (GB50917-2013) and use the Neo4j graph database to construct and apply a knowledge graph of bridge design codes.

2　Related Works

A knowledge graph stores knowledge in a structured triplet form, which can represent the subject, predicate, and object respectively. The knowledge graph can be represented as (1):

$$G = \{E, R, T, F_k\} \tag{1}$$

where G represents the knowledge graph, E is the set of entities, R is the set of relations, T is the set of triples in the knowledge graph, and F_k represents the set of rules formed by background knowledge [5]. T can be expressed as (2):

$$T = \{(h, r, t) | h, t \in E\} \tag{2}$$

where T is the set of triples, h is the head entity, t is the tail entity, r is the relation, and E is the set of entities [6]. Triples are represented in the form of (s, p, o) triples consisting of subject, predicate, and object [7]. They are typically presented in the form of node-edge-node in graph databases.

With the development of deep learning, pretrained models represented by BERT [8] are widely used for the construction and application of knowledge graph. Although knowledge graphs have made progress in many fields such as biomedicine, practical experience and electronic databases are still primarily relied on in the fields of construction and transportation [9]. There are several studies on building knowledge graph based on code files. For example, [10] proposed a semi-automatic construction method for design code knowledge graphs to analysis the structure and content of code texts. The code text review method based on ontology proposed in [11] can obtain the SPARQL query based on NLP and national fire prevention code text parsing, and use it for the review of building model. In [12], a knowledge graph construction method based on mandatory fire safety regulations was proposed, which defined the pattern layer using the IFC standard as a reference.

These works have achieved diverse results in the semantic parsing of specifications and evaluation of BIM. On the other hand, the above research mainly focuses on civil residential buildings and building fire safety review, while the research on transportation infrastructure and review of other specifications is not enough. In terms of knowledge graph construction, research focuses on semantic parsing and normative review methods, which is not suitable for direct application in the field of knowledge graph construction in transportation facilities. Therefore, this article will implement the construction of ontology and knowledge graph based on the parsing of transportation facility related code documents.

3   Construction Method of Knowledge Graph

3.1   Construction and Application Process of Knowledge Graph for Bridge Design Code

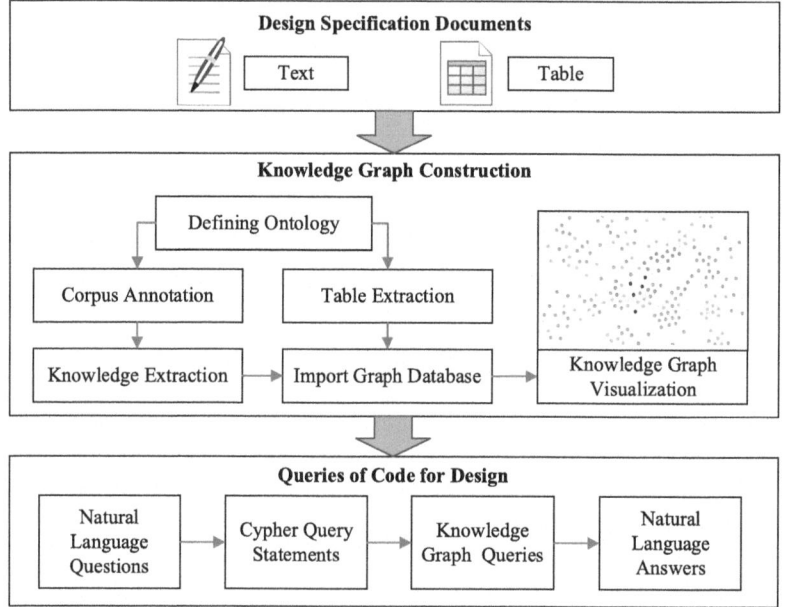

Fig. 1. Construction and application framework of knowledge graph for bridge design codesy.

Design Code Information Acquisition. The construction and application of a knowledge graph for bridge design codes require obtaining information from bridge design specification documents. The information in the file mainly exists in the form of text and tables, and needs to be extracted separately. The text information needs to be annotated with corpus, and knowledge extraction should be carried out based on the annotated entities and relations (Fig. 1).

Knowledge Graph Construction Based on Ontology. Firstly, define the ontology of the knowledge graph for bridge design codes. Determine the logical structure of entities, concepts, and relations in the knowledge graph based on the professional domain knowledge in the design specifications and construct the pattern layer of the knowledge graph. Next, determine the annotated objects, annotate text corpus and extract information based on the ontology. Then construct the data layer, import entities, relations, and attributes into Neo4j graph database to achieve the construction and visualization of the knowledge graph.

Knowledge Graph Query Based on NLP. Build a bridge design code query system based on NLP. Firstly, parse the input statements by intent recognition and named entity recognition. Then convert natural language questions into Cypher statements, and query in the bridge design code knowledge graph. Finally, based on the knowledge graph query results and matching with the question answering template to generate answers.

3.2   Knowledge Graph Construction Process Based on Ontology

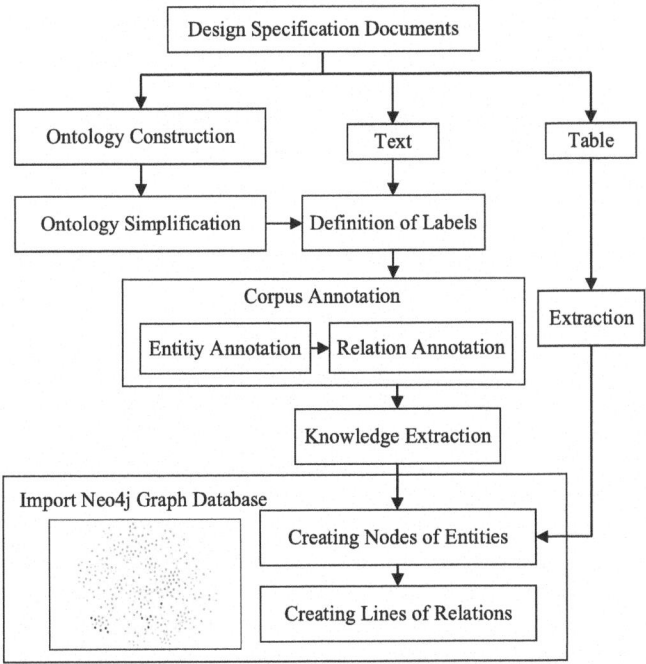

Fig. 2. The construction process of knowledge graph for bridge design codes.

The flowchart for constructing the knowledge graph is shown in Fig. 2. Firstly, construct an ontology based on the content of design codes and establish the pattern layer of the knowledge graph. Next, simplify the ontology and annotate the text with entities and relations based on the simplified ontology. Then extract entity and relations between table and text separately, import the entity and relations of the knowledge graph into the graph database to form nodes and lines, and complete the construction of the data layer. Finally, optimize and adjust the structure of the knowledge graph to complete the construction of the knowledge graph.

Firstly, build an ontology based on the content of bridge design codes and sort out a reasonable logical structure to complete the preliminary construction of the ontology. It is necessary to extract appropriate entity concept relationships from them and sort out a reasonable logical structure to complete the preliminary construction of the design specification ontology. Next, simplify the ontology in order to better adapt to the storage structure of the knowledge graph. The ontology obtained in the first step that meets the professional requirements of the bridge industry is relatively complex, and after simplification, it can better adapt to the storage structure of the knowledge graph. Then, annotate corpus based on defined ontology. Due to the lack of relevant professional datasets, this article needs to annotate the corpus of bridge design codes in order to obtain the data sources of the knowledge graph. After that, extract the entity relationships contained in the text based on the annotated corpus, and extract knowledge by reading the entity concepts and relationships in the text. Then build the knowledge graph based on the extracted entities and relations, and import the created entities, concepts, relations, and attributes into the Neo4j graph database. Import the extracted entities into the graph database in the form of nodes firstly, and then import the relationships in the form of edges. Finally, implement the visualization of knowledge graph based on Neo4j. In the Neo4j graph database, knowledge graphs are presented in the form of a network of nodes and edges, where nodes represent entities or concepts, and edges represent relationships between nodes.

3.3 Construction and Simplification of the Knowledge Graph Pattern Layer

The logical structure of a knowledge graph is divided into a pattern layer and a data layer. The pattern layer defines the logical structure and stores conceptual knowledge, the data layer stores instance data [13]. The process of constructing a knowledge graph is usually divided into two ways: top-down and bottom-up. Top-down refers to building the pattern layer first, and then building the data layer under the guidance of the pattern layer; bottom-top constructs the data layer first, and then extract conceptual information from the data layer to obtain the pattern layer [14]. In these two methods, the former is mostly used for constructing domain knowledge graphs, while the latter is mainly used for constructing general knowledge graphs. The knowledge graph of bridge design codes belongs to domain knowledge graphs in transportation field, which involves many professional concepts that are not suitable for automated extraction and definition methods. Therefore, to ensure professionalism and rigor, top-down approach is selected. Firstly, construct an ontology based on the content of code file and establish the pattern layer. Then, extract entities and relations from table information and text information separately based on the requirements, import the entities and relations into the graph database to form nodes and relations, and complete the construction of the knowledge graph data layer.

This article uses Protégé as a tool to construct a knowledge graph pattern layer ontology. It constructs an ontology based on the logical layering and association between

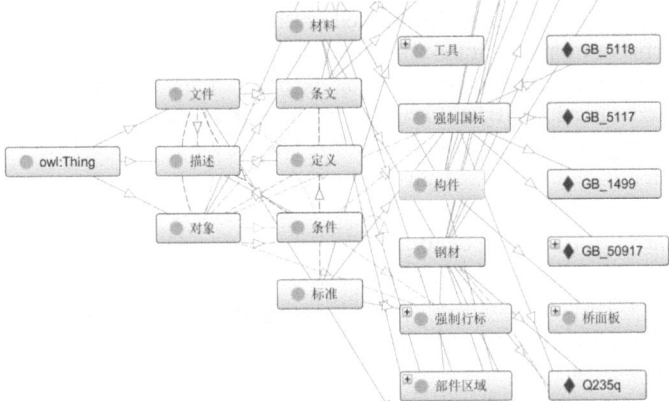

Fig. 3. A part of the preliminary constructed ontology.

different concepts. A part of the preliminary constructed ontology is shown in the Fig. 3. By defining the hierarchical relationships between concept classes, the reference relationships between concept classes and object attributes, and data attributes, it can express multiple semantics more rigorously.

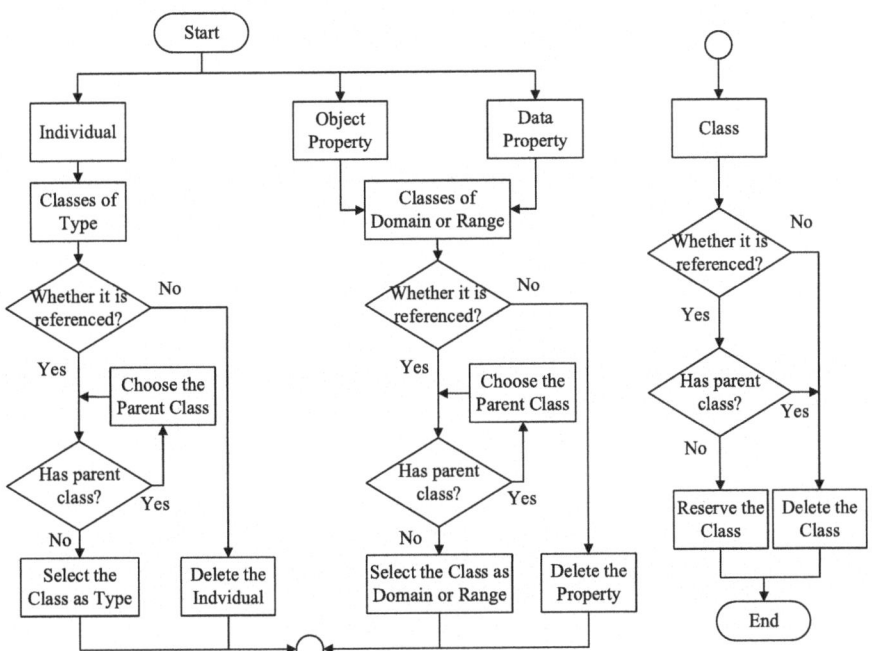

Fig. 4. Simplified process of ontology.

The concept ontology constructed based on the terminology in the normative document can achieve more accurate semantic parsing of normative text content, so the pattern ontology can be constructed based on the professional terminology concepts and relationships contained in GB50917. In GB50917, the requirements that need to be followed in the design of steel-concrete composite bridges are explained. After completing the construction of the bridge design code ontology in Protégé, rule reasoning can be carried out based on the inference machine and rules, achieving query and preliminarily completing the construction of the bridge design code knowledge graph. However, due to the complexity of the ontology structure and the use of SPARQL for queries, the function of the knowledge graph at this stage needs improvement. Thus, it is necessary to construct the data layer and achieve visualization of the knowledge graph through Neo4j graph database.

The logical structure of the original ontology constructed by Protégé is relatively complex, so it is necessary to transform the ontology based on a more suitable logical form for the Neo4j graph database, and then build a knowledge graph based on the transformed ontology. The process of ontology simplification is shown in Fig. 4, it requires merging conceptual classes, object attributes, and data attributes, and subsequently changing the attributes of entities. In this study, the top-level concepts in the original ontology are used as concept labels in the graph database, while concepts at other levels are used as nodes which connects to instance nodes through relations. The concepts and attributes are defined based on the simplified ontology.

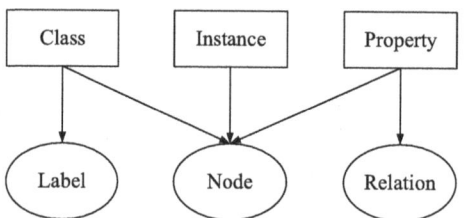

Fig. 5. Conversion logic for importing graph databases.

The conversion logic for constructing a knowledge graph based on Neo4j graph database storage based on the Protégé ontology is shown in Fig. 5. The concepts defined in the ontology are mainly divided into classes, instances, and property attributes. Through strict rule constraints and complex relation transmission, concepts and entities in knowledge are layered and classified, which can rigorously reflect the hierarchical relations between concepts. However, cumbersome relations and complex nesting are redundant and difficult to understand, while Neo4j based storage forms are relatively intuitive. In the Neo4j graph database, data is stored in the form of Labels, Nodes, and Relation edges. Labels represent classes and concepts to which a node belongs. Nodes generally represent entities, but can also represent concepts and attributes. Relations are connections between entities.

3.4   Construction of the Data Layer and Visualization of the Knowledge Graph

After the construction of the pattern layer, it is necessary to import entities and relations into the Neo4j graph database to construct the data layer of the knowledge graph. During the construction process, nodes and edges are created using Cypher statements. There are still some problems of missing entities and relations, as well as logical confusion, in the knowledge graph based on ontology. Therefore, it is necessary to optimize the concepts and unreasonable logical structures in the knowledge graph. Compared with the emphasis on concepts, constraints, and logical structures in ontology, knowledge representation in a knowledge graph emphasizes more on the triple formed by entity-relation-entity, which is mainly presented in the form of node-edge-nodes in the Neo4j graph database. Finally, optimize and adjust the structure of the knowledge graph.

4   Construction and Application Experiment of Knowledge Graph

4.1   Data Sources and Corpus Annotation

The knowledge graph of bridge design codes is based on national standard documents in the field of transportation facilities. This article uses the Chinese national standard GB50917 "Code for design of steel and concrete composite bridges" as the main data source to construct the knowledge graph of bridge design codes. GB50917 consists of 8 chapters with 128 clauses. This study annotates entities and relations for the code, and uses the annotated files as data for constructing the knowledge graph. It combines relevant domain data on the network with the parameters of some transportation facility models to construct the data layer. There are contents such as text, tables, and images in the specification files. This study mainly focuses on text and tables, and has not yet considered the extraction of image information. For structured table data, information extraction can be directly achieved through the extraction and transformation process. For text, data annotation is required first, followed by entity relation extraction and other operations to import the data into the knowledge graph.

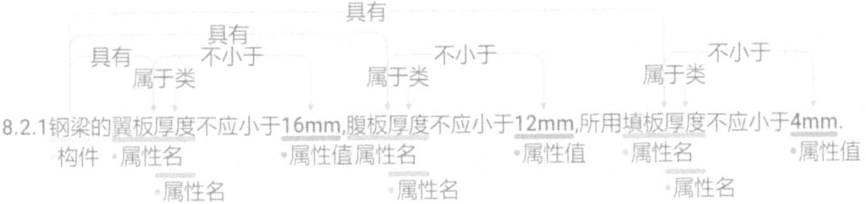

Fig. 6. Example of entity relationship annotation for GB50917.

This article uses the Doccano annotation tool for entity and relation annotation, and then imports the entities and relations into the graph database through a Python program. The example of entity relationship annotation is shown in Fig. 6, selected from clause 8.2.1 of GB50917-2013. It says that *"The thickness of the wing plate of the steel beam should not be less than 16 mm, the thickness of the web plate should not be less than*

12 mm, and the thickness of the filling plate used should not be less than 4 mm", where the entity "steel beam" belongs to the "component" class, the entity "thickness of the web" belongs to the "attribute name" class, and the entity "steel beam" has a "has" relationship pointing to the entity "thickness of the web". According to the transformed pattern layer logic, entities are divided into seven categories: bridges, definitions, standards, components, materials, attribute names and values. Other concepts and instances are classified into these six categories. When performing semantic annotation, the first step is to annotate the entities that belong to the six categories in the text. Then, annotate relation to establish relations between the entities. The annotation uses the BIO annotation method, using a single Chinese character as the basic element for annotation. After annotation, the JSONL file is obtained. Use a Python program to parse the JSONL file and import the annotated entities and relations into Neo4j to obtain a preliminary constructed knowledge graph.

4.2 Construction and Application Process of Knowledge Graph for Bridge Design Code

After importing entities and relations into the Neo4j graph database, entities and relations are stored in the graph database in the form of nodes and edges. The visualization results of the bridge design specification knowledge graph in the Neo4j graph database are divided into 7 categories with a total of 308 nodes and 412 edges. The information and some examples of nodes are shown in the Table 1.

Table 1. Information of nodes in the Neo4j graph database.

Label of Nodes	Example	Number
Bridge	Beam bridge, large bridge, composite bridge	11
Specification codes	GB50917-2013, JTGD62, GB/T700	17
Definition	Definition of material strength standard values	18
Component	Beam, continuous beam, bridge deck	37
Material	Steels, prestressed concrete, Q235 steel	31
Attribute Name	Geometric parameters, section form, tensile strength	132
Attribute Value	40mm, Class A, 95%, 100 years, 2.5, C30	62

4.3 Knowledge Graph Query System Based on NLP

In the bridge design code query system, the input question is converted into a Cypher query statement, realizing a query based on knowledge graph, and outputting the query results. The system flowchart is shown in Fig. 7.

It uses the BERT model for intent recognition of input statements. Sequence vectors are input into a bidirectional Transformer for feature extraction. Based on self-attention

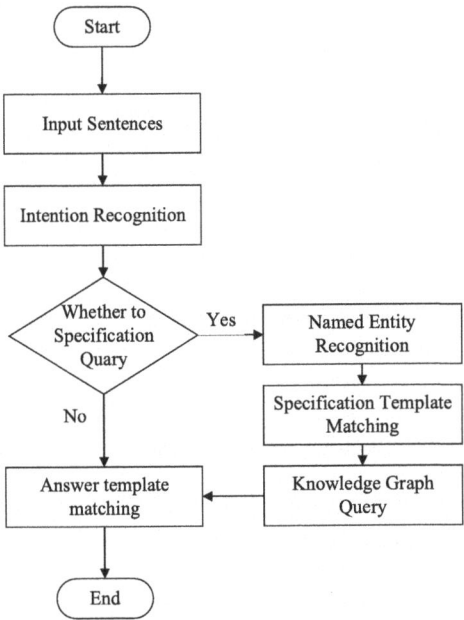

Fig. 7. Query process of knowledge graph for bridge design codes.

mechanism, the Transformer can adjust weights according to the association of words, as shown in (3), where Q is the query matrix, K is the key matrix, and V is the value matrix.

$$Attention(Q, K, V) = softmax(\frac{QK^T}{\sqrt{d_k}})V \tag{3}$$

Classify question types based on the attention score of the CLS identifier corresponding to the token in the statement, obtain scores for different categories of the question, and select the category corresponding to the highest score as the candidate category. When the score exceeds the threshold, select an answer template based on the candidate class question and query the knowledge graph for answers. If the score does not exceed the threshold, it is considered that the problem has a low correlation with the bridge field and no knowledge graph query is performed.

This study is based on BiLSTM for entity extraction of text, and CRF is used to constrain the results. BiLSTM can simultaneously process contextual information and reduce the problems of gradient vanishing and exploding that RNN may encounter. The output of the BiLSTM model serves as the input of the CRF layer, which can effectively constrain the label results and avoid improper label sequences as the output. The CRF layer score is calculated as (4), for the input text sequence X, the score S for the label sequence Y is:

$$S(X, Y) = \sum_{i=1}^{n} P_{iy_i} + \sum_{i=0}^{n} T_{y_i y_{i+1}} \tag{4}$$

P is the emission score matrix, and T is the transition score matrix. For the input sequence X, select the label with the highest score S as the output for named entity recognition.

4.4 Knowledge Graph Query Process and Results

When querying, select answers based on the question and answer template of design codes, and use Cypher statements based on the Neo4j to query the knowledge graph. Combine the query results with the selected answers as the final reply. Firstly, the input natural language question is segmented, and then the segmentation result is matched with the domain terms stored in the dictionary to obtain the query entity. Select the query template based on the type of question, and use the query entity as a node to query in the bridge design code knowledge graph to obtain entities related to answer. Finally, the answer is generated by combining the entity with the question type. An example of a Cypher query is shown below.

Question:

What is the range of concrete strength grade for reinforced concrete components?

Cypher query:

MATCH(p: concrete components)-[r: has]-> (q: property_name) -[r1]-> (s: property_value) WHERE q.name = 'concrete strength grade' RETURN type(r1), s.name

During the query process, Cypher statements are obtained based on the input natural language questions. After querying in the knowledge graph, the query results are combined with the answer template to generate natural language answers.

In specific professional field question and answer, the method based on knowledge graph can ensure the accuracy of answers compared to ChatGPT's possible fabricated answers. In our test, the query based on the bridge code knowledge graph can query and answer the references of codes, the definitions of components, and the attribute constraints of entities. On the contrary, ChatGPT mistakenly stated the meaning of GB50917-2013, and fabricated its referenced code number and name, which may be misleading in the engineering. Moreover, professional queries based on knowledge graphs do not answer questions unrelated to the specific field. Therefore, the answer based on the knowledge graph of bridge specifications has higher reliability.

5 Conclusion

This article proposes a method for constructing a knowledge graph of bridge design code based on ontology and specification parsing. It constructs a knowledge graph through specification parsing and ontology construction, realizing the structuring and integration of information for bridge design codes. Based on triplets and graph databases, it realizes the storage and display of conceptual and semantic information. By converting natural language text into data stored in the graph database, it structures the text of codes. And the difficulty of querying specifications in bridge design and construction processes is reduced by queries based on knowledge graph. By using the method proposed in this article, a knowledge graph of bridge design codes can be constructed. Applying the constructed knowledge graph to the design and review process of bridges can reduce the possibility of non-compliant design schemes and provide guidance for the modification and optimization of design schemes. The research deficiency lies in mainly using manual ontology construction methods. If combined with automatic generation methods and large language models, and integrated with BIM, it is possible to generate larger and more practical knowledge graphs for different codes and applying them to engineering.

Acknowledgments. This work is supported by the National Key Research and Development Program of China under grants 2021YFB2600302.

Disclosure of Interests. The authors have no competing interests to declare that are relevant to the content of this article.

References

1. Wu, T., Qi, G., Li, C., Wang, M.: A survey of techniques for constructing Chinese knowledge graphs and their applications. Sustainability **10**(9), 3245 (2018)
2. Lin, J., Zhao, Y., Huang, W., Liu, C., Pu, H.: Domain knowledge graph-based research progress of knowledge representation. Neural Comput. Appl. **33**(2), 681–690 (2021)
3. Pan, S., Luo, L., Wang, Y., Chen, C., Wang, J., Wu, X.: Unifying large language models and knowledge graphs: a roadmap. IEEE Trans. Knowl. Data Eng. (2024). https://doi.org/10.1109/TKDE.2024.3352100
4. Lonsdale, D., Embley, D.W., Ding, Y., Xu, L., Hepp, M.: Reusing ontologies and language components for ontology generation. Data Knowl. Eng. **4**(69), 318–330 (2010)
5. Zhong, L., Wu, J., Li, Q., Peng, H., Wu, X.: A comprehensive survey on automatic knowledge graph construction. ACM Comput. Surv. **56**(4), 94 (2024)
6. Ji, S., Pan, S., Cambria, E., Marttinen, P., Yu, P.S.: A survey on knowledge graphs: representation, acquisition, and applications. IEEE Trans. Neural Netw. Learn. Syst. **33**(2), 494–514 (2022)
7. Nickel, M., Murphy, K., Tresp, V., Gabrilovich, E.: A review of relational machine learning for knowledge graphs. Proc. IEEE **104**(1), 11–33 (2016)
8. Devlin, J., Chang, M., Lee, K., Toutanova, K.: BERT: pre-training of deep bidirectional transformers for language understanding, arXiv preprint arXiv:1810.04805 (2018)
9. Hu, Z., Leng, S., Lin, J., Li, S., Xiao, Y.: Knowledge extraction and discovery based on BIM: a critical review and future directions. Arch. Comput. Methods Eng. **29**(1), 335–356 (2022)
10. Yang, M., et al.: Semi-automatic representation of design code based on knowledge graph for automated compliance checking. Comput. Ind. **150**, 103945 (2023)
11. Zheng, Z., Zhou, Y., Lu, X., Lin, J.: Knowledge-informed semantic alignment and rule interpretation for automated compliance checking. Autom. Constr. **142**, 104524 (2022)
12. Peng, J., Liu, X.: Automated code compliance checking research based on BIM and knowledge graph. Sci. Rep. **13**(1), 7065 (2023)
13. Wang, N., Haihong, E., Song, M., Wang, Y.: Construction method of domain knowledge graph based on big data-driven. In: 2019 5th International Conference on Information Management (ICIM), Cambridge, UK, pp. 165–172 (2019)
14. Liu, Q., Li, Y., Duan, H., Liu, Y., Qin, Z.: Knowledge graph construction techniques. Comput. Res. Dev. **53**(3), 582–600 (2016)

Data-Driven Dynamic Decision-Making Strategy for Gear-Shaft Robotic Assembly Process

Ruizhang Wang[1,2(✉)], Wenjun Xu[1,2], Jiayi Liu[1,2], Ping Lou[1,2], Yi Zhong[1,2], Quan Liu[1,2], and Zude Zhou[1]

[1] School of Information Engineering, Wuhan University of Technology, Wuhan 430070, China
{wrz098039,xuwenjun,jyliu,louping,zhongyi,quanliu,
zudezhou}@whut.edu.cn
[2] Hubei Key Laboratory of Broadband Wireless Communication and Sensor Networks, Wuhan University of Technology, Wuhan 430070, China

Abstract. Gear-shaft assembly is a common process in assembling process of products. It can be divided into rough alignment process, precise alignment and insertion process. Inspired by operator's gear-shaft assembly process, visual data is utilized to detect the relative pose between the shaft and the gear, while force/torque data is utilized to handle their interaction. In this paper, Proximal Policy Optimization (PPO) algorithm is employed to make dynamic decisions during the gear-shaft robotic assembly. First, the framework of the proposed method is studied. Afterwards, Convolutional Neural Network (CNN) and PPO are employed to make dynamic decisions in rough alignment process, while PPO is employed to make dynamic decisions in precise alignment and insertion process. Finally, case study is carried out to verify the proposed method. The results show the converged PPO model could dynamically generate optimal strategy to complete the gear-shaft robotic assembly.

Keywords: Proximal policy optimization · gear-shaft robotic assembly · dynamic decisions

1 Introduction

Robotic assembly plays a crucial role in manufacturing by efficiently integrating individual components into assembled products. The assembly process mainly includes peg-in-hole assembly, gear-shaft assembly and so on. Gear-shaft assembly is a typical step in product's assembly process. The gear-shaft is defined as the purpose of making the shaft transmit power to the gear using keyway system. Keyways with square channels can provide a strong connection between the shaft and the gear when the key is inserted between them [1]. The current challenge of gear-shaft robotic assembly is to ensure high precision to avoid damaging the individual components.

Humans can easily complete contact-rich assembly tasks by adequately combining visual and force/torque data [2].Gear-shaft assembly process operated by humans can be generally divided into four stages. (1) Rough alignment: the operator observes the relative pose between the shaft and the gear, and then adjusts the shaft pose at a certain height until the shaft is roughly aligned with the gear. (2) Reaching: the operator moves the shaft vertically until operator senses that the shaft is in contact with the gear. (3) Precise alignment: the operator senses the contact force/torque between the shaft and the gear, and then adjusts the shaft pose to reduce contact force/torque until the shaft is precisely aligned with the gear. (4) Insertion: the operator inserts the shaft into the gear vertically with small contact force/torque until the height of the shaft is less than the threshold. Inspired by the operator's gear-shaft assembly process, this paper proposes data-driven dynamic decision-making strategy for gear-shaft robotic assembly process. The data used in this paper includes visual and force/torque data. Like the operator's assembly process, for the gear-shaft robotic assembly process, the (1) is defined as rough alignment, and then the (2) (3) (4) is defined as precise alignment and insertion.

In rough alignment process, CNN and deep reinforcement learning (DRL) are employed. CNN has made remarkable achievements in image classification, object detection, semantic segmentation and so on [3–5]. In this paper, to handle with visual data, CNN is utilized to transform low-level image representations into higher-level vector representations. DRL can make robots achieve human-like intelligence [6]. In this paper, PPO algorithm, which performs excellently in continuous tasks for its simplicity, effectiveness and high sample efficiency, is employed to make dynamic decisions according to extracted vector representations [7]. Then, the optimal strategy optimized by PPO is utilized to adjust the pose of robotic end-effector.

In precise alignment and insertion process, PPO is employed to make dynamic decisions. PPO takes the assembly force/torque data and the pose of robotic end-effector as inputs. Then, the optimal strategy optimized by PPO is utilized to adjust the pose of robotic end-effector.

The rest of this paper is described as follows. In Sect. 2, related works are summarized. In Sect. 3, framework of data-driven dynamic decision-making strategy for gear-shaft robotic assembly process is proposed. In Sect. 4, the gear-shaft robotic assembly employs both CNN and PPO to make dynamic decisions. In Sect. 5, the proposed method is verified through a case study. Finally, Sect. 6 concludes this paper.

2 Related Works

Robotic assembly has always been a challenging task. Researchers solved robotic assembly based on the perception data, which are visual data and force/torque data.

Vision-based approaches focus on detecting the pose of individual components. Litvak et al. proposed a high-accuracy two-stage pose estimation procedure employing CNN to solve a generic assembly task, with the networks trained on simulated depth images to ensure successful application to the real robot [8]. Chang et al. proposed cartesian-based visual servoing with finite-time control employing deep learning and iterative closest point algorithm for automated robotic picking and assembly [9]. Solorzano et al. employed PPO algorithm to solve the positioning of printed circuit boards, with the sensed image serving as the current environment [10]. Zhuang et al. proposed a novel vision-based semantic human-robot-collaboration system for a real-world assembly task, which integrated all sub-modules by utilizing implicit semantic information from visual perception [11]. Liu et al. proposed a hybrid visual servoing controller based on deep deterministic policy gradient (DDPG) algorithm to complete the autonomous robotic assembly [12]. The controller takes current observed image features and camera pose as inputs. Bartyzel et al. proposed a stereo-view method based on Soft Actor-Critic (SAC) algorithm for the robust assembly of electronic parts, and the proposed method was robust to applied perturbations [13]. Nevertheless, vision-based approaches require high-resolution images, making it difficult to detect the pose of individual components when the image resolution is low.

Many researchers utilized the force/torque data to handle the interaction between components. Zou et al. proposed a variable impedance controller employing the fuzzy Q-learning algorithm to yield compliant behavior from the robot during the peg-in-hole assembly inserting stage [14]. Luo et al. proposed a learning approach utilizing Recurrent Distributed DDPG to solve lap-joint and peg-in-hole assembly tasks. Force/torque data in task space was only observation [15]. Oikawa et al. proposed a method for online selection of non-diagonal stiffness matrices for admittance control using deep Q-learning (DQN), which was evaluated on peg-in-hole and gear-insertion assembly tasks [16]. Shi et al. proposed a novel contact force sensing and control method for the inserting operation during precise assembly process, which relied on a micromanipulator integrated with force sensor [17]. Vartanov et al. proposed a method for determining the shaft position and sleeve using a force/torque sensor, which is utilized to estimate a three-point contact position [18]. Markert et al. proposed a DQN framework to solve the peg-in-hole insertion task, and chosen force/torque data as inputs [19]. Nevertheless, the above work does not utilize visual data, and requires the rough assembly positions as prior knowledge.

Recently, many researchers have effectively combined visual and force/torque data to complete robotic assembly. Chen et al. proposed a method to reduce positioning errors in inserting a nut onto a screw without threading by combing visual and force/torque data [20]. Ming et al. proposed a vision-guided two-stage approach with force/torque data to achieve high-precision and flexible gear assembly [21]. The approach integrated YOLO to coarsely localize the target workpiece in a searching stage and employed DRL to complete the insertion stage. Zhang et al. proposed a residual reinforcement learning policy including a vision-based strategy and a force/torque-based strategy for peg-in-hole assembly via combing visual and force/torque data [2]. The vision-based strategy focused on spatial search, while the force/torque-based strategy handled the interactive behaviors. From the related works, it is obvious that, few researchers studied the gear-shaft robotic assembly. In this paper, the visual and force/torque data are utilized to make dynamic decisions during the gear-shaft robotic assembly.

3 Framework

The framework of data-driven dynamic decision-making strategy for gear-shaft robotic assembly process is proposed in Fig. 1.

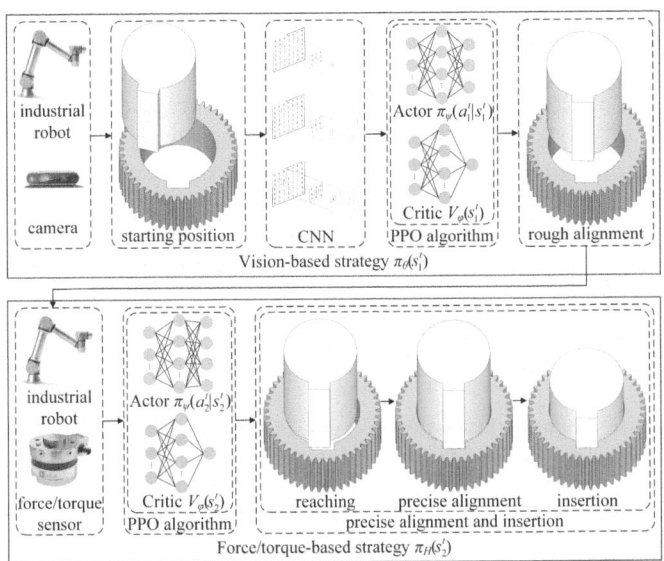

Fig. 1. Framework of data-driven dynamic decision-making strategy for gear-shaft robotic assembly process.

The gear-shaft robotic assembly is divided into two processes, which are rough alignment, precise alignment and insertion. Furthermore, the assembly model can be formulated as a Markov decision process (MDP) by a tuple (S, A, P, R, γ), where S denotes the environment's state space, A represents the agent's action space, $P(s_{t+1}|s_t, a_t)$ indicates transition function, $R(s_t, a_t)$ refers to the reward function and γ signifies the discount factor of reward. The goal of the agent's strategy learning is to maximize the expected total reward, and the optimal strategy π^* is calculated by (1)

$$\pi^* = \arg\max_{\pi} \mathbb{E}_{(s_t,a_t)\sim\rho_\pi} \sum_{t=0}^{T-1} \gamma^t r(s_t, a_t) \tag{1}$$

where ρ_π denotes the probability density of (s_t, a_t) with the strategy π and T is the total time steps of an episode. In this paper, a camera and a force/torque sensor are utilized to acquire visual and force/torque data. Then, a vision-based strategy $\pi_\theta(s_1^t)$ and a force/torque-based strategy $\pi_H(s_2^t)$ are proposed, which both employ PPO to learn gear-shaft robotic assembly skills. Furthermore, PPO consists of an actor network and a critic network.

The gear-shaft robotic assembly process is described as follows. The Cartesian translation and rotation of robotic end-effector at step t are represented as $[x_t, y_t, z_t, rx, ry, rz_t]$.

The shaft is attached to robotic end-effector which is aligned with the pose of the shaft. The gear is put on a flat surface, so rx and ry remain unchanged during the assembly process. Furthermore, the pose of the gear is represented as $[g_x, g_y, g_z, g_{rx}, g_{ry}, g_{rz}]$.

In rough alignment process, the relative pose between the shaft and the gear at step t is defined by (2)

$$d_t = \left[d_x^t, d_y^t, d_z^t, 0, 0, d_{rz}^t \right] = \left[|x_t - g_x|, |y_t - g_y|, z_t - g_z, 0, 0, |rz_t - g_{rz}| \right] \quad (2)$$

Robotic end-effector starts from a random pose, and relative pose d_0 can be represented as $[d_x^0, d_y^0, d_z^0, 0, 0, d_{rz}^0]$. d_x^0, d_y^0 and d_{rz}^0 are, respectively, less than the constant α_x, α_y and α_{rz}. d_z^0 represents the vertical distance from the lower surface of the shaft to the lower surface of the gear. During training, the camera, which is mounted on robotic end-effector, is utilized to capture images. Furthermore, CNN is employed to extract higher-level vector representations of images. In the strategy $\pi_\theta(s_1^t)$, CNN is shared between actor and critic networks. To avoid ineffective exploration, d_x^t, d_y^t and d_{rz}^t should be, respectively, limited to be less than β_x, β_y and β_{rz}. The aim of the trained optimal strategy $\pi_\theta(s_1^t)$ is to guarantee that d_x^t, d_y^t and d_{rz}^t, are, respectively, less than δ_x, δ_y and δ_{rz}.

In precise alignment and insertion process, the relative pose between the shaft and the gear at step t is defined by (3)

$$e_t = \left[e_x^t, e_y^t, e_z^t, 0, 0, e_{rz}^t \right] = \left[|x_t - g_x|, |y_t - g_y|, z_t - g_z, 0, 0, |rz_t - g_{rz}| \right] \quad (3)$$

Robotic end-effector starts from a random pose, and relative pose e_0 can be represented as $[e_x^0, e_y^0, e_z^0, 0, 0, e_{rz}^0]$. e_x^0, e_y^0 and e_{rz}^0 are, respectively, less than the constant η_x, η_y and η_{rz}. The value of e_z^0 is equal to d_z^0. During training, the force/torque sensor is utilized to perceive force/torque data. The force/torque data at step t is represented as $[F_x^t, F_y^t, F_z^t, T_x^t, T_y^t, T_z^t]$. F_z^t needs to filter out the gravity of the shaft. The resultant force and torque are, respectively, represented as F_t and T_t. Then, the strategy $\pi_H(s_2^t)$ takes the force/ torque data and the pose of robotic end-effector as inputs. The trained optimal strategy $\pi_H(s_2^t)$ is employed to adjust the pose of robotic end-effector. Furthermore, the strategy $\pi_H(s_2^t)$ consists of reaching, precise alignment and insertion. $F_{max}^1/T_{max}^1, F_{max}^2/T_{max}^2, F_{max}^3/T_{max}^3$, respectively, represent the maximum resultant force and torque during reaching, precise alignment and insertion. F_{min}/T_{min} represent the minimum resultant force and torque. During reaching, robotic end-effector moves the shaft vertically. When F_t is within the range of $[F_{min}, F_{max}^1]$ or T_t is within the range of $[T_{min}, T_{max}^1]$, the strategy $\pi_H(s_2^t)$ enters precise alignment. If e_z^t is less than 0, the gear-shaft robotic assembly is completed. During precise alignment, robotic end-effector adjusts the pose of the shaft to reduce contact force/torque. To avoid ineffective exploration, e_x^t, e_y^t and e_{rz}^t should be, respectively, limited to be less than μ_x, μ_y and μ_{rz}. If F_t is less than F_{min} and T_t is less than T_{min}, the strategy $\pi_H(s_2^t)$ will enter insertion. During insertion, robotic end-effector inserts the shaft into the gear vertically. The gear-shaft robotic assembly is completed when e_z^t is less than 0.

4 Gear-Shaft Robotic Assembly Process Dynamic Decision-Making Strategy Based on CNN and PPO

4.1 Dynamic Decision-Making Strategy in Rough Alignment Process

CNN and PPO are employed to make dynamic decision in rough alignment process. As shown in Fig. 2, it consists of CNN and PPO. PPO includes state, action, reward, environment and trajectory memory.

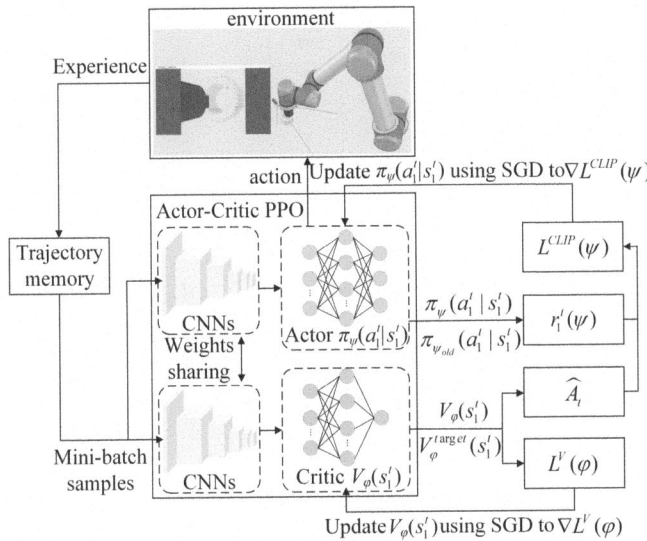

Fig. 2. Workflow of the vision-based strategy in rough alignment process.

CNN is employed to handle with visual data. As shown in Fig. 3, to obtain the relative pose d_t, a 5-layer CNN architecture is proposed to extract visual data. The input of the CNN architecture is an image with dimensions of Channel × Height × Width ($C \times H \times W$), and the output is a one-dimensional vector with dimensions of K. The filter sizes of the convolutional layer are, respectively, $64 \times 8 \times 8$, $25 \times 4 \times 4$, $16 \times 4 \times 4$, $9 \times 2 \times 2$ and $4 \times 2 \times 2$. The strides of the convolutional layer are, respectively, 4, 2, 2, 1 and 1. There is no padding in each convolutional layer. The default activation function after each convolution layer is a rectified linear unit (ReLU) nonlinearity function. Then, the image features extracted by 5 convolutional layers are flatten into one-dimensional vector, which generates 128 features through a fully connected layer. The 128 features represent the relative pose d_t.

The state s_1^t, action a_1^t and reward 4_1^t are, respectively, defined by (4)–(7)

$$s_1^t = image \tag{4}$$

$$a_1^t = [\Delta x, \Delta y, \Delta rz] \tag{5}$$

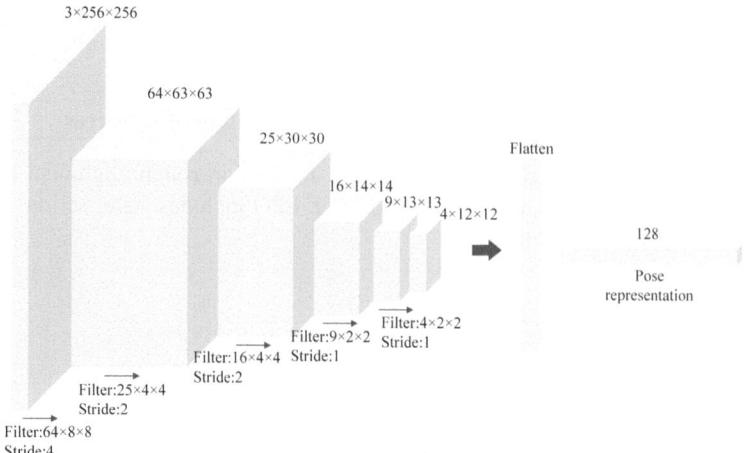

Fig. 3. The proposed CNN architecture for obtaining relative pose between the shaft and the gear.

$$r_1^t = (d_x^{t-1} - d_x^t)/\alpha_x + (d_y^{t-1} - d_y^t)/\alpha_y + (d_{rz}^{t-1} - d_{rz}^t)/\alpha_{rz} - 1/m_1 + r_{end}^1 \quad (6)$$

$$r_{end}^1 = \begin{cases} 2 & d_x^t > \delta_x \text{ and } d_y^t > \delta_y \text{ and } d_{rz}^t > \delta_{rz} \\ -2 & d_x^t > \beta_x \text{ or } d_y^t > \beta_y \text{ or } d_{rz}^t > \beta_{rz} \\ 0 & others \end{cases} \quad (7)$$

The state s_1^t is an image, which represents the relative pose d_t. For action a_1^t, it includes Δx, Δy and Δrz that are the Cartesian translation and rotation increments of robotic end-effector. For action r_1^t, the maximum number of steps per episode is m_1.

In DRL, the agent interacts with the environment. The agent choose action a_1^t according to the current state s_1^t, and then the environment provides reward r_1^t and the next state s_1^{t+1} for the agent. In this paper, the environment consists of a gear, a shaft, a robotic end-effector, a camera and a force/torque sensor.

For each step t, the experience from the environment, which is represented by tuple $<s_1^t, a_1^t, r_1^t, s_1^{t+1}>$, is stored into the trajectory memory. Furthermore, trajectory memory randomly choose a finite mini-batch of samples to update the parameters of actor network $\pi_\psi(a_1^t|s_1^t)$ and critic network $V_\varphi(s_1^t)$.

PPO is a model-free, online, on-policy, policy gradient reinforcement learning method. It is built upon the actor-critic architecture, and the objectives for actor network $\pi_\psi(a_1^t|s_1^t)$ and critic network $V_\varphi(s_1^t)$ are, respectively, described by (8)–(9)

$$L^{CLIP}(\psi) = \widehat{E}_t\left[\min(r_1^t(\psi)\widehat{A}_t, \text{clip}(r_1^t(\psi), 1 - \varepsilon, 1 + \varepsilon)\widehat{A}_t)\right] \quad (8)$$

$$L^V(\varphi) = \widehat{E}_t\left[(V_\varphi(s_1^t) - V_\varphi^{t\,arget}(s_1^t))^2\right] \quad (9)$$

$\widehat{E}_t$ denotes the empirical average over a finite batch of samples. $rt\ 1(\psi)$ denotes the probability ratio $r_1^t(\psi) = \pi_\psi(a_1^t|s_1^t)/\pi_{\psi old}(a_1^t|s_1^t)$, and ψ_{old} is the vector of strategy parameters before updating process. $\hat{A}_t$ indicates an estimator of the advantage function, which is calculated by the generalized advantage estimation. The clipping function clip() is utilized to clip the gradients of strategy updates to prevent actions from becoming too high or too low. Furthermore, the clipping function clip() improves the stability and convergence of PPO algorithm. The critic network $V_\varphi(s_1^t)$ gives value function for time-difference (TD) error. $V_\varphi(s_1^t)$ denotes the estimated state value, and the target state value is calculated by $V_\varphi^{target}(s_1^t) = r_1^t + \gamma\, V_\varphi(s_1^t)$. Then, the parameters of $L^{CLIP}(\psi)$ and $L^V(\varphi)$ are updated by Stochastic Gradient Descent (SGD) algorithm [22].

4.2 Dynamic Decision-Making Strategy in Precise Alignment and Insertion Process

PPO is employed to make dynamic decisions in precise alignment and insertion process. As shown in Fig. 4, PPO includes state, action, reward, environment and trajectory memory. Environment and trajectory memory have been introduced in the previous section.

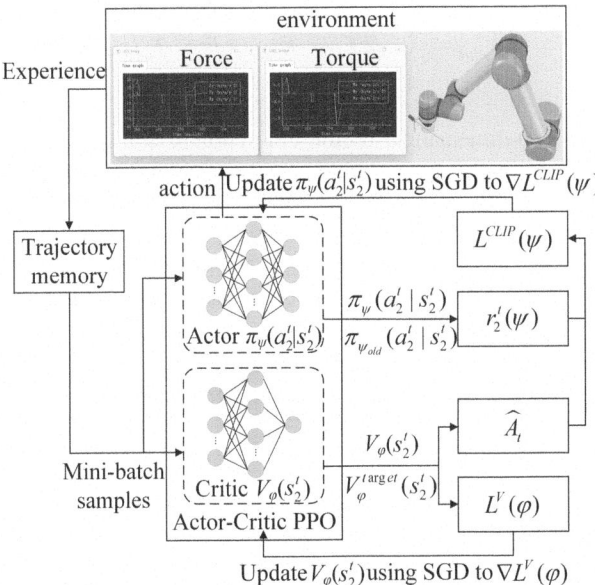

Fig. 4. Workflow of the force/torque-based strategy in precise alignment and insertion process.

The state s_2^t, action a_2^t and reward r_2^t are, respectively, defined by (10)–(18)

$$s_2^t = \left[x_t, y_t, z_t, rz_t, F_x^t, F_y^t, F_z^t, T_x^t, T_y^t, T_z^t\right] \tag{10}$$

$$a_2^t = \left[\Delta x, \Delta y, \Delta z, \Delta rz\right] \tag{11}$$

$$r_2^t = k_1 I_1 + k_2 I_2 + k_3 I_3 \tag{12}$$

$$I_1 = (z_{t-1} - z_t)/\lambda_1 - 1/m_2 + r_{end}^2 \tag{13}$$

$$r_{end}^2 = \begin{cases} 2 & F_{min} \leq F_t \leq F_{max}^1 or T_{min} \leq T_t \leq T_{max}^1 or e_z^t < 0 \\ -2 & F_t > F_{max}^1 or T_t > T_{max}^1 \\ 0 & others \end{cases} \tag{14}$$

$$I_2 = \frac{e_x^{t-1} - e_x^t}{\mu_x} + \frac{e_y^{t-1} - e_y^t}{\mu_y} + \frac{e_{rz}^{t-1} - e_{rz}^t}{\mu_{rz}} + \frac{F_{t-1} - F_t}{F_{max}} + \frac{T_{t-1} - T_t}{T_{max}} - \frac{1}{m_2} + r_{end}^3 \tag{15}$$

$$r_{end}^3 = \begin{cases} 2 & F_t < F_{min} and T_t < T_{min} \\ -2 & F_t > F_{max}^2 or T_t > T_{max}^2 or e_x^t > \mu_x or e_y^t > \mu_y or e_{rz}^t > \mu_{rz} \\ 0 & others \end{cases} \tag{16}$$

$$I_3 = (z_{t-1} - z_t)/\lambda_2 - \xi F_t/F_{max}^3 - \xi T_t/T_{max}^3 - 1/m_2 + r_{end}^4 \tag{17}$$

$$r_{end}^4 = \begin{cases} 2 & e_z^t < 0 \\ -2 & F_t > F_{max}^3 or T_t > T_{max}^3 \\ 0 & others \end{cases} \tag{18}$$

The state s_2^t is a 10-dimensional vector. For action at 2, it contains Δx, Δy, Δz and Δrz that are the Cartesian translation and rotation increments of robotic end-effector. For reward rt 2, it consists of Reaching (I_1), Precise alignment (I_2) and Insertion (I_3) which have the same maximum step m_2. k_1, k_2 and k_3 represent the scale factor. During reaching, k_1 is 1 while both k_2 and k_3 are 0. Meanwhile, Δx, Δy and Δrz are 0. λ_1 represents the vertical distance from the lower surface of the shaft to the upper surface of the gear during rough alignment process. During precise alignment, k_2 is 1 while both k_1 and k_3 are 0. Meanwhile, Δz is 0. During insertion, k_3 is 1 while both k_1 and k_2 are 0. λ_2 represents the height of the gear. ξ represents the scale factor.

5 Case Study

5.1 Experimental Setup

The gear-shaft robotic assembly process is simulated using Coppeliasim software. Furthermore, physics engine, simulation time step and friction are ODE, 50 ms and 0.15, respectively. All experiments were taken on PC with 5.80 GHz Inter Core i9-13900K CPU, 128 GB memory and NVIDIA GeForce RTX 4090 GPU. The industrial robot used in this paper is UR5. The lengths are measured in millimeters (mm), while angles are measured in radians (rad). Force is measured in Newtons (N), while torque is measured in Newton-meters(N·m). As shown in Fig. 5, the gear has an outer radius of 30 mm, an inner radius of 20 mm, a height of 30 mm, and a keyway length of 12 mm and width of 4.2 mm. The shaft has a radius of 19.5 mm, a height of 60 mm, and a keyway length of 11.4 mm and width of 4.2 mm.

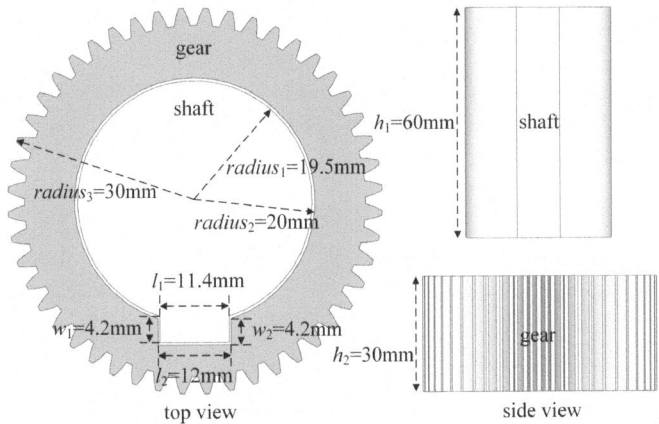

Fig. 5. Dimensions of gear and shaft.

5.2 Experimental Analysis of Rough Alignment

Comparative experiments (Ex) are conducted to validate the efficacy of the proposed method under low image resolution conditions. Ex1 takes images with dimensions of $3 \times 128 \times 128$ as inputs, while Ex2 takes images with dimensions of $3 \times 256 \times 256$ as inputs. Both Ex 1 and Ex2 employ CNN and PPO, and share the same parameters and network architecture. The parameters of PPO are listed as follows: learning rate is 0.0003, discount factor is 0.99, advantage estimation dicounting factor is 0.95, clipping parameter is 0.2, batch size is 64, the number of steps to run per update is 2048, number of epoch when optimizing the surrogate loss is 10, optimizer is Adam. In this paper, stable-baselines3 is utilized to implement the actor-critic architecture for PPO [23]. Actor network $\pi_\psi(a_1^t|s_1^t)$ and critic network $V_\varphi(s_1^t)$ share the same network architecture, which is characterized by having three fully connected layers, with each layer consisting of 64 units. The activation function is ReLU. The maximum values for Δx, Δy and Δrz are 2 mm, 2 mm and $\pi/90$ rad, respectively. α_x, α_y, α_{rz}, β_x, β_y, β_{rz}, δ_x, δ_y, δ_{rz}, d_z^0 and m_1 are, respectively, 30 mm, 30 mm, $\pi/6$ rad, 40 mm, 40 mm, $2\pi/9$ rad, 2 mm, 2 mm, $\pi/90$ rad, 50 mm and 64. The training process of PPO is described in Fig. 6(a). It is obvious that PPO gradually converges with the step increases.

To evaluate the assembly quality in rough alignment process, five evaluation metrics are proposed: the maximum (Max), minimum (Min), mean(Mean) and standard deviation (Std) of d_x^t, d_y^t and d_{rz}^t, successful rate (SR). The units for d_x^t, d_y^t and d_{rz}^t are, respectivsly, mm, mm and rad. The trained optimal strategy $\pi_\theta(s_1^t)$ is tested over 100 episodes and the result is described in Table 1. Furthermore, the box plots of d_x^t, d_y^t and d_{rz}^t are described in Fig. 6(b)–(d). The result shows that the both Ex1 and Ex2 can achieve 100% accuracy. Therefore, the proposed method maintains good performance even when the image resolution is low.

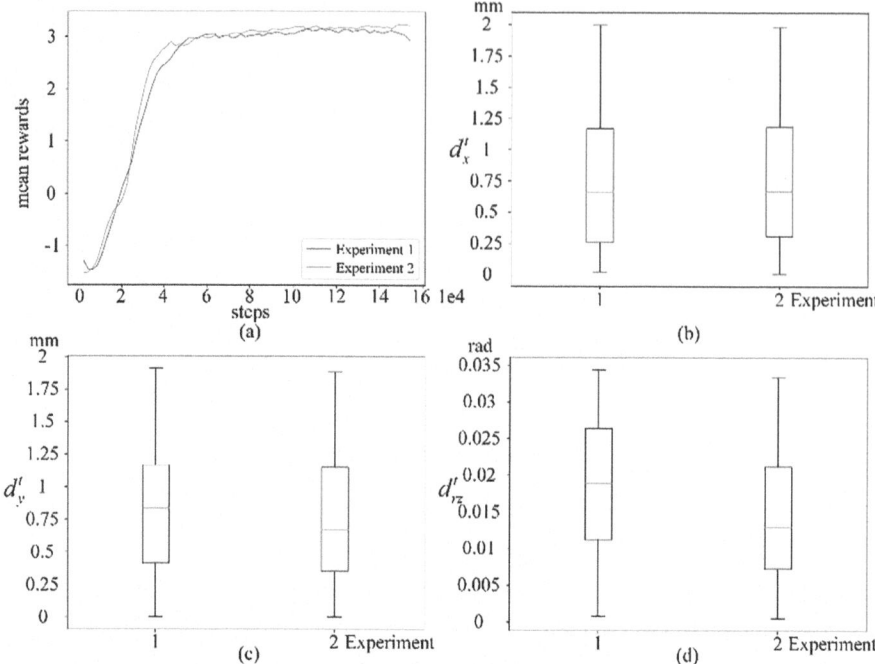

Fig. 6. Training and testing process of PPO. (a) mean rewards in rough alignment. (b) the box plot of $d\,t\,x$. (c) the box plot of $d\,t\,y$. (d) the box plot of $d\,t\,rz$.

Table 1. Performance of converged model in rough alignment process.

Ex	Max (d_x^t, d_y^t, d_{rz}^t)	Min (d_x^t, d_y^t, d_{rz}^t)	Mean (d_x^t, d_y^t, d_{rz}^t)	Std (d_x^t, d_y^t, d_{rz}^t)	SR
1	1.997, 1.912, 0.0344	0.023, 0.002, 0.0008	0.749, 0.816, 0.0184	0.534, 0.502, 0.0094	100%
2	1.981, 1.885, 0.0334	0.007, 0.003, 0.0005	0.773, 0.756, 0.0145	0.540, 0.491, 0.0091	100%

5.3 Experimental Analysis of Precise Alignment and Insertion

Bartyzel et al. employed SAC algorithm to complete the robust assembly of electronic parts [13]. SAC is a model-free reinforcement learning method characterized by its off-policy, actor-critic architecture, which leverages entropy maximization to enhance exploration [24]. In this paper, SAC algorithm is employed to compare with the proposed method. Furthermore, Ex3 employ PPO, while Ex4 employ SAC.

Ex3 and Ex4 share the same parameters as follows: learning rate is 0.0003, discount factor is 0.99, batch size is 64, optimizer is Adam. Furthermore, the specific parameters for Ex3 are listed as follows: advantage estimation dicounting factor is 0.95, clipping parameter is 0.2, the number of steps to run per update is 2048, number of epoch when optimizing the surrogate loss is 10. The specific parameters for Ex4 are listed as follows: buffer size is 1000000, the soft update coefficient is 0.005, entropy regularization coefficient is auto, target entropy is auto. Both Ex 3 and Ex4 share the same network architecture. In this paper, stable-baselines3 is utilized to implement the actor-critic architecture for PPO and SAC. Actor network and critic network share the same network architecture, which is characterized by having two fully connected layers, with each layer consisting of 32 units. The activation function is ReLU. The maximum values for Δx, Δy, Δz and Δrz are 0.2 mm, 0.2 mm, 2 mm and $\pi/900$ rad. η_x, η_y, η_{rz}, μ_x, μ_y, μ_{rz}, F^1_{max}, T^1_{max}, F^2_{max}, T^2_{max}, F^3_{max}, T^3_{max}, F_{min}, T_{min}, λ_1, λ_2 and ξ are, respectively, are 2mm, 2mm, $\pi/90$ rad, 3mm, 3mm, $\pi/60$ rad, 40N, 2 N·m, 120N, 6 N·m, 40N, 2 N·m, 0.1 N, 0.01 N·m, 20mm, 30mm and 0.5. The training process of PPO and SAC is described in Fig. 7(a). It is obvious that PPO and SAC gradually converge with the step increases. F_1/T_1, F_2/T_2 and F_3/T_3, respectively, represent the resultant force and torque during reaching, precise alignment and insertion. To evaluate the assembly quality in precise alignment and insertion process, five evaluation metrics are proposed: Max, Min, Mean and Std of F_1/T_1, F_2/T_2, F_3/T_3 and z_t, SR. The units for F_1/T_1, F_2/T_2, F_3/T_3 and z_t are, respectively, N/N·m, N/N·m, N/N·m and mm. The trained optimal strategy $\pi_H(s^t_2)$ is tested over 100 episodes and the result is described in Table 2.

Table 2. Performance of converged model in precise alignment and insertion process.

Ex	Max (F_1/T_1, F_2/T_2, F_3/T_3, z_t)	Min (F_1/T_1, F_2/T_2, F_3/T_3, z_t)	Mean (F_1/T_1, F_2/T_2, F_3/T_3, z_t)	Std (F_1/T_1, F_2/T_2, F_3/T_3, z_t)	SR
3	30.11/0.49, 181.98/4.68, 23.18/1.09, 31.43	0.10/0.002, 0.10/0.002, 0/0, 0.04	11.44/0.14, 17.60/0.57, 1.52/0.06, 2.72	6.80/0.11, 20.44/0.73, 3.51/0.14, 7.22	94%
4	19.54/0.30, 186.51/4.15, 67.15/2.62, 31.74	0.11/0.005, 0.11/0.001, 0/0, 0.14	5.35/0.07, 18.73/0.43, 2.06/0.08, 4.02	3.73/0.05, 25.40/0.58, 5.85/0.23, 8.49	86%

Furthermore, the box plots of z_t, F_1, T_1, F_2, T_2, F_3, and T_3 are described in Fig. 7(b)–(h). The result shows that the proposed method exhibits a higher success rate compared to SAC algorithm. In Ex3, the reason for the failure of the gear-shaft robotic assembly is that F_2 exceeds 120 N. In Ex4, the reason for the failure of the gear-shaft robotic assembly is that F_2 exceeds 120 N or F_3 exceeds 40 N or T_3 exceeds 2 N·m.

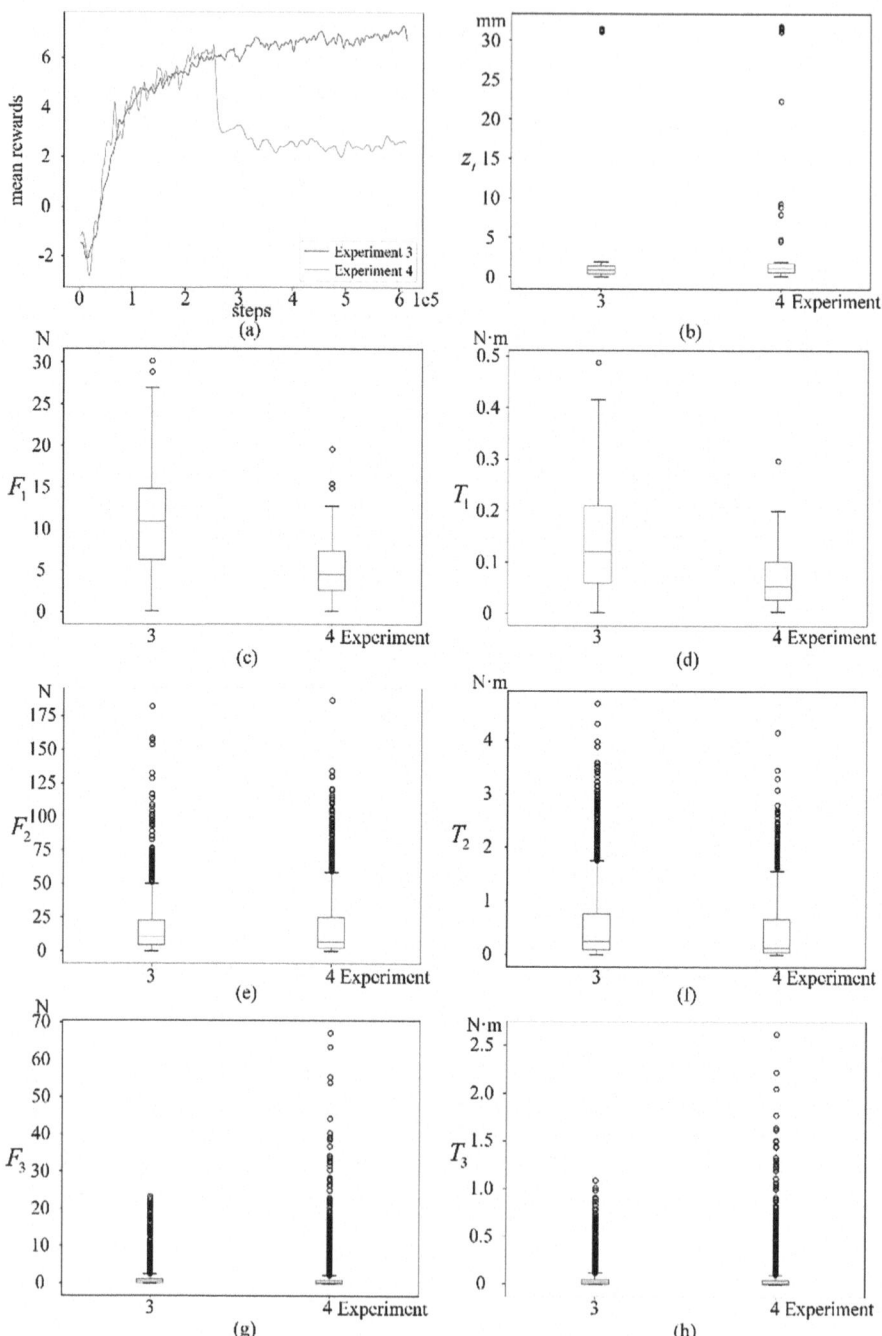

Fig. 7. Training and testing process of PPO and SAC. (a) mean rewards in precise alignment and insertion. (b) the box plot of z_t. (c) the box plot of F_1. (d) the box plot of T_1. (e) the box plot of F_2. (f) the box plot of T_2. (g) the box plot of F_3. (h) the box plot of T_3.

6 Conclusion

In this paper, visual and force/torque data were used to realize gear-shaft robotic assembly. CNN and PPO are employed to make dynamic decisions during the assembly. First, framework of data-driven dynamic decision-making strategy for gear-shaft robotic assembly process was proposed. Then, CNN and PPO were employed to make dynamic decisions in rough alignment process, while PPO was employed to make dynamic decisions in precise alignment and insertion process. Finally, case study was conducted to verify the proposed method. The results show the converged PPO model can maintain good performance in rough alignment process, even when the image resolution is low. In precise alignment and insertion process, the converged PPO model can complete the gear-shaft robotic assembly, and exhibit a higher success rate compared to SAC algorithm.

Nevertheless, the proposed method performs badly due to excessively force and torque. In the future, compliance controller will be designed to reduce the contact force/torque between the shaft and the gear. Furthermore, the proposed method did not generalize to gears of different sizes and shapes. The generalizability of the proposed method will be validated in the future.

Acknowledgments. This work is supported by National Natural Science Foundation of China (Grant Nos. 92367108 and 52205538), Hubei Provincial Key Research and Development Program of China (Grant No. 2023BAB012), and Wuhan East Lake High-tech Development Zone's Leading the Charge with Open Competition Project (Grant No. 2023KJB215).

Disclosure of Interests. It is confirmed that no competing interests are declared.

References

1. Rao, A., Shinde, V., Upasani, R., Gadhia, D., Devasagayam, D.: Design and fabrication of keyway cutting fixture for gears, pp. 1115–1119 (2016)
2. Zhang, Z., Wang, Y., Zhang, Z., Wang, L., Huang, H., Cao, Q.: A residual reinforcement learning method for robotic assembly using visual and force information. J. Manuf. Syst. **72**, 245–262 (2024)
3. He, K., Zhang, X., Ren, S., Sun, J.: Deep residual learning for image recognition. In: Proceedings of the IEEE Conference on Computer Vision and Pattern Recognition, pp. 770–778 (2016)
4. Redmon, J., Divvala, S., Girshick, R., Farhadi, A.: You only look once: unified, real-time object detection. In: Proceedings of the IEEE Conference on Computer Vision and Pattern Recognition, pp. 779–788 (2016)
5. Long, J., Shelhamer, E., Darrell, T.: Fully convolutional networks for semantic segmentation. In: Proceedings of the IEEE Conference on Computer Vision and Pattern Recognition, pp. 3431–3440 (2015)
6. Mnih, V., et al.: Human-level control through deep reinforcement learning. Nature **518**(7540), 529–533 (2015)
7. Schulman, J., Wolski, F., Dhariwal, P., Radford, A., Klimov, O.: Proximal policy optimization algorithms. arXiv preprint arXiv:1707.06347 (2017)

8. Litvak, Y., Biess, A., Bar-Hillel, A.: Learning pose estimation for high-precision robotic assembly using simulated depth images. In: 2019 International Conference on Robotics and Automation (ICRA), pp. 3521–3527 (2019)
9. Chang, W.C., Lin, Y.K., Pham, V.T.: Vision-based flexible and precise automated assembly with 3D point clouds. In 2021 9th International Conference on Control, Mechatronics and Automation (ICCMA), pp. 218–223 (2021)
10. Solorzano, C., Tsai, D.M.: Environment-adaptable printed-circuit board positioning using deep reinforcement learning. IEEE Trans. Components Packag. Manuf. Technol. **12**(2), 382–390 (2022)
11. Zhuang, Z., Ben-Shabat, Y., Zhang, J., Gould, S., Mahony, R. GoferBot: a visual guided human-robot collaborative assembly system. In: 2022 IEEE/RSJ International Conference on Intelligent Robots and Systems (IROS), pp. 8910–8917 (2022)
12. Liu, Z., Wang, K., Liu, D., Wang, Q., Tan, J.: A motion planning method for visual servoing using deep reinforcement learning in autonomous robotic assembly. IEEE/ASME Trans. Mechatron. 3513–3524 (2023)
13. Bartyzel, G., Półchłopek, W., Rzepka, D.: Reinforcement learning with stereo-view observation for robust electronic component robotic insertion. J. Intell. Rob. Syst. **109**(3), 57 (2023)
14. Zou, P., Zhu, Q., Wu, J., Xiong, R.: Learning-based optimization algorithms combining force control strategies for peg-in-hole assembly. In: 2020 IEEE/RSJ International Conference on Intelligent Robots and Systems (IROS), pp. 7403–7410 (2020)
15. Luo, J., Li, H.: A learning approach to robot-agnostic force-guided high precision assembly. In: 2021 IEEE/RSJ International Conference on Intelligent Robots and Systems (IROS), pp. 2151–2157 (2021)
16. Oikawa, M., Kusakabe, T., Kutsuzawa, K., Sakaino, S., Tsuji, T.: Reinforcement learning for robotic assembly using non-diagonal stiffness matrix. IEEE Robot. Autom. Lett. **6**(2), 2737–2744 (2021)
17. Shi, B., Wang, F., Huo, Z., Tian, Y., Cong, R., Zhang, D.: Contact force sensing and control for inserting operation during precise assembly using a micromanipulator integrated with force sensors. IEEE Trans. Autom. Sci. Eng. 2147–2155 (2022)
18. Vartanov, M.V.: Identify the position of the shaft and hole using a force-torque sensor in three-point contact assembly operations. In: 2022 International Ural Conference on Electrical Power Engineering, pp. 295–300 (2022)
19. Markert, T., Hoerner, E., Matich, S., Theissler, A., Atzmueller, M.: Robotic peg-in-hole insertion with tight clearances: a force-based deep q-learning approach. In 2023 International Conference on Machine Learning and Applications (ICMLA), pp. 1045–1051 (2023)
20. Chen, Y., Yu, J., Shen, L., Lin, Z., Liu, Z.: Vision-based high-precision assembly with force feedback. In: 2023 9th International Conference on Control, Automation and Robotics (ICCAR), pp. 399–404 (2023)
21. Ming, J., Bargmann, D., Cao, H., Caccamo, M.: Flexible gear assembly with visual servoing and force feedback. In: 2023 IEEE/RSJ International Conference on Intelligent Robots and Systems (IROS), pp. 8276–8282 (2023)
22. Kingma, D.P., Ba, J.: Adam: a method for stochastic optimization. arXiv preprint arXiv:1412.6980 (2014)
23. Raffin, A., Hill, A., Gleave, A., Kanervisto, A., Ernestus, M., Dormann, N.: Stable-baselines3: reliable reinforcement learning implementations. J. Mach. Learn. Res. **22**(268), 1–8 (2021)
24. Haarnoja, T., Zhou, A., Abbeel, P., Levine, S.: Soft actor-critic: off-policy maximum entropy deep reinforcement learning with a stochastic actor. In: International Conference on Machine Learning (PMLR), pp. 1861–1870 (2018)

Simulation for Science, Industry and Society

A Multiview Approach for Pedestrian 3D Pose Detection and Reconstruction

Kai Chen[1], Xiaodong Zhao[1(✉)], Yujie Huang[1], and Pengfei Wang[2]

[1] Nanjing University of Aeronautics and Astronautics, Nanjing 210016, China
xdzhao@nuaa.edu.cn
[2] Nanjing Research Institute of Electronic Engineering, Nanjing 210007, China

Abstract. Aiming at the problems that most of the existing methods of constructing 3D models of human body based on 2D human body surface pose points will lead to continuous modeling jitter and local distortion of the modeling results, we propose a 3D pose point detection method based on the skinned multiplayer linear model (SMPL) in the human body, which maps the 2D pose points of the body to 3D pose points of the real scene in the multiview perspective through a clustering algorithm, and introduces Kalman filtering to de-noise the human body pose points. The Kalman filter is introduced to denoise the human body posture points. In the process of constructing a 3D model of the human body based on 3D pose points, we construct an end-to-end human 3D modeling network (SMPL-VAE) based on the correction of gradient descent regression network by the automatic variational approach (VAE), which is more in line with the local modeling of the human body's motion structure while maintaining the overall proportion. The results on open dataset Shelf show that our methods improve the quality of human post point detection and modeling.

Keywords: Multi-Ocular Vision · Skinned Multi-Person Linear Modeling (SMPL) · 3D Pose Detection

1 Introduction

With the continuous development of science and technology and the significant improvement of urban informatization level, future robots and intelligent systems in the real world need to be able to perceive and understand human beings from visual inputs and interact with the real world, such as digital twin modeling for intelligent factories and panoramic imaging system for in-vehicle AVMs, and other technologies. Therefore, 3D modeling of human body in three-dimensional space is of great significance.

Multi-target detection and 3D modeling generally refer to the detection of pedestrians under the viewpoint of the device when the number of unknown targets is unknown, and modeling in 3D space in order to carry out the subsequent trajectory prediction and analysis. In practical application scenarios, multi-target detection not only needs to be real-time, but also needs to be accomplished between different targets in each frame, which puts forward higher requirements for detection and modeling due to the respective

© The Author(s), under exclusive license to Springer Nature Singapore Pte Ltd. 2024
S. Saito et al. (Eds.): AsiaSim 2024, CCIS 2170, pp. 87–100, 2024.
https://doi.org/10.1007/978-981-97-7225-4_7

differences between the targets as well as the intricate correlation situation between the pose points. Most of the existing multi-target detection algorithms [1–4] start from the surface of human body, and mainly improve on accuracy, computational resources and algorithms [5–7]. However, they are unable to solve the problems of target occlusion and target matching with pose points well. In terms of modeling, although the existing methods [8–10] have achieved better modeling results in terms of reconstruction error, floating point operations per second (FLOPs), etc., they are still prone to local pose distortion and jitter in continuous modeling when there is more occlusion and the target cascade relationship is more complex.

Aiming at the existing problem of detecting and modeling the 2D body surface pose points of dense pedestrians under binocular vision, we propose a 3D pose point detection method inside the human body based on the Skinned Multi-Person Linear Model [11] (SMPL), and then construct an end-to-end SMPL-VAE human body regression model by noise reduction and smoothing of 3D pose points inside the body through Kalman Filtering: the step-by-step training and the joint training are combined, and the VAE training results are used to correct the modeling results of the SMPL, to take into account the overall proportion and at the same time to ensure that the local modeling results conform to the structure of the human body movement more closely.

2   Related Works

Thanks to the wide application of deep learning technology, the field of 2D human pose detection has made remarkable development. DeepPose [4] extends the AlexNet [14] network, adopts DNN [15] to learn the human pose point feature representations, and optimizes the results of the previous processing step through a cascade structure; OpenPose proposes a PAF+ confidence-associated detection network that performs feature extraction on a single RGB image, which effectively solves problems such as low accuracy of key human pose detection due to partial occlusion. However, associating only the surface pose of 2D images cannot accurately reflect the complex body poses of the human body, and the final 3D skeleton projection and body modeling will lose the stereoscopic sense because of neglecting the thickness of the human body itself; moreover, with the increase in the number of targets, the detection performance and efficiency of 2D pose detection methods are poor in dealing with problems such as multi-person occlusion.

Based on the inadequacy of 2D human pose detection techniques and the lack of depth information, 3D human pose detection techniques [16–18] were developed.Tekin [16] et al. established a deep learning regression framework by fusing convolutional neural network CNN with recurrent neural network RNN; Moon G et al. [17] proposed a camera distance sensing framework to extract the target from a 2D image, using RootNet to localize the root node of the human body and PoseNet to localize the relative root node of the human body to maintain spatial consistency and obtain 3D human pose detection; VoxelPose [18] operates directly in the 3D space, avoiding the case of misdetection for each camera viewpoint. Although these methods achieve better detection results, they still cannot effectively solve the problems of target occlusion as well as correlation and matching between pose points when facing dense scenes.

As 3D modeling of the human body continues to emerge as a promising application in various fields, this technology is gradually attracting scholars to invest in research: SMPLify [19] uses a CNN network to estimate the posture points on the 2D body surface, and then uses a statistical model to fit them to the 2D joints; HMR, an end-to-end human posture detection network proposed by Kanazawa et al. [20] adopts a regressive approach, where images are processed through an encoder and then fed into a 3D regression module as well as a discriminator to minimize the reprojection error before determining the authenticity of the human pose; MVPOSE [21] introduces a recursive Bayesian filter function to detect each pedestrian individually as a way of reducing the problem of correlation determination between the pedestrian's 3D pose points. Although these methods have also achieved fruitful results, when facing the situation of human detection modeling in a huge state space, due to the poor treatment of local poses of individuals in a dense state, there are still obvious distorted modeling results in some wrist joints and other places.

3 Proposed Method

To address the problems of mis-correlation and matching of pose points, poor local poses, and jitter in continuous modeling results caused by the complex cascading relationship between dense pedestrians occluding each other under binocular vision in related work. In this section, the method we proposed in this will be introduced in detail, which mainly focuses on two major aspects of 3D intra-personal pose detection as well as modeling.

3.1 Multi-target Intra-personal 3D Pose Detection

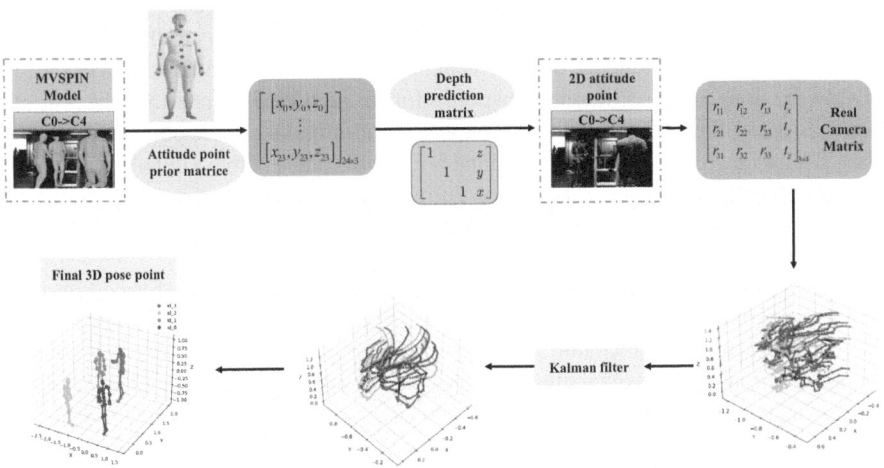

Fig. 1. 3D pose point detection framework.

In the actual multi-view pedestrian detection applications, there are differences in the parameter models, shooting angles, and viewing angles of different cameras, as well as possible occlusion problems between different viewing angles, so it is difficult to obtain accurate pose point data, which increases the complexity of tasks such as pedestrian trajectory prediction and 3D modeling. Therefore, it is particularly important to further improve the human posture detection methods to realize more realistic and effective multi-view pedestrian association.

As shown in Fig. 1, we propose a 3D pose point detection method based on the Skinned Multi-Person Linear Model (SMPL) for the interior of the human body, which obtains the SMPL rendering model of a single view angle and the joint pose a priori matrices through the MVSPIN model, and then computes the 3D predicted pose points that contain the depth prediction information. In order to integrate the information of each viewpoint and enhance the correlation, this paper adopts the camera matrix generated by the weak perspective model to map the 3D predicted pose points, and obtains the 2D pose points under the corresponding viewpoints; and then adopts the real camera matrix to correlate the 2D pose points of each viewpoint, and returns to the 3D space; at the same time, a Kalman filter is introduced to smooth the human pose points and solve the jitter problem of continuous modeling, and finally realizes the multiview multiple pose points under the spatial scene.

2D Image Pose Point Detection Based on MVSPIN. We use the MVSPIN pose dete-ction model to obtain the rendering model information of each target in the planar im-age, and extract 6890 surface pose point coordinates, while obtaining the joint regres-sion prior matrices from the SMPL basic model, which in turn computes the coordina-tes of 24 3D predicted pose points in the generalized human body model.

$$pose\_pre\_result = J\_regressor\_prior \times v\_posed \tag{1}$$

where $J\_regressor\_prior$ is the joint regression prior matrix, v_{posed} is a matrix containing 24 3D pose points.

Since the MVSPIN model is based on a single view for 3D pose point detection, the limitations that are present in the single view angle make errors in the longitudinal depth direction. In addition, due to camera angle, occlusion, etc., there is also a deviation between the images from different viewpoints for the predicted 3D pose and the original image. To solve this problem, this paper considers eliminating the depth data predicted for each view and restoring the 3D coordinates to the 2D plane of the original image based on the depth prediction matrix generated during the construction of each target model. In this way, preparation is made for a more accurate restoration of the 3D information on the image.

As shown in Fig. 2, (a) the camera prediction parameters are transformed into a 3 × 4 camera prediction matrix by a unit matrix; (b) the 3D pose points predicted by the SMPL model are transformed into a 1 × 24 × 3 × 1 pose matrix; and (c) the two matrices are multiplied together to obtain the coordinates of the planar 2D pose point 1 × 24 × 2 × 1. In this paper, the predicted 3D pose points are matrix transformed by constructing a 4 × 4 unit matrix and multiplied with the camera prediction matrix to obtain the coordinates of the pose points in the camera direction of the image prediction.In order to eliminate the error of depth prediction, we eliminate the depth component in the camera direction to obtain the 2D pose point coordinates. This method ensures the completeness of individual target pose points and takes the thickness of pedestrians into account, which provides a good foundation for the following regression from 2D to 3D poses.

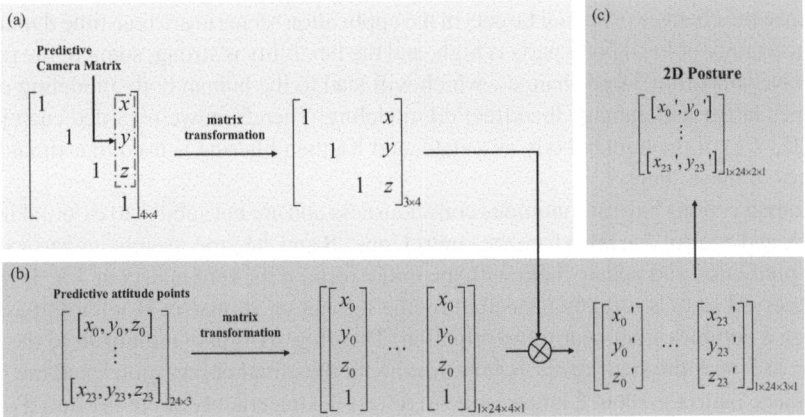

Fig. 2. 2D pose point restoration under various viewing.

3D Pose Point Regression and Filtering Processing. We use the real camera matrix to reproject the 2D pose points of all views into 3D spatial.Since 3D pose point detec-tion is performed in multiple views, there is no need to obtain depth data. It is suffici-ent to regress the 3D coordinates through the vertically projected 2D coordinates of each view and the real camera matrix.

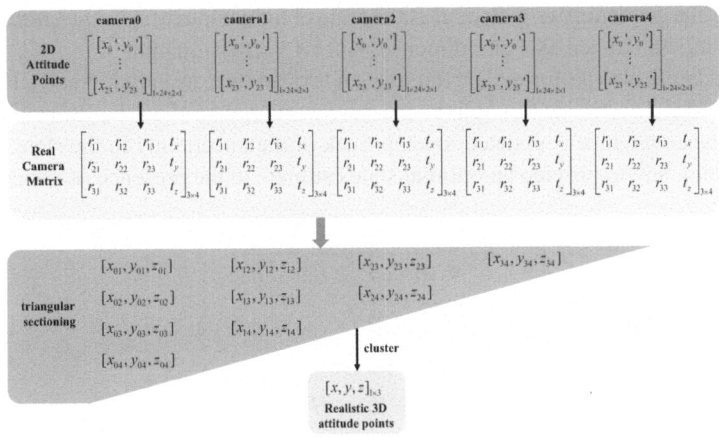

Fig. 3. 3D coordinate regression.

As in Fig. 3, based on the 2D pose point coordinates calculated above, each view is processed with a real camera matrix, and according to the monoclinic constraints existing between the planes [22], the matching and reconstruction between the corresponding points of the two planes are realized by triangular dissection [23] to obtain the 3D pose point coordinates containing the thickness of the human body, in order to better characterize the real position of the pedestrian in space.

Since the changes of spatial targets in the application scenario are real-time dynamic, the uncertainty of local body parts is high, and the flexibility is strong, some of the poses may have short-time large changes, which will lead to the human body modeling error becomes larger and damage the effect od modeling. Therefore, we regarded each pose point (x, y, z) of the human body as a state, and Kalman filtering is used to estimate the changes of these states.

Human actions have autonomous consciousness and are not subject to external intervention and control, we introduce the control input **B** and the process noise covariance **Q** in the prediction and update them with the initial form of the zero matrix of 3×3, while the observed state is directly measured by the measurement matrix **H** without passing through a complex nonlinear transformation. Therefore **H** is also initialized to the unit matrix of 3×3, the state vector is initialized with the initial observations, and the state covariance matrix is set to a larger value to reflect the uncertainty about the initial state.

The whole filtering process is carried out in two major steps, prediction and update [21], and the system model is used to predict the state $\widehat{X}_k$ at the next time step as in Eq. 2. The state covariance $\widehat{P}_k$ is also predicted to determine the uncertainty of the prediction as in Eq. 3.

$$\widehat{X}_k = F \times \widehat{X}_{k-1} + B \times u_k \tag{2}$$

$$P_k = F \times P_{k-1} \times F^T + Q \tag{3}$$

where F is the state transfer matrix, B is the control input matrix, u_k is the control input, for which it is initialized as a one-dimensional row vector of length 3.

The update is also performed on the state and covariance, as in Eqs. 5 and 6 to correct the predicted state estimate through the measurements z_k as well as the Kalman gain K_k, and then the covariance is updated to keep track of the uncertainty in the state estimate of the system, and to provide an optimal state estimate in the presence of uncertainty.

$$K_k = \frac{P_k \times H^T}{H \times P_k \times F^T + R} \tag{4}$$

$$\widehat{X}_k = \widehat{X}_k + K_k \times \left(z_k - H \times \widehat{X}_k \right) \tag{5}$$

$$P_k = (I - K_k \times H) \times P_k \tag{6}$$

where I is the unit matrix, R is the measurement noise covariance matrix.

As shown in Fig. 4 for the comparison of Kalman filtering effect, we select 0-pelvis, 10-left foot, 11-right foot, 15-chin, 18-left elbow and 19-right elbow with a larger range of activity for comparison of the pose points, which can be seen that the filtered trajectory of the movement is smoother.

After the acquisition of 2D pose points from each viewpoint and 3D pose regression, and then Kalman filtering, the 3D human pose points with good poses and smooth continuous modeling are finally obtained, which lays a good foundation for 3D modeling of the human body.

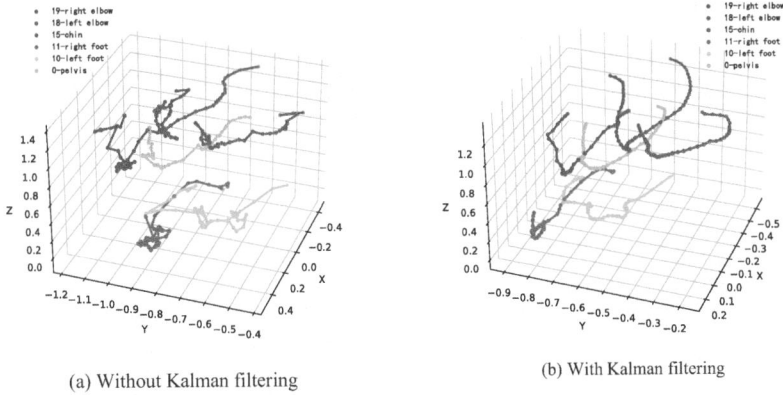

(a) Without Kalman filtering

(b) With Kalman filtering

Fig. 4. Comparison of Kalman filter processing effect

3.2   3D Modeling of Human Body Based on SMPL-VAE

Since the work in the section of 3.1 is global processing for each target as a whole and no local processing is applied, if these pose points are directly used for 3D modeling of the human body, the modeling of local poses of human joints will be distorted.

Therefore, for the problem of 3D modeling of the human body with processed pose points, we refer to the idea of VAE [13] and propose the SMPL-VAE model, as shown in Fig. 6, which combines step-by-step training with joint training to generate 3D modeling results of the human body. The two-way step-by-step training processes the input pose points in different ways, and then the final modeling results are obtained by joint training.

Step-by-step training I: The 3D pose point X finally obtained in the section of 3.1 is taken as input, and the SMPL parameter $\Theta_S(\theta, \beta)$ is trained by the iterative optimization method of gradient descent [24] using the ground truth as a criterion, and the data is continuously fitted to make it closer to the ground truth. Due to the high degree of autonomy and irregularity of the target activity, a loss function that is robust to outliers is needed, so we choose to use Smooth L1 Loss [25], whose strong robustness provides more stable convergence and more reliable error gradients than other loss functions. This loss function strikes a balance between the mean square error and the absolute error, thus reducing the impact of large error values and ensuring a smoother and more gradual optimization process as in Eq. 7.

$$L_{Smpl} = \begin{cases} \frac{1}{2}(X_{gt} - J)^2, & if \left|X_{gt} - J\right| < 1 \\ \left|X_{gt} - J\right| - \frac{1}{2}, & if \left|X_{gt} - J\right| \geq 1 \end{cases} \tag{7}$$

where J is the coordinates of the human joint points output by the model and X_{gt} is the ground truth of the human joint points (Fig. 5).

In this way, we are able to obtain the human body's mesh vertices *verts* and human body joints coordinates J, where *verts* size is $n \times 6890 \times 3$, n is the number of targets in the scene during modeling, and J size is $n \times 24 \times 3$. The essence of this method lies in its ability to refine the model parameters during iteration, and more closely and accurately represent the human body pose 3D step-by-step ground-truthing values. Up

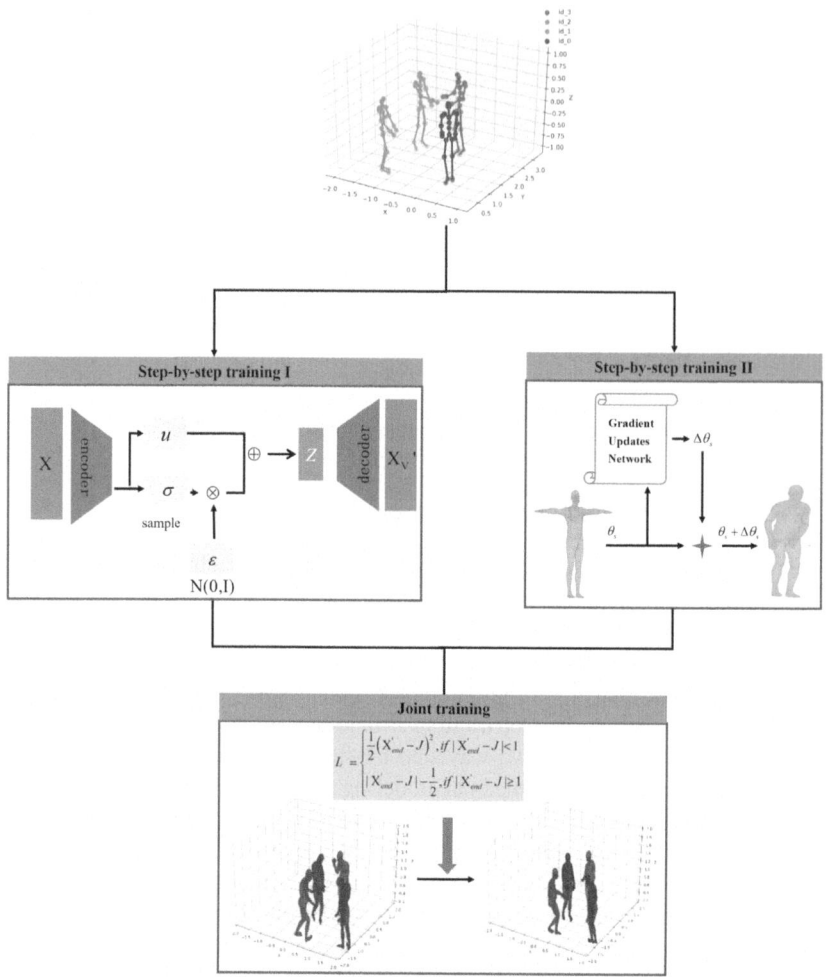

Fig. 5. Human body 3D modeling framework

to this point, step-by-step training one obtains preliminary human 3D modeling results, but in the local joints of the human body, there is still a violation of the laws of human kinematics, which still does not meet the final modeling needs.

Step-by-step training II: We use a Variable Auto-Encoder (VAE) framework to encode and decode the filtered dataset. The VAE obtains the generated 3D human pose points X_v by learning a latent space representation of the input data X, and then reconstructing the inputs using this latent space. This process introduces a relaxation mechanism designed to learn the substeps on the latent space, allowing for generation of new samples, thus accommodating human pose transformations that conform to kinematic constraints.

The encoding phase plays an important role by compressing X into a lower dimensional space that captures the stepwise characteristics of the potential variables needed for reconstruction, and the decoding phase attempts to reconstruct the input data from this compressed representation, preserving the critical structure without compromising the global and local fidelity of the 3D modeling of the human body, as in Eqs. 8 and 9 respectively.

$$z \sim q(z|X) \tag{8}$$

$$X'_V \sim p(X|z) \tag{9}$$

where $q(z|X)$ is the latent variable stepping defined by the encoder; $p(X|z)$ is the conditional stepping defined by the generator.

In the training process, then $\Theta_V(\phi_{en}, \phi_{de})$ is trained, which is done by maximizing the likelihood of the data. However, solving the likelihood function directly is difficult because the latent variables in VAE are continuous. To solve this problem, VAE uses a regularization term that learns the distribution of latent variables to simplify solving the likelihood function. The loss function consists of two parts: reconstruction loss and regularization loss. The reconstruction loss is the difference between the output sample and the original sample in VAE when the input sample is transformed into latent variables by the encoder and then the output sample is generated by the decoder. The smaller the reconstruction loss, the better the performance of the generator. Regularization loss, on the other hand, refers to the difference between the substeps of the latent variables and the a priori substeps in VAE by introducing the KL divergence [26]. This difference is quantified by the KL divergence, with a smaller KL divergence indicating that the latent variable is closer to the a priori substep. Taken together, the loss function of the VAE can be tabulated as:

$$L_{Vae} = -\mathrm{E}\big[\log p(X|z)\big] + \alpha \cdot KL(q(z|X)||p(z)) \tag{10}$$

where α is the tradeoff of the parameters, and $p(z)$ is the prior step. Finally, the results of step-by-step training I and step-by-step training II are fused for joint training. Using the results of step-by-step training I as a starting point, the output of step-by-step training I is corrected according to the results of step-by-step training II to better converge to the actual situation. And the trained parameters Θ (Θ_S, Θ_V) in the step-by-step training are still used for the second training using Smooth L1 Loss, as in Eq. 11, to obtain the final trimmed human pose points X'_{end}.

$$L = \begin{cases} \frac{1}{2}\big(X'_{end} - J\big)^2, & if \big|X'_{end} - J\big| < 1 \\ \big|X'_{end} - J\big| - \frac{1}{2}, & if \big|X'_{end} - J\big| \geq 1 \end{cases} \tag{11}$$

where X'_{end} is the final human joint point coordinates obtained after joint training.

This joint training model allows for knowledge transfer between the two methods. The iterative refinement of the parameters is enhanced by the relaxed constraints imposed by the VAE, and the excellent results of the overall model benefit from the ability to adjust local modeling effects. This combined training approach not only improves the accuracy of the pose estimation, but also ensures that the model adapts to the diversity of human movements, contributing to a more natural and intuitive interaction between humans and the digital environment.

4 Experiments

4.1 Introduction to Public Datasets and Evaluation Metrics

The Shelf [27] dataset was used in this study as a benchmark for model testing. The dataset contains an indoor scene with four people carrying shelves in close proximity and is equipped with five calibrated cameras. Each view presents a different degree of occlusion and the pedestrians may appear to be added, completely disappeared, reappeared, and other complexities, which provides favorable conditions for algorithm testing. The dataset provides a more realistic representation of the pedestrians' action postures, taking into account factors that may have an impact on the tasks of detection, tracking and feature association.

In the process of human 3D pose detection and modeling, we take the ground truth of the complete 3D pedestrian pose point coordinates as the standard in the training process, and due to the problems of missing targets and ID errors in the real values provided by Shelf, we re-labeled 1500 images of the dataset in 5 views based on the 24 bit-pose points of the SMPL model.

Considering that the mean positional error per joint [28] (MPJPE) is expressed in terms of the physical positional errors of the joints, which makes the results more intuitive and easy to understand, the error per joint can help to analyze the model's performance in different parts, thus providing more detailed feedback. Therefore, the quantitative evaluation metric we take is MPJPE, the smaller its value, the better the performance of the model in estimating the joint positions, as in Eq. 12.

$$MPJPE = \frac{1}{N} \sum_{i=1}^{N} \|p_i - \hat{p}_i\| \tag{12}$$

where $\| \cdot \|$ is the Euclidean distance, i.e., the distance between the true joint position and the model-predicted joint position, N is the 24 joint points of the human body, p_i is the true position of each joint point; and $\hat{p}_i$ is the position predicted by the model.

Percentage of Correct Keypoints [29] (PCK) focuses on the location of keypoints, provides an intuitive percentage representation of keypoint accuracy, applies to different thresholds, is flexible and versatile, and facilitates the comparison of the performance of different models on the same task, so we select PCK as an evaluation metric as in Eq. 13.

$$PCK = \frac{1}{N} \sum_{i=1}^{N} \left(\frac{\|p_i - \hat{p}_i\|}{D} \leq T \right) \times 100\% \tag{13}$$

where T is the threshold value, which is selected as 160mm,D is the reference length, which is set as the normalization factor of a single target, δ is the indicator function, which takes the value of 1 if the condition is true and 0 otherwise.

As shown in Table 1, the method proposed in this paper has demonstrated ideal test results. It improved by 3.88 on the MPJPE metric compared to the well-performing Shape-Aware, and it improved by 3.50 on the PCK metric compared to the well-performing SMPLify.

Table 1. Comparison with previous work based on Shelf dataset.

method	MPJPE	PCK
SMPL	107.12	86.24
SMPLify	109.30	88.76
HMR	114.62	83.26
Shape-aware	105.62	85.35
our	101.74	92.26

4.2 Human Posture Detection and Modeling Experiment

The relative positions and occlusion patterns of targets vary greatly between different viewpoints, and there are more ID matching errors and omissions in the detection algorithms. In the context of dense pedestrians under binocular vision, we addresses the issues of pedestrian pose point detection and association matching, jitter in continuous modeling, and local pose distortion. Therefore, a few more critical images in the video stream are extracted to demonstrate the operation effect of our method.

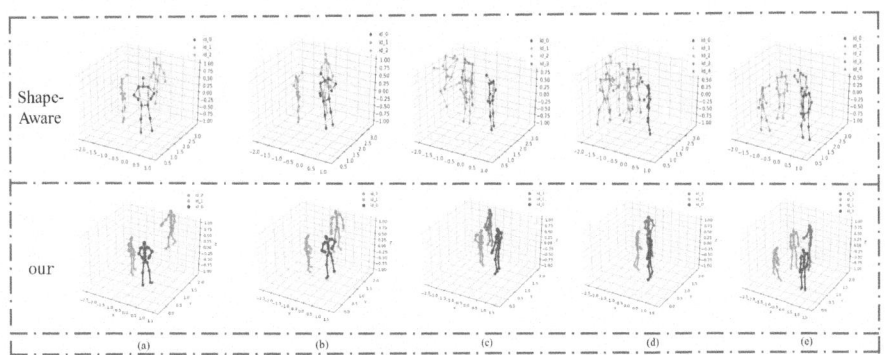

Fig. 6. 3D pose detection result in key frame

As shown in Fig. 6, comparing the algorithm we proposed with Shape-Aware, the framework is able to stably and accurately detect human 3D pose points when dealing with complex human interactions, and the algorithm still achieves better results even when there is an occlusion or overlapping of pedestrians, but Shape-Aware may produce different types of errors, such as (c) in which Shape-Aware gets a distorted pose detection that is not realistic when dealing with the person in the border, and (d) and (e) both of which appear to match the detected pose points to the wrong pedestrian target.

As shown in Fig. 7, the results of comparative and ablation experiments in modeling are presented. In subfigure (a), the 3D human body models constructed using the Shape-Aware and SPEC methods exhibited localized postural distortions, such as upper body deformation, leg shape distortion, and abnormal twisting of the feet. Although the

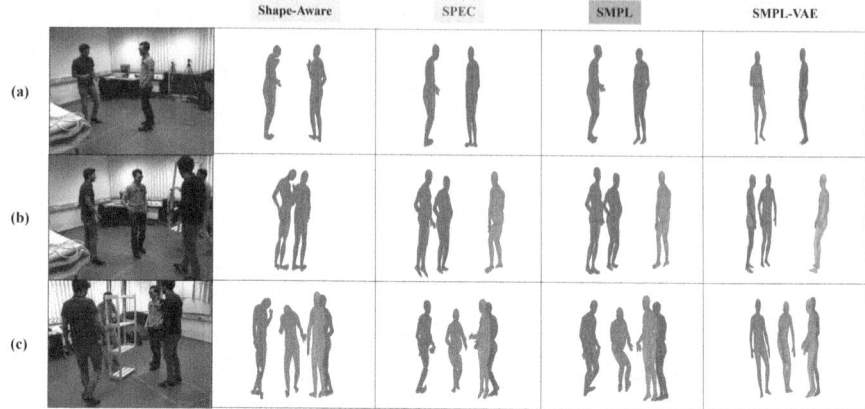

Fig. 7. Human body 3D modeling results

SMPL method improved the modeling results to some extent, significant corrections were achieved when incorporating the VAE module into the SMPL-VAE method. In subfigure (b), the Shape-Aware method exhibited omissions, and the SPEC and SMPL methods resulted in unnatural twists in the foot and head regions, respectively. These issues were effectively optimized in the modeling process using SMPL-VAE. In subfigure (c), problems in the detection of human posture points using Shape-Aware led to erroneous modeling results; SPEC and SMPL exhibited various degrees of modeling distortions (such as the orange model's abdomen and the blue pedestrian's upper body) that did not align with the actual human form, all of which were improved in the SMPL-VAE approach.

5 Conclusion

The 3D pose point detection method for the human body interior based on the skinned multiplayer linear model (SMPL) achieves excellent performance, with a value of 101.74 in the average per-joint position error metric and 92.26 in the percentage of correct keypoint metric, and also matches accurate pose points for different targets. For the noise generated by the clustering algorithm that maps the 2D pose points inside the human body from multiple viewpoints to the 3D pose points in the real scene, we remove the noise by means of Kalman filtering, which effectively solves the problem of jitter in the continuous modeling of pedestrians. The algorithm is based on the consideration of global optimization modeling results on the basis of focusing on the optimization of the local posture, through the automatic variational enconder (VAE) to the gradient descent regression network to correct the construction of the end-to-end human body 3D modeling network (SMPL-VAE), in order to maintain the overall proportion at the same time, more in line with the human body movement structure of the local modeling.

Acknowledgement. This work was supported by the National Natural Science Foundation of China (52202417); China Postdoctoral Science Foundation (2022TQ0155,2022M721605); Open

Project Program of State Key Laboratory of Virtual Reality Technology and Systems, Beihang University (VRLAB2023A02); Young Elite Scientists Sponsorship Program by CAST (2023QNRC001); Young Elite Scientists Sponsorship Program by JSTJ (JSTJ-2023-XH032). Authors thank reviews for their valuable comments.

References

1. Cao, Z., Hidalgo, M.G., Simon, T.: OpenPose: realtime multi-person 2D pose estimation using part affinity fields. IEEE Trans. Pattern Anal. Mach. Intell. **43**, 172–186 (2021)
2. Zhang, W.Q., Fang, J., Wang, X.G.: EfficientPose: efficient human pose estimation with neural architecture search. Comput. Vis. Media **7**, 335–347 (2021)
3. Sun, K., Xiao, B., Liu, D.: Deep high-resolution representation learning for human pose estimation. In: Proceedings of the IEEE/CVF Conference on Computer Vision and Pattern Recognition, NJ, pp. 5693–5703. IEEE (2019)
4. Toshev, A., Szegedy, C.: DeepPose: human pose estimation via deep neural networks. In: Proceedings of the IEEE Conference on Computer Vision and Pattern Recognition, NJ, pp. 1653–1660. IEEE (2014)
5. Sun, Y., Ye, Y., Liu, W.: Human mesh recovery from monocular images via a skeleton-disentangled representation. In: Proceedings of the IEEE/CVF International Conference on Computer Vision, NJ, pp. 5349–5358. IEEE (2019)
6. Pavlakos, G., Choutas, V., Ghorbani, N.: Expressive body capture: 3D hands, face, and body from a single image. In: Proceedings of the IEEE/CVF Conference on Computer Vision and Pattern Recognition, NJ, pp. 10975–10985 (2019)
7. Liu, S.C., Saito, S., Chen, W.K.: Learning to infer implicit surfaces without 3D supervision. In: Advances in Neural Information Processing Systems, vol. 32 (2019)
8. Fang, H.S., Xie, S., Tai, Y.W.: RMPE: regional multi-person pose estimation. In: Proceedings of the IEEE International Conference on Computer Vision, NJ, pp. 2334–2343. IEEE (2017)
9. Li, W.B., Wang, Z., Yin, B.Y., et al.: Rethinking on multi-stage networks for human pose estimation. arXiv preprint arXiv:1901.00148 (2019)
10. Dong, Z.J., Song, J., Chen, X.: Shape-aware multi-person pose estimation from multi-view images. In: Proceedings of the IEEE/CVF International Conference on Computer Vision, NJ, pp. 11158–11168. IEEE (2021)
11. Loper, M., Mahmood, N., Romero, J.: SMPL: a skinned multi-person linear model. Seminal Graph. Pap.: Push. Bound. **2**, 851–866 (2023)
12. Li, Z., Oskarsson, M., Heyden, A.: 3D human pose and shape estimation through collaborative learning and multi-view model-fitting. In: Proceedings of the IEEE/CVF Winter Conference on Applications of Computer Vision, NJ, pp. 1888–1897. IEEE (2019)
13. Diederik, P.K., Max, W.: Auto-encoding variational bayes. Comput. Sci. (2013)
14. Krizhevsky, A., Sutskever, I., Hinton, G.E.: ImageNet classification with deep convolutional neural networks. Commun. ACM **60**(6), 84–90 (2017)
15. Ishwarya, K., Nithya, A.A.: Squirrel search optimization with deep convolutional neural network for human pose estimation. Comput. Mater. Continua **74**(3) (2023)
16. Tekin, B., Katircioglu, I., Salzmann, M.: Structured prediction of 3D human pose with deep neural networks. arXiv preprint arXiv:1605.05180 (2016)
17. Moon, G., Chang, J.Y., Lee, K.M.: Camera distance-aware top-down approach for 3D multi-person pose estimation from a single RGB image. In: Proceedings of the IEEE/CVF International Conference on Computer Vision, NJ, pp. 10133–10142. IEEE (2019)

18. Tu, H., Wang, C., Zeng, W.: VoxelPose: towards multi-camera 3D human pose estimation in wild environment. In: Vedaldi, A., Bischof, H., Brox, T., Frahm, JM. (eds.) ECCV 2020. LNCS, vol. 12346, pp. 197–212. Springer, Cham (2020). https://doi.org/10.1007/978-3-030-58452-8_12

19. Bogo, F., Kanazawa, A., Lassner, C.: Keep it SMPL: automatic estimation of 3D human pose and shape from a single image. In: Leibe, B., Matas, J., Sebe, N., Welling, M. (eds.) ECCV 2016, Part V. LNCS, vol. 9909, pp. 561–578. Springer, Cham (2016). https://doi.org/10.1007/978-3-319-46454-1_34

20. Kanazawa, A., Black, M.J., Jacobs, D.W.: End-to-end recovery of human shape and pose. In: Proceedings of the IEEE Conference on Computer Vision and Pattern Recognition, NJ, pp. 7122–7131. IEEE (2018)

21. Kwon, O.-H., Tanke, J., Gall, J.: Recursive Bayesian filtering for multiple human pose tracking from multiple cameras. In: Ishikawa, H., Liu, C.-L., Pajdla, T., Shi, J. (eds.) ACCV 2020. LNCS (LNAI and LNB), vol. 12623, pp. 438–453. Springer, Cham (2021). https://doi.org/10.1007/978-3-030-69532-3_27

22. Tulsiani, S, Efros, A.A., Malik, J.: Multi-view consistency as supervisory signal for learning shape and pose prediction. In: Proceedings of the IEEE conference on computer vision and pattern recognition, NJ, pp. 2897–2905. IEEE (2018)

23. Ke, S.R., Zhu, L.J., Hwang, J.N.: Real-time 3D human pose estimation from monocular view with applications to event detection and video gaming. In: 2010 7th IEEE International Conference on Advanced Video and Signal Based Surveillance, NJ, pp. 489–496. IEEE (2010)

24. Shi, J.R., Wang, D., Shang, F.H.: Research progress of stochastic gradient descent algorithm. Acta Autom. Sinica 47(9), 2103–2119 (2019)

25. Wei, L., Zheng, C., Hu, Y.: Oriented object detection in aerial images based on the scaled smooth L1 loss function. Remote Sens. 15(5), 1350 (2023)

26. Ramakrishna, V., Munoz, D., Hebert, M.: Pose machines: articulated pose estimation via inference machines. In: Fleet, D., Pajdla, T., Schiele, B., Tuytelaars, T. (eds.) ECCV 2014, Part II. LNCS, vol. 8690, pp. 33–47. Springer, Heidelberg (2014). https://doi.org/10.1007/978-3-319-10605-2_3

27. Belagiannis V, Amin S, Andriluka M.: 3D pictorial structures for multiple human pose estimation. In: 2014 IEEE Conference on Computer Vision and Pattern Recognition, Los Alamitos, CA, USA, vol. 1, pp. 1669–1676. IEEE Computer Society (2014)

28. Tian, W, Gao, Z, Tan, D.: Single-view multi-human pose estimation by attentive cross-dimension matching. Front. Neurosci. 17 (2023)

29. Rani, C.J., Devarakonda, N., Kumari, K.W.S.N., Malavath, P.: A monadic and effective frame work for single human pose estimation of 2D images and videos. In: Chen, J.I.Z., Tavares, J.M.R.S., Iliyasu, A.M., Du, K.L. (eds.) ICIPCN 2021. LNNS, vol. 300, pp. 254–268. Springer, Cham (2022). https://doi.org/10.1007/978-3-030-84760-9_23

Simulation of Space Radiation Model for Multilayer Electret Based Low Frequency Antennas

Zhihong Yuan, Yong Cui(✉), Xiao Song, Zhi Cui, Yiming Li, and Wenjie Qu

Beihang University, Beijing 100191, China
cuiyong@buaa.edu.cn

Abstract. Low frequency electromagnetic waves, characterized by strong anti-interference capability and high penetrative power, are recognized as excellent carriers for communication. Multilayer FEP/THV based antennas, emerging as an advanced low frequency antennas, offer advantages such as compactness, lightweight design, and high transmission efficiency. Different from other simulations that only focus on increasing the number of layers to enhance radiation intensity, this work utilized COMSOL simulation software to establish a finite element model of multilayer electret based low frequency antennas to further understand the spatial radiation field variations. The simulation results provide the spatial radiation model of the antenna and the primary magnetic induction direction is obtained. Additionally, the investigation explores variations in the radiation intensity of the antenna with differing numbers of layers while maintaining constant total charge density. These findings offer valuable insights for the realization of multilayer multilayer electret based low frequency antenna in the future.

Keywords: Low frequency · Multilayer electret · COMSOL

1 Introduction

Low frequency (LF, between 3 Hz and 300 kHz) electromagnetic (EM) wave possesses significant advantages in robustness and penetration, which means it maintains strong capability to transmit information even within complex media such as seawater, soil, and coal beds [1–3]. Thus, the LF EM signal is of great potential to be applied in underwater communication, pipeline monitoring, through-the-earth communication, and so on [4–6]. However, traditional LF antennas have drawbacks of large size, low radiation efficiency, and high-power consumption which severely limit LF EM signal's application.

In recent years, a new type of low frequency antennas called mechanical antenna has been proposed to solve the above-mentioned problems. This type of antenna does not require a huge impedance matching network and can realize the miniaturization of low-frequency communication equipment.Depending on the different core materials,

S. Saito et al. (Eds.): AsiaSim 2024, CCIS 2170, pp. 101–113, 2024.
https://doi.org/10.1007/978-981-97-7225-4_8

they can be broadly classified into three categories: permanent magnet [1, 7–9], electret [10–13] and piezoelectric [14–17].

Among them, a rotating unipolar electret antenna based on Fluorinated ethylene propylene (FEP) and Terpolymer of tetrafluoroethylene, hexafluoropropylene and vinylidene fluoride (THV) becomes the focus of attention [18]. It possesses several advantages including low near-field attenuation rate, broad frequency modulation range, high transmission efficiency, and lightweight design, which endow it with significant potential for various applications.

Benefiting from the flexibility of electrets, the layers of electrets has become the primary method for enhancing the radiation intensity of antennas. However, at present, the main focus of FEP/THV based antennas simulation remains on exploring methods to increase radiation intensity through layer stacking, while lacking micromesh research into spatial radiation characteristics and the influence of interlayer spacing on radiation intensity. Therefore, to further understand the radiation characteristic of multilayer electret based low frequency antennas and expand their application potential, this work established a finite element model using COMSOL simulation software and conducted a systematic simulation analysis.

2   Model Building

2.1   Theoretical Model

The mechanism of electromagnetic field radiation from a rotating unipolar polarized low-frequency antenna can be equivalently represented as the radiation generated by a charged particle undergoing uniform circular motion.

The charge carried by the particle is denoted as q

$$q = n\rho_{ele}S \qquad (1)$$

where, n is the number of layers of polarized bodies, ρ_{ele} is the surface charge density of the polarized bodies, and S is the surface area of the polarized bodies.

Based on the Liénard-Wiechert potentials and considering the retardation effect, we can determine the electromagnetic scalar $\varphi(x, t)$ and vector $A(x, t)$ produced by a charge moving along a specific trajectory at time t, while the distance from the center of circular motion to the observation point is x.

$$\varphi(x, t) = \frac{n\rho_{ele}S}{4\pi\varepsilon_0\left(r - \frac{1}{c}v \cdot r\right)} \qquad (2)$$

$$A(x, t) = \frac{\mu_0 n\rho_{ele}Sv}{4\pi\left(r - \frac{1}{c}v \cdot r\right)} \qquad (3)$$

where ε_0 is the vacuum relative permittivity, μ_0 is the vacuum permeability, v is the velocity of the charge, r is the distance from the charge to the observation point at time t, and c is the velocity of light.

Because the velocity v of charge motion during the operation of the mechanical antenna is much smaller than the speed of light c, Eqs. (2) and (3) can be rewritten as:

$$\varphi(x, t) = n\rho_{\text{ele}}S\left(\frac{1}{4\pi\varepsilon_0 r} + \frac{v \cdot r}{4\pi\varepsilon_0 c r^2}\right) \tag{4}$$

$$A(x, t) = \frac{n\rho_{\text{ele}}S}{4\pi\varepsilon_0 c^2 r} \tag{5}$$

Combining the Maxwell's equations, we can obtain the magnetic flux density and electric field strength generated by the moving charge as:

$$\begin{cases} B = \nabla \times A = \frac{n\rho_{\text{ele}}S}{4\pi\varepsilon_0}\left(\frac{\hat{v}\times\hat{r}}{c^2 r^2} + \frac{\hat{a}\times\hat{r}}{c^3 r}\right) \\ E = -\frac{\partial A}{\partial t} - \nabla\varphi = \frac{n\rho_{\text{ele}}S}{4\pi\varepsilon_0}\left(\frac{\hat{r}}{r^3} + \frac{\hat{r}\times(\hat{r}\times\hat{a})}{c^2 r}\right) \end{cases} \tag{6}$$

where $\hat{a}$ is the angular acceleration of charge

Due to the charge undergoing uniform circular motion, v and a can be rewritten as:

$$v = \omega R \tag{7}$$

$$a = \omega^2 R \tag{8}$$

where ω is the angular velocity of the charge's rotation, and R is the radius of the charge's circular motion. Combing with Eqs. (7) and (8), Eqs. (6) can be rewritten as:

$$\begin{cases} B = \frac{n\rho_{\text{ele}}S\omega R}{4\pi\varepsilon_0 c^2 r^2}\hat{v}\times\hat{r} + \frac{n\rho_{\text{ele}}S\omega^2 R}{4\pi\varepsilon_0 c^3 r}\hat{a}\times\hat{r} \\ E = \frac{n\rho_{\text{ele}}S}{4\pi\varepsilon_0 r^3}\hat{r} + \frac{n\rho_{\text{ele}}S\omega^2 R}{4\pi\varepsilon_0 c^2 r}\hat{r}\times(\hat{r}\times\hat{a}) \end{cases} \tag{9}$$

According to the theoretical model, the magnetic flux density generated by a rotating unipolar electret based low frequency antennas is directly proportional to factors such as the number of polarized body layers, the charge density per layer, the surface area of the polarized bodies, and the radius of the cylinder support structure. Moreover, higher antennas' rotation frequencies correspond to increased magnetic flux density production.

Because the angular velocity of charge rotation ω is much smaller than the speed of light c, the second term in the expression for magnetic flux density and electromagnetic intensity in Eq. (9) is significantly smaller than the first term. Therefore, it can be approximated that for multilayer electret based low frequency antenna, the magnetic flux density decreases inversely proportional to the square of the distance, and the electric field intensity decreases inversely proportional to the cube of the distance. Therefore, considering practical applications, this paper mainly focuses on the spatial distribution of the magnetic flux density from the transmitting source.

2.2 Simulation Model

In theoretical models, the external shape and volume of unipolar electret based low-frequency antennas are disregarded when considering their electromagnetic radiation characteristics. To further understand the spatial electromagnetic radiation properties of low-frequency antennas based on the multilayer unipolar electret, this work incorporated the shape and volume of the antenna into our analysis. Utilizing COMSOL simulation software, we constructed models of the antenna and its radiation space, and conducted simulation analyses on the electromagnetic field intensity it generates.

To reduce the complexity of the model, certain adjustments were made to the antenna model. Firstly, multilayer electret based low frequency antenna typically employ motors as driving structures, which introduce interference signals at the same frequency but different phase as the dipole. To mitigate this interference, reducers were employed to alter the frequency of the interfering signals generated by the motor drive. Consequently, the influence of the motor on the magnetic induction intensity was disregarded, and the antenna was simplified to a cylindrical shape with a radius of 4 cm and a height of 10 cm.

To streamline the representation of the uneven charge distribution on the dipole, we replaced it with a uniform charge density σ. Additionally, since the mechanical antenna lacks a rotation mechanism, we substituted the magnetic field generated by the movement of charges on the dipole with a current-induced magnetic field. The formula for calculating the current is as follows:

$$J_{s0} = \vec{n} * \left(\vec{H_1} - \vec{H_2}\right) \tag{10}$$

Finally, we constructed a simplified model as shown in Fig. 1, where the negative polarity dipole (blue portion) occupies 50% of the surface area relative to the cylinder. Under this configuration, the magnetic induction intensity generated by the low-frequency antenna is maximized.

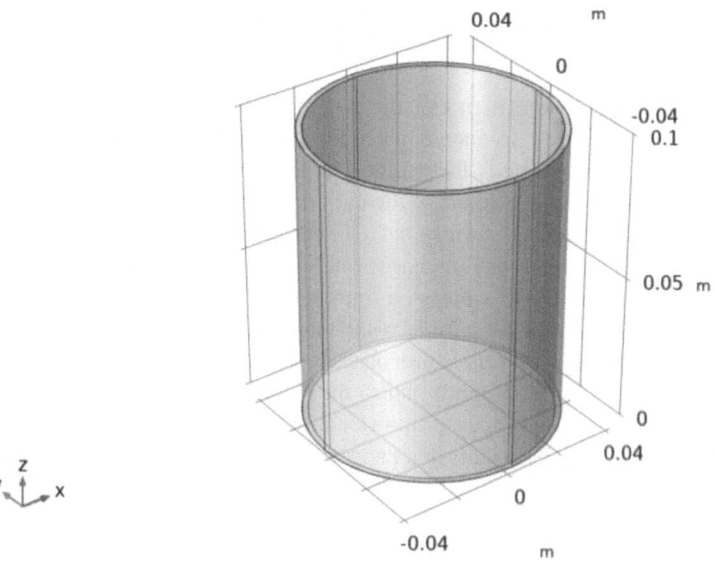

Fig. 1. Simplified model of multilayer electret based low frequency antenna

In order to investigate the spatial radiation characteristics of the multilayer electret based low frequency antenna, it is imperative to establish an electromagnetic wave propagation medium outside the transmitting source. Given that the material properties of the propagation medium has significant influence on the attenuation of electromagnetic waves, it is essential to judiciously design the material parameters. The magnetic permeability μ, electrical conductivity σ, and relative dielectric ε constant of the propagation medium are critical parameters affecting the magnetic induction intensity. Additionally, the size of the medium domain represents the computational distance for finite element simulation of the magnetic field radiation characteristics. Therefore, based on the physical properties of air, we set $\mu = 1$, $\sigma = 1 \times 10^{-14}$ S/m, $\varepsilon = 1$ and the radius of the spherical medium domain is 1m. The holistic model is shown in Fig. 2(a).

Finite element simulation entails discretizing the physical system into numerous small elements and employing finite element methods for solution using simple functions. Therefore, it is necessary to meticulously mesh the model in order to ensure accurate finite element simulation results. To guarantee uniformly dense computational

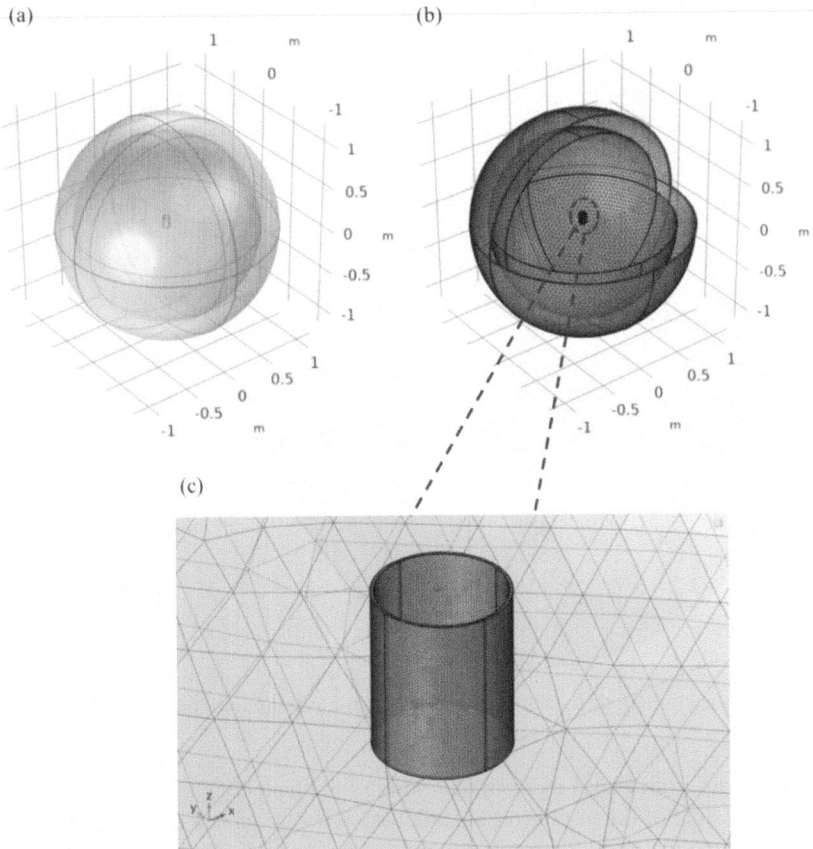

Fig. 2. Model with media domain and meshing

elements, which have a direct impact on the accuracy of the finite element simulation results, mesh generation is conducted in different mesh-sizes for the antenna and the propagation medium domain separately. Tetrahedral meshes exhibit excellent geometric adaptability, thus in this work, the tetrahedral meshes are employed for both the propagation medium domain and the antenna. The partitioning results are depicted in Fig. 2(b) and 2(c). The results indicate that the model has been adequately partitioned.

3 Simulation Results and Discussion

3.1 Space Radiation Characteristic

The surface charge density of the charge attached to the cylinder is equivalent to three layers of unipolar polarized bodies. Each layer has a charge density of 0.56 mC/m^2, resulting in a total charge density of 1.68 mC/m^2 across the three layers. The antenna operates at a frequency of 20 Hz. Utilizing the transient electromagnetic wave module, the Maxwell equation group is solved to conduct a finite element simulation of the magnetic field radiation characteristics of the low-frequency antenna. The resulting magnetic field distribution in the plane (perpendicular plane) containing the rotation axis is depicted in Fig. 3. In Fig. 3, the black rectangle represents the side view projection of the cylinder support structure, while the white dashed line indicates the rotation axis.

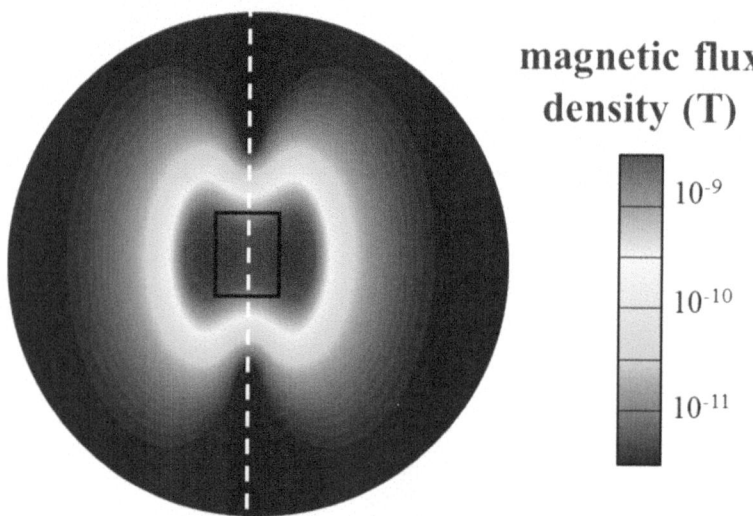

Fig. 3. Space radiation characteristic of multilayer electret based low frequency antenna

The results from Fig. 3 indicate that within the plane containing the rotation axis, the radiation field of the multilayer electret based low frequency antenna exhibits axial symmetry along the rotation axis. Furthermore, the magnetic flux density generated by the low frequency antenna is weakest at the rotation axis and strongest at a position passing through the center of the cylinder and perpendicular to the rotation axis.

Furthermore, to compare the components of the magnetic flux density generated by a multilayer electret based low frequency antenna in different directions, we establish a cartesian coordinate system as shown in Fig. 4. In Fig. 4, the direction of the antenna's rotation axis is considered as the Z- axis, the rotation plane as the XY-plane, and the origin O as the center of rotation.

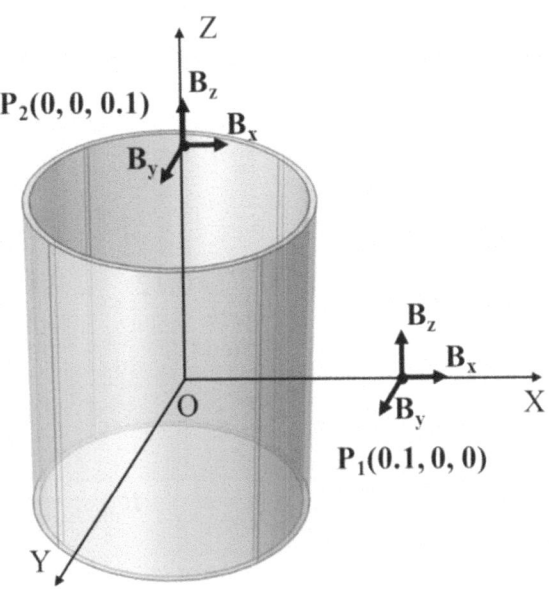

Fig. 4. Cartesian coordinate system

Due to the axial symmetry of the multilayer electret based low frequency antenna, we select th epoint P_1 (0.1, 0, 0) as observation point located within the rotating plane at the center of the cylinder and 0.1 m away from the antenna. The temporal variations of the magnetic flux density in the X, Y, and Z directions at this point are observed. The simulated results are shown in Fig. 5(a), and the enlarged temporal variations of the magnetic induction intensity in the x and y directions are shown in Fig. 5(b).

On the rotating surface, the magnetic flux density in the X, Y, and Z directions at point P_1 exhibits periodic variations that synchronize with the rotation frequency of the antenna.Notably, the magnetic flux density along the Z-axis significantly surpasses that of the X-axis and Y-axis, reaching up to 1.2×10^{-8} T, which constitut the primary component. Meanwhile, the magnetic flux density along the X-axis and Y-axis are comparatively smaller, approximately 6.0×10^{-11} T and 8.5×10^{-11} T respectively, representing secondary components.

Similarly, select point P_2 (0, 0, 0.1) located 0.1 m away from the antenna on the rotation axis. The time-domain variation curves of the magnetic induction intensity in the X, Y, and Z directions at this point are shown in Fig. 6. In Fig. 6, the components of the low-frequency electromagnetic wave along the X-axis and Y-axis are sinusoidal signals, which are the main components, with the peak value of approximately $5 \times$

(a)

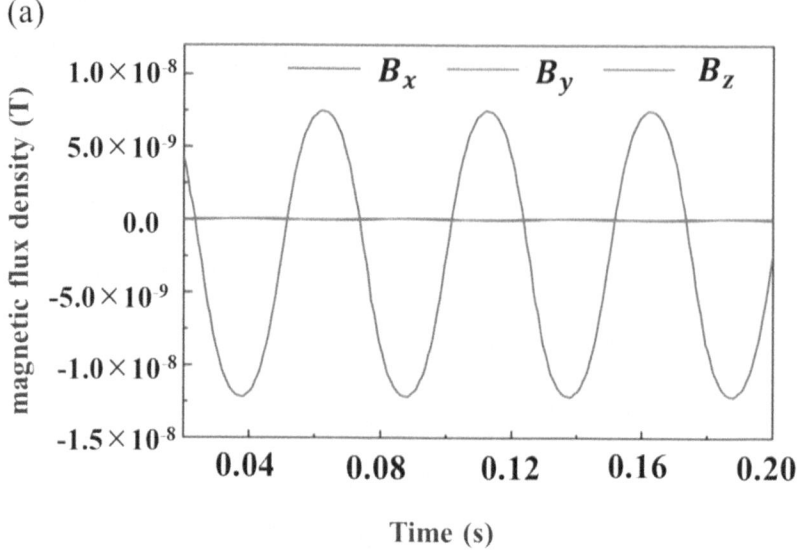

(b)

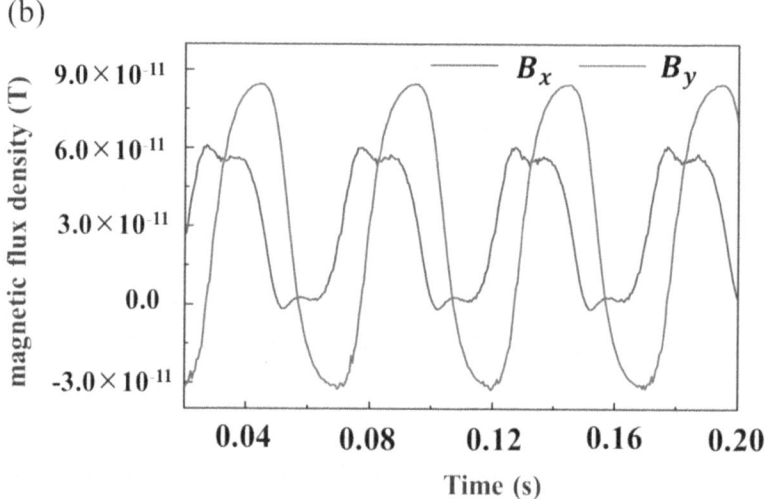

Fig. 5. The magnetic flux density alone X, Y, Z-axis on the rotating surface

10^{-9} T. Different form the rotating surface, the magnetic field along the Z-axis on the rotation axis remains a constant static magnetic field, while magnetic flux density is unchangable with the antennas' rotaing frequency.

In summary, multilayer electret based low frequency antenna operates within the rotation plane where its center is situated, with the vertical component serving as the predominant magnetic induction intensity within this plane.

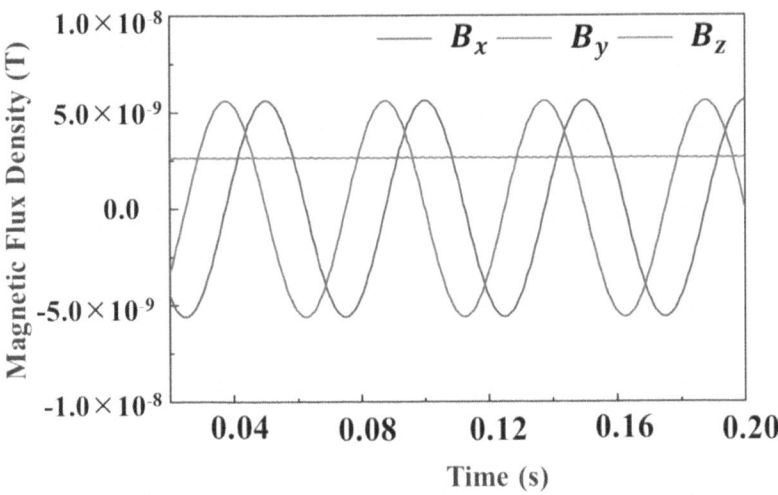

Fig. 6. The magnetic flux density alone X, Y, Z-axis on the rotating axis

3.2 Simulation Results Considering the Number of Layers

In the above simulation, the issue of interlayer spacing for electrets was overlooked. When considering the actual stacking of electret layers, the small interlayer spacing can result in breakdown in the air. According to the Paschen's law, the breakdown voltage V_{bre} between electret layers can be calculated as the product of the gas pressure P and the gap distance d between the layers.

$$V_{bre} = \frac{APd}{ln(BPd) - ln\left[ln\left(1 + \frac{1}{\gamma_{se}}\right)\right]} \tag{11}$$

where A and B are experimentally determined constants, and their values are determined by the ratio of electric field intensity to gas pressure. γ_{se} is the secondary electron emission coefficient, which characterizes the ratio of the number of electrons leaving the material surface under the action of incident electrons to the number of incident electrons. It means, to avoid breakdown, when the charge density of the electrets is higher,, the spacing between layers should be larger. Therefore, this paper considers using multi-layer electrets with low charge density to replace electrets with high charge density, which can reduce the requirement for interlayer spacing.

In order to investigate the impact of varying layer numbers on the magnetic flux density of multilayer electret based low frequency antenna, two models are constructed as depicted in Fig. 7. Model 1 (Fig. 7(a)) incorporates three layers of unipolar electrets with a charge density of 0.56 mC/m², while Model 2 (Fig. 7(a)) comprises five layers with a charge density of 0.336 mC/m². The interlayer spacing in both models is set to $d = 3$ mm, which ensures the absence of breakdown. Due to comparable surface areas for the electrets in both models, the total charge remains consistent. The relationship between simulated mechanical antenna radiation intensity and distance variation is delineated in Fig. 8.

(a)   (b)

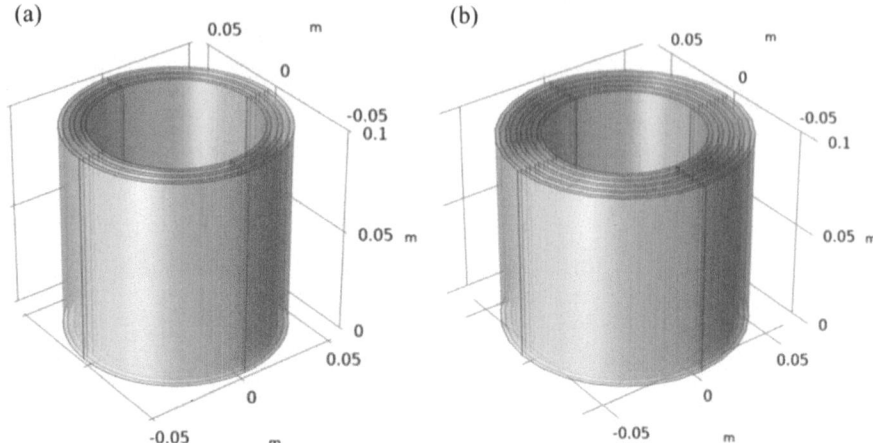

Fig. 7. Antennas model with different numbers of electret layers

In Fig. 8, the simulation results indicate that the near-field radiation intensity of electret-based mechanical antennas decreases inversely with distance. When the distance to the center of the antenna is relatively close, the size of the antenna cannot be neglected. Due to the fact that the antenna's radius is the sum of the supporting structure radius and the interlayer spacing, while Model 2 has more layers, the overall radius of the antenna in Model 2 is larger than that in Model 1, resulting in a greater radiation intensity. Comparing the results obtained in Sect. 3.1, the enhancement in radiation intensity when considering interlayer spacing is relatively minor, which further corroborates the validity of the simplified model.

Extracting the magnetic field intensity data at a distance of 25 cm from the antennas' center, where the size of the antennas can be approximately neglected, data retrieved from COMSOL indicates 1.7×10^{-9} T for Model 1 and 1.8×10^{-9} T for Model 2, with a difference of only 5.8%. Therefore, as the distance increases, the influence of the radius diminishes, and the radiation intensities of the two models become approximately equal.

Therefore, by increasing the number of layers of electrets, multilayer electre based low frequency antennas can generate strong radiation intensity even when the net charge density of the electrets is low. It implies that we can utilize this approach to alleviate the constraints on interlayer spacing for electrets.

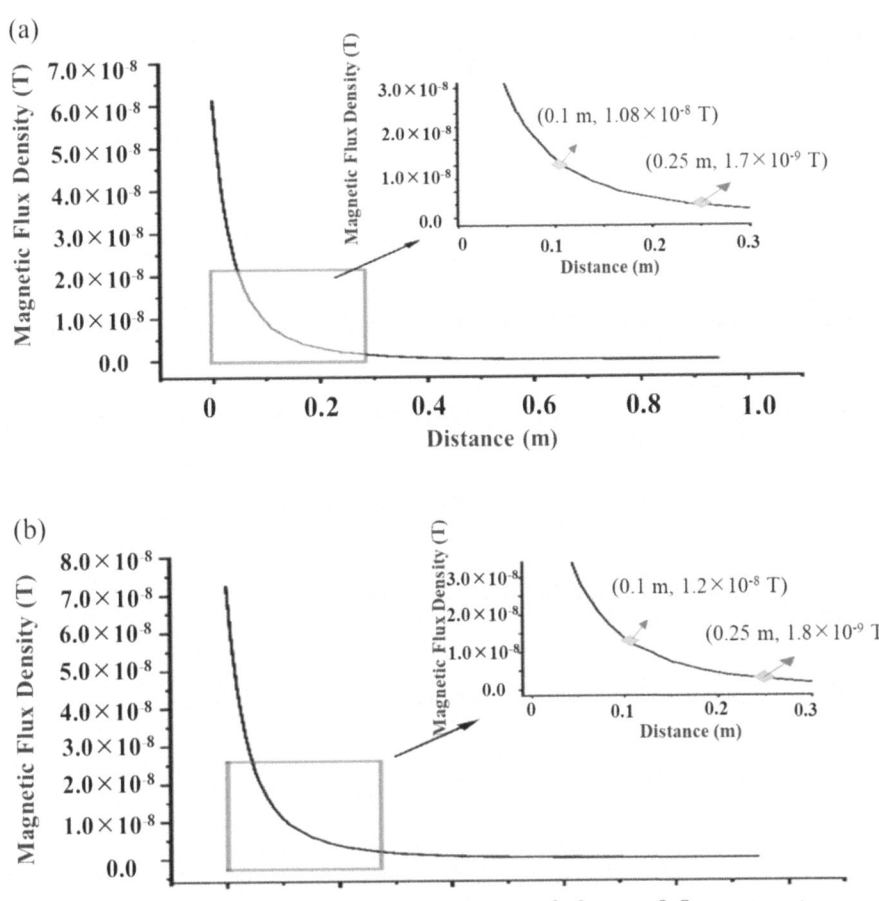

Fig. 8. The magnetic flux density changing with distance in different number of electret layers

4 Conclusion

This work employs COMSOL simulation software to construct a model of the multi-layer electret based low frequency antenna. The simulation results reveal that the antenna operates predominantly within the rotation plane centered on its location, with the vertical component serving as the primary magnetic induction intensity within this plane. Meanwhile, this work also considers the influence of the number of electret's layers on the radiation intensity. When the total charge density remains constant, increasing the number of layers while reducing the charge density of individual electret layers can mitigate the requirements for interlayer spacing and remain the same magnetic flux density strength. This approach facilitates the realization of multilayer electret based antennas in the future and enhances its spatial utilization efficiency.

Acknowledgments. This work was supported by National Natural Science Foundation of China (62271029).

References

1. Yang, S., et al.: Long-range EM communication underwater with ultracompact ELF magneto-mechanical antenna. IEEE Trans. Antennas Propag. **71**(3), 2082–2097 (2022)
2. Zhang, W., Cao, Z., Wang, X., Quan, X., Sun, M.: Design, array, and test of super-low frequency mechanical antenna based on permanent magnet. IEEE Trans. Antennas Propag. (2023)
3. Han, J., Geng, J., Wu, H., Wang, K., Zhou, H., Ren, C., et al.: The ultra-compact ELF magneto-mechanical transmission antenna with the speed modulated EM signal based on three-phase induction motor. IEEE Trans. Antennas Propag. **69**(9), 5286–5296 (2021)
4. Cui, Y., et al.: A novel positioning method for UAV in GNSS-denied environments based on mechanical antenna. IEEE Trans. Ind. Electron. (2024)
5. Chen, Y.-L., Liu, Z.-C., Li, A., Ning, D.-Y., Hou, J.-Y., Zhang, Z.-M., et al.: 3-D positioning method of pipeline robots based on a rotating permanent magnet mechanical antenna. IEEE Sens. J. **23**(7), 7095–7104 (2023)
6. Yenchek, M.R., Homce, G.T., Damiano, N.W., Srednicki, J.R.: NIOSH-sponsored research in through-the-earth communications for mines: a status report. IEEE Trans. Ind. Appl. **48**(5), 1700–1707 (2012)
7. Song, W., Shuai, C., Zhai, Q.: Magnetic field model of radially magnetized cylindrical permanent magnets in a self-biased magnetic pendulum array. Heliyon. **9**(11) (2023)
8. Liu, Y., Liu, Q., Gong, S., Hou, M., Xu, R.: Chirp-rate shift keying modulation for mechanical antenna based on rotating dipoles. IEEE Trans. Antennas Propag. **71**(4), 2989–2999 (2023)
9. Liu, Y., Gong, S., Liu, Q., Hou, M.: A mechanical antenna for undersea magnetic induction communication. IEEE Trans. Antennas Propag. **69**(10), 6391–6400 (2021)
10. Barani, N., Sarabandi, K.: A frequency multiplier and phase modulation approach for mechanical antennas operating at super low frequency (SLF) band. In: IEEE International Symposium on Antennas and Propagation and USNC-URSI Radio Science Meeting. Atlanta, GA, USA, pp. 2169–2170 (2019)
11. Barani, N., Sarabandi, K.: Mechanical antennas: emerging solution for very-low frequency (VLF) communication. In: IEEE Antennas and Propagation Society International Symposium on Antennas and Propagation & Usnc/Ursi National Radio Science Meeting. Boston, MA, USA, pp. 95–96. IEEE (2018)
12. Cui Y, et al.: The optimization of multilayer spacing for miniaturization of mechanical antenna based on unipolar electrets. IEEE Trans. Dielectr. Electr. Insul. (2023)
13. Zhang, W., et al.: An ultra-compact mechanical antenna based on PTFE highly charged electret for extremely low frequency communications. Heliyon (2024)
14. Cao, J., Yao, H., Pang, Y., Xu, J., Lan, C., Lei, M., et al.: Dual-band piezoelectric artificial structure for very low frequency mechanical antenna. Adv. Compos. Hybrid Mater. **5**(1), 410–418 (2022)
15. Xu, J., Cao, J., Guo, M., Yang, S., Yao, H., Lei, M., et al.: Metamaterial mechanical antenna for very low frequency wireless communication. Adv. Compos. Hybrid Mater. **4**(3), 761–767 (2021)
16. Zaeimbashi, M., et al.: Ultra-compact dual-band smart NEMS magnetoelectric antennas for simultaneous wireless energy harvesting and magnetic field sensing. Nat. Commun. **12**(1), 1–11 (2021)

17. Xiao, N., Wang, Y., Chen, L., Wang, G., Wen, Y., Li, P.: A 2D electro-magneto-elastic model and experimental verification for nonvolatile pattern reconfigurable magnetoelectric antenna. IEEE Trans. Antennas Propag. (2024)
18. Cui, Y., Wu, M., Li, Z., Song, X., Wang, C., Yuan, H., et al.: A miniaturized mechanical antenna based on FEP/THV unipolar electrets for extremely low frequency transmission. Microsyst. Nanoeng. **8**(1), 58 (2022)

Reduction of Knocking Sound Annoyance Through Engine Sound Order Simulation

Tomoyuki Hakozaki[1]([✉]), Shunsuke Ishimitsu[1], Ryuhei Suzuki[1], Kiyoaki Iwata[2], Satoshi Fujikawa[2], Mitsunori Matsumoto[2], Masakazu Kikuchi[2], and Naoki Kishikawa[2]

[1] Hiroshima City University, 3-4-1, Ozukahigashi, Asaminami-ku, Hiroshima-shi, Hiroshima 731-3194, Japan
hakozaki@sd.info.hiroshima-cu.ac.jp
[2] Mazda Motor Corporation, 3-1 Shinchi, Fuchu-cho, Aki-gun, Hiroshima 730-8670, Japan

Abstract. The objectives of this study were to mitigate the annoyance caused by engine noise identify the underlying factors contributing to knocking noise during engine vehicle acceleration, and explore methods for its reduction. A time-frequency analysis was employed to examine the characteristics of knocking sound, revealing a peak frequency between 1 and 5 kHz at the onset of acceleration, which was identified as the knocking sound. Given the proximity of the frequencies in the low-order components at the onset of acceleration, it was hypothesized that annoyance arises from the simultaneous occurrence of low-frequency humming or fluctuations alongside the knocking sound. To address this, stimulus sounds were generated by isolating the order components and knocking sound during acceleration. A sound synthesis method, called the Sinusoidal Model, was used to generate the stimulus sounds with high and low-order components. The knocking sound was isolated by applying a 1 kHz high-pass filter to the acceleration sound. Subsequently, the isolated order components were combined with the knocking sound to create the stimulus. Auditory evaluations were conducted using these stimulus sounds, and the results were analyzed in conjunction with psychoacoustic indices such as loudness and roughness.

Keywords: Knocking sound · Order components · Psychoacoustical indices

1 Introduction

Sound is an ever-present physical phenomenon in our daily lives. The same sound can evoke different auditory impressions, depending on the listener. For instance, the driver of an engine vehicle may have a positive opinion of engine sounds and perceive them as sporty. In contrast, residents living near roads often express negative opinions and find the same sounds noisy. Owing to this noise problem, there is an on-going transition from internal combustion engine vehicles to electric vehicles. Nevertheless, efforts have been made to integrate loudspeakers into electric vehicles to simulate the acceleration sound characteristics of internal combustion engines. A previous study [1], showed that

S. Saito et al. (Eds.): AsiaSim 2024, CCIS 2170, pp. 114–124, 2024.
https://doi.org/10.1007/978-981-97-7225-4_9

the "sportiness" and "power" of sound increased when the order component of the engine vehicle acceleration sound was adjusted. This suggests that although electric vehicles are becoming more popular, a certain number of gasoline and diesel engine vehicles and users will remain. However, the problem of engine-caused noise and its annoyance remains. Although various studies have been conducted on noise control, few have focused on noise annoyance; and, previous studies [1] have been inefficient in this regard. Therefore, in this study, With the aim of constructing a system that reduces the annoyance in engine sound and leaves the desirable components, the effect of annoyance on knocking sound due to engine sound order components was investigated.

2   Sinusoidal Model [2]

The order components of the engine sound can be easily estimated and reproduced based on the RPM and the number of cylinders of the engine, but it is difficult to estimate and reproduce the instantaneous power of the order components. In this study, a sinusoidal model was adopted as a synthesis method. A flowchart for the Sinusoidal Model is shown in Fig. 1.

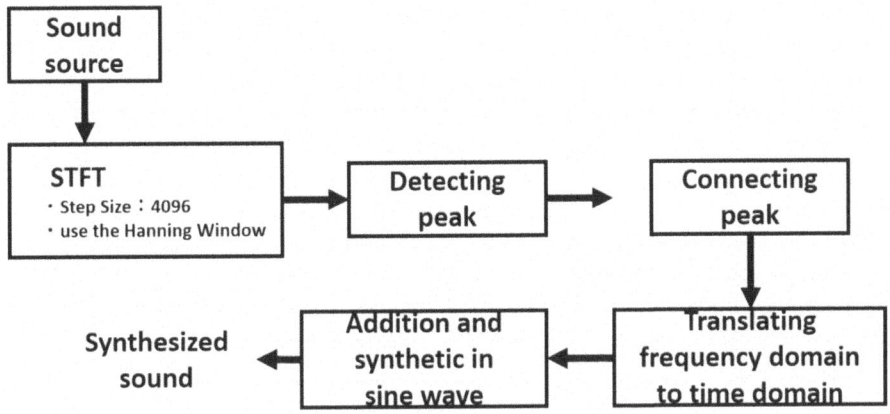

Fig. 1. Flowchart of the Sinusoidal Model

The procedure is as follows: First, a short-term Fourier transform (STFT) is applied to the source data for analysis. For this process, a step size of 4096, an overlap of 50%, and a Hanning window were utilized. Subsequently, peaks were identified in the spectrogram of each frame. From these peaks, instantaneous amplitude, frequency, and phase information were extracted, synthesized, and subjected to inverse fast Fourier transform (IFFT). Then, a constant amplitude signal was generated via resynthesis with a sinusoidal waveform. In this study, amplitude information was considered; therefore, no additional sine waves were added at the end of the flow diagram. Instead, the time-domain signal created by the IFFT of the peak-connected data was used as the stimulus material.

3 Psychoacoustical Index

Various sensations arise when people hear sounds [3]. However, sound can be physically analyzed in terms of frequency and sound pressure. Numerous studies have analyzed sound in relation to human psychological factors. These studies have often attempted to link the psychological quantities, that quantify human sensations, with the physical quantities of sound. However, no linear relationship was observed between them. Therefore, based on the results of studies examining the relationship between the psychological and physical quantities of sound, as well as studies attempting to elucidate the information processing process of hearing, methods have been proposed to calculate the psychological quantities of loudness, sharpness, and roughness of human sensory sounds from the physical quantities of sound. The current describes the loudness and roughness of human sensory sounds, as detailed next.

3.1 Loudness

Loudness is a subjective measure of sound intensity that is influenced by factors such as sound pressure, frequency, and bandwidth [3]. This scale has undergone several revisions, and the one proposed by Steven [4] has often been used recently. This measurement is frequently used for sonication. Zwicker and Moore–GlasBerg [5] proposed a loudness calculation model that has been standardized in the international standards ISO 532-1 and 532-2. In this study, the model standardized in ISO 532-1 was utilized to analyze unsteady sounds [5]. In the Zwicker model, the first step is to consider the physical spectrum of sound. In the auditory system, the physical spectrum of sound is transformed into an excitation pattern corresponding to its spectral representation. The excitation level in each critical band of the excitation pattern is converted into a loudness function. The specific loudness of each critical band is calculated from its excitation level in the excitation pattern. For convenience, the audible frequency band up to 16 kHz is divided into 24 non-overlapping critical bands [5]. Loudness is summed over these 24 critical bands to calculate the overall loudness of the sound as follows:

$$N = \int_0^{24Bark} N'(z)dz[sone]$$

where $N'(z)$ is the loudness per critical bandwidth and z is the critical bandwidth number. "Bark" used in the integral interval refers to the Bark Scale, a psycho-acoustic scale proposed by Eberhard Zwicker in 1961 [6].

3.2 Roughness

Roughness is a measure of the perceived roughness of a sound caused by fluctuations in its amplitude envelope and frequency [3]. At carrier frequencies (f_c) below 1 kHz, the sensation of roughness is influenced by the frequency selectivity of hearing. Conversely, at carrier frequencies above 2 kHz, the sensation of roughness is not related to the critical bandwidth; instead, it is primarily influenced by the temporal resolution of the auditory system. Generally, the modulation frequency at which the sensation of roughness occurs is approximately 15–300 Hz, with a maximum roughness of approximately 70 Hz [5].

The effect of sound pressure level on subjective roughness is not significant, with a 20-dB increase resulting in a doubling of the subjective roughness. Roughness is calculated as follows:

$$R = 0.3f_{mod} \int_0^{24Bark} \Delta L(z)dz[asper]$$

where f_{mod} is the modulation frequency and $\Delta L(z)$ is the difference between the peaks and valleys of the excitement level for each critical band.

4   Knocking Sound Analysis

First, the knocking sound characteristics were examined. The sound source used in this study is the operational sound of a six-cylinder diesel vehicle during acceleration. A 30-Hz high-pass filter was applied to enhance the clarity of the order components. The analytical parameters are listed in Table 1 (Fig. 2).

Table 1. Analytical conditions

Car model	Six-cylinder diesel-powered vehicle
Running conditions	Fifth gear driving
Recording location	Front center (between driver and passenger seat)
Sampling frequency	8 kHz

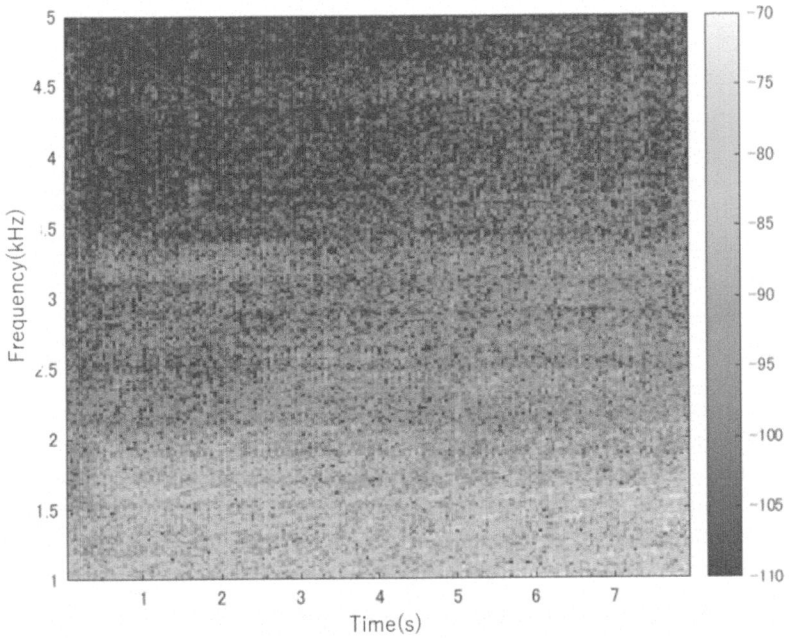

Fig. 2. Spectrogram of the knocking sound analyzed in this study

According to [7], a peak is observed in the high-frequency range of approximately 2 kHz at the onset of knockdown, which indicates that the frequency peak appears over a wide range at the onset of acceleration, especially around 1.8 s. Given that the engine speed is lower during the initial phase of acceleration compared to the latter half, it is anticipated that the knocking sound is influenced by order components such as humming and fluctuations.

5   Stimulus Creation

Stimulus sounds were generated to investigate the effect of knocking sounds on order components. The first–sixth-order components were extracted using a sinusoidal model for 6 s of acceleration-running sounds. The results are shown in Figs. 3, 4, 5 and 6.

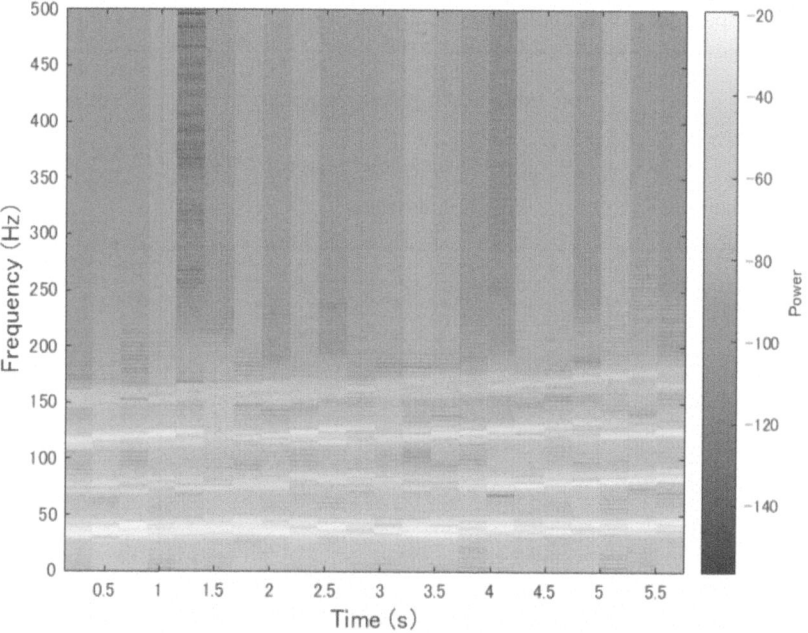

Fig. 3. 1st- to 4th-order components

These order components were combined with the knocking sound extracted by applying the 1-kHz high-pass filter. Additionally, to investigate the change in auditory impression resulting from order-component reduction, the first- to fourth-order components; odd-order components of the first, third, and fifth orders; and even-order components of the second, fourth, and sixth orders were extracted using the Sinusoidal Model. Then, the extracted components were synthesized with the knocking sound.

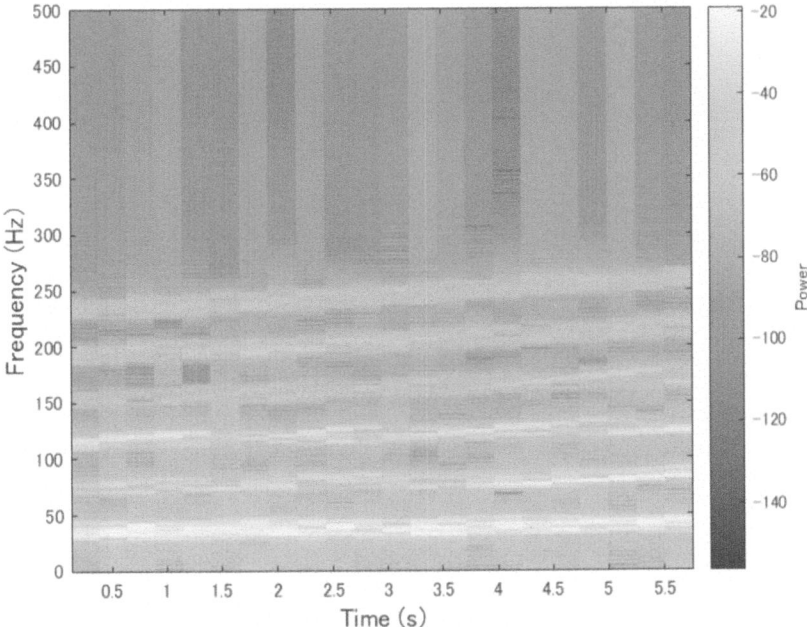

Fig. 4. 1st- to 6th-order components

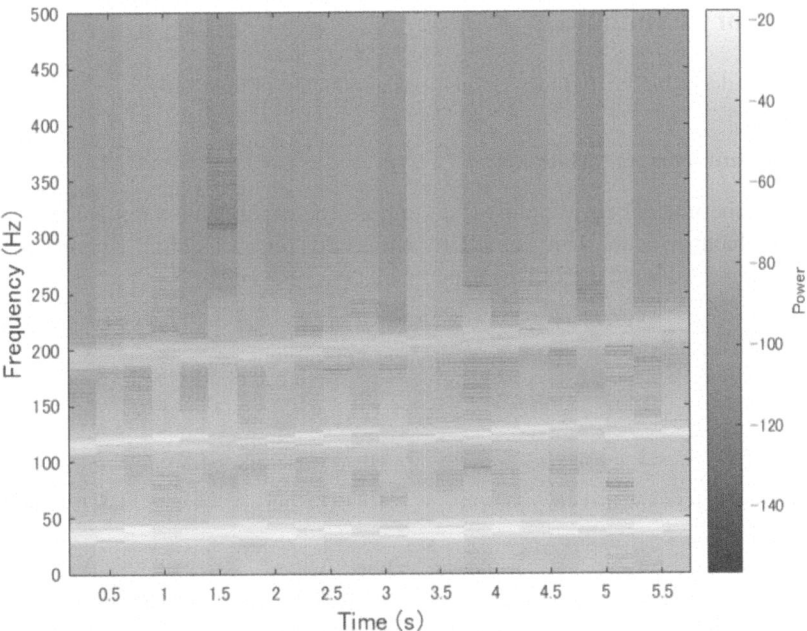

Fig. 5. 1st, 3th, 5th-order components

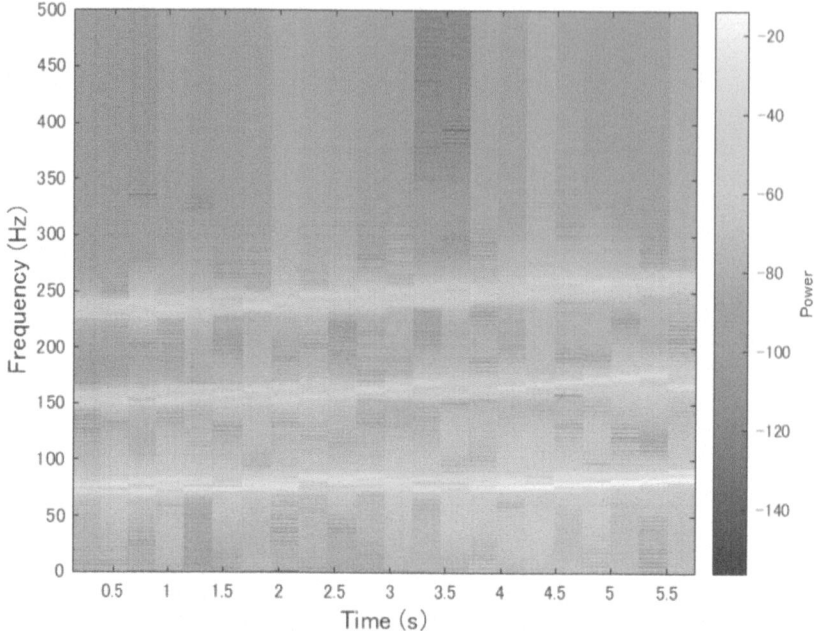

Fig. 6. 2nd, 4th, 6th-order components

6 Auditory Evaluations

Next, let us describe the auditory evaluations conducted in this study.

6.1 Environments and Methods

First, the experimental environment is described. The experiment was conducted in an anechoic chamber at the Hiroshima City University, Japan. Table 2 summarizes the equipment and software used in the experiments.

Table 2. Equipment and software

Headphone	AKG K812
Frontend	HEAD acoustics lab2-V1
Stimulus sound reproduction environment	HEAD acoustics ArtemiS SUITE 10.0

First, six stimulus pairs were created by combining four stimulus sounds created before the start of the experiment. To familiarize the participants with the knocking sound, they were asked to listen to the acceleration sounds several times. Then, they were asked to evaluate the knocking sounds in the areas where they heard them. Two evaluation criteria were selected: the sound's "dryness" (defined as dry if the knocking sound is easily discernible) and its "unpleasantness". Furthermore, two forced-choice questions were used to determine the sounds that were drier and more unpleasant. Figure 7 shows the evaluation format.

1st stimulus sound

1.   Which was drier sound?

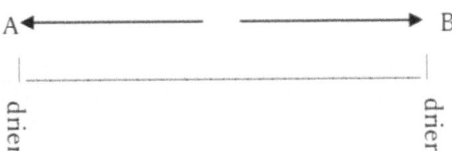

2.   Which was more unpleasant sound?

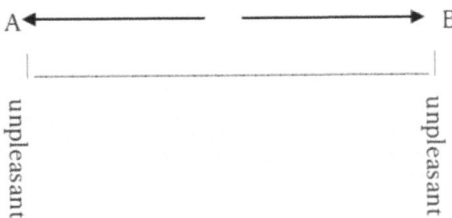

Fig. 7. Evaluation form

6.2   Result and Discussion

Thurston's pairwise comparison method was used to analyze the auditory evaluation results. Figures 8 and 9 show the interval scales for dryness and annoyance, respectively. The participants were Japanese, consisting of nine males and four females with an average age of 22.4 years.

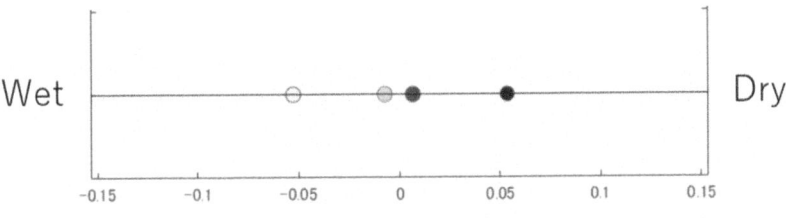

Fig. 8. Dryness interval scale

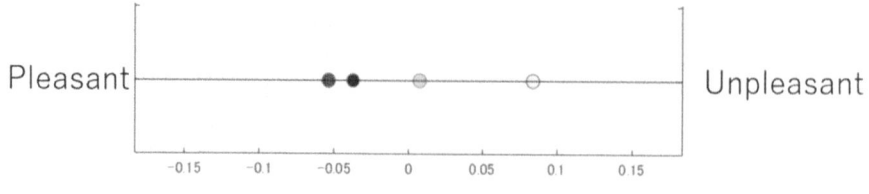

Fig. 9. Annoyance interval scale

Red dot : 1~4 order components
light blue dot : 1~6 order components
Blue dot : 1,3,5 order components
Yellow dot : 2,4,6 order components

An analysis was conducted based on driving frequency. Subjects who drove more than once a week were classified as frequent drivers. Six participants fell into this category, whereas seven were classified as infrequent drivers. The analyses for frequent and infrequent drivers are shown in Figs. 10 and 11, respectively. Notably, the results shown in Fig. 10(a) show an overlap between the evaluation of the first- to fourth-order components (represented by the red dots) and the first- to sixth-order components (represented by the light-blue dots) of the stimulus sound.

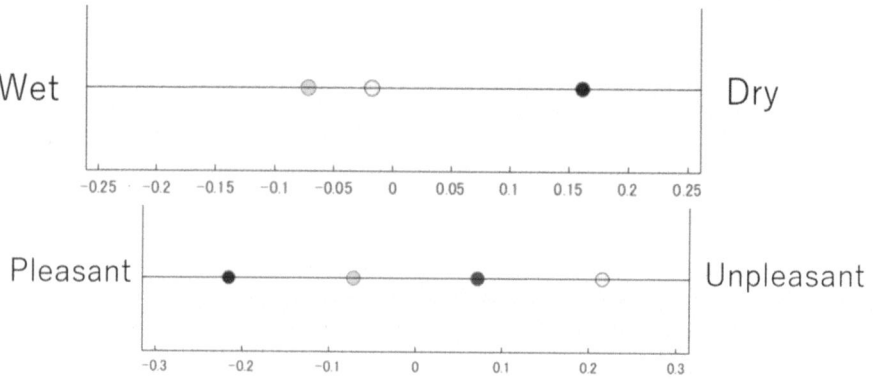

Fig. 10. (a) Dryness and (b) annoyance interval scales for the frequent-driver group

These results confirmed that when the driving frequency was high, stimulus sounds with fewer order components were evaluated as less unpleasant than those with more order components. Conversely, when the driving frequency was low, stimulus sounds with more order components were evaluated as less unpleasant. Next, we discuss how these sounds are evaluated using psychoacoustic indices, such as loudness and roughness. The maximum loudness, maximum roughness, final roughness, and time at which the maximum roughness occurred for each stimulus sound are listed in Table 3.

These psychoacoustic index results, along with the annoyance scale values according to driving frequency, indicate that the lower the roughness, the lower the annoyance

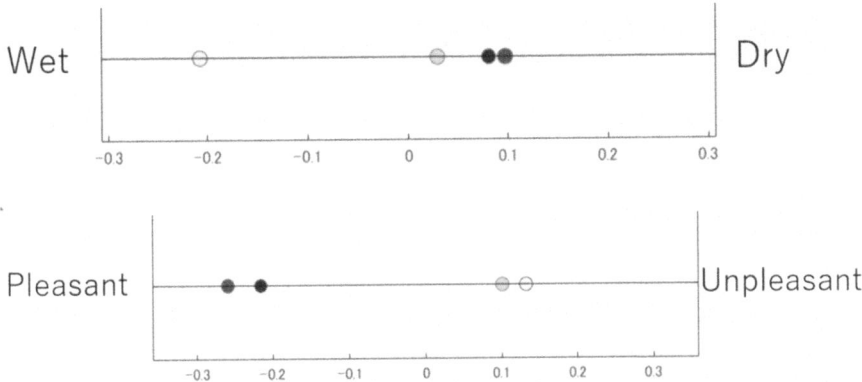

Fig. 11. (a) Dryness and (b) annoyance interval scales for the infrequent-driver group

Table 3. Psychoacoustic indices

Stimulus	Maximum loudness (sone)	Final roughness (asper)	Maximum roughness (asper)	Time of maximum roughness (s)
1 to 4	30.5	0.157	0.157	6.0
1 to 6	31.0	0.082	0.124	5.9
1, 3, 5	30.9	0.026	0.053	0.83
2, 4, 6	37.6	0.026	0.065	5.44

when the driving frequency is high. Conversely, the higher the roughness, the lower the annoyance when the driving frequency is low. The loudness of the second-, fourth-, and sixth-order components was approximately seven sones louder than that of the other stimulus sounds, suggesting that they were evaluated as unpleasant regardless of the driving frequency. No relationship was found between the psychoacoustic index and the degree of dryness. We plan to further study the dryness index in future work.

7 Conclusions

Knocking sound, which is a type of noise, was investigated in this study to determine the factors that contribute to annoyance and methods for its reduction. We considered the simultaneous occurrence of knocking sounds and low-order components of humming and shaking as factors contributing to annoyance and conducted auditory evaluation experiments. The results demonstrated that annoyance varied with driving frequency. The psychoacoustic index analysis results revealed a relationship between roughness and driving frequency. However, the degree of dryness was not sufficiently considered. Nevertheless, the order component intervals indicate that they were approximately 50 Hz apart, suggesting that the investigation of the effects of humming may be insufficient for this stimulus sound. In future work, we intend to explore the changes in auditory perception under various conditions, including loudness adjustments, the presence of humming in the stimulus sound, and the reduction of order components using adaptive control.

References

1. Shibatani, N., Ishimitsu, S., Yamamoto, M.: Command filtered-x LMS algorithm and its application to car interior noise for sound quality control. Int. J. Innov. Comput. Inf. Control (IJICIC) **14**(2), 647–656 (2018)
2. Depalle ftal. Proceedings of 1993 International Computer Music Conference. Computer Music Association, San Fransisco, pp. 94–97 (1993)
3. Iwamiya, S.: Sensitivity Studies of Sound Tone. Corona Publishing (2011)
4. Stevens, S.S.: The measurement of loudness. J. Acoust. Soc. Am. **27**, 815–829 (1995)
5. Fastl, H., Zwicker, E.: Psychoacoustics: Facts and Models, 3rd edn. Springer, Heideelberg (2006). https://doi.org/10.1007/978-3-540-68888-4
6. Zwicker, E.: Subdivision of the audible frequency range into critical bands (Frequenzgruppen). J. Acoust. Soc. Am. **33** (1961)
7. Kasahara (Ono Sokki). Estimation of knocking sound and cylinder pressure from engine radiation sound by deep learning. J. Soc. Autom. Eng. Jpn. **52**(2) (2021)

Inference of a Multi-server Queue Using External Observations

Jinsoo Park[1] and Yun Bae Kim[2]([✉])

[1] Yong In University, Yongin, Korea
[2] Sungkyunkwan University, Suwon, Korea
kimyb@skku.edu

Abstract. Since the autocorrelated arrivals and services complicate queue analysis, numerous modeling and analysis methods for the autocorrelated cases have been developed. This study aims to overcome the limits of the existing methods in some situations with a time-dependent structure in arrival or service process and trouble to observe the internal operations. With only external observations, the proposed method identifies the internal operation of a multi-server queue via a non-parametric approach even when the service times are autocorrelated. The efficiency and demonstration of this suggested inference method is measured by queue delays and service times of each customer. Mathematical proof and numerical examples are provided to verify the proposed inference method.

Keywords: queue inference · multi-server queue · queueing time · autocorrelated arrivals · autocorrelated service times

1 Introduction

The autocorrelation structure of the arrival processes and the service times have prompted the development of various stochastic process models, such as the Markovian arrival process (MAP) [1], the Pareto distribution for long-tailed probability density, and so on. In addition, the diffusion process approximations such as Brownian motion [2, 3], which make facilitate analysis of self-similarity, have been used to model Internet traffic.

Since the analytical solutions of queueing systems have limited application for real system modeling, inference methods have been developed as alternative approaches. A typical method is the "queue inference engine (QIE)" developed by Larson [4–6]. Starting from the QIE, various studies have inferred the queueing systems [7–17]. In queueing theory, Baccelli et al. [18] introduced the inverse problem, which unites all of the previous studies on inference methods. Recently, Bayesian inference [19, 20] and maximum likelihood [21] have been applied to some queueing systems. Those inference methods, however, necessarily rely on distributional or parametric assumptions regarding the arrival processes. Due to the lack of information or data for the fulfillments the requirement of the assumptions, difficulties arise when these methods are applied to model real systems. Our method differs from these methods in that we make neither distributional nor parametric assumptions. Using external explanations of the arrival

S. Saito et al. (Eds.): AsiaSim 2024, CCIS 2170, pp. 125–136, 2024.
https://doi.org/10.1007/978-981-97-7225-4_10

and departure times, we are able to infer the internal performance measures such as the queueing delays and service times. Park *et al.* [22] introduced this method that limited the service times to be independent. Similarly, Keith *et al.* [23] also proposed a robust method to infer the queue from the arrival and departure times. Our study focuses on the method of Park *et al.* [22]. We relax the independent condition to the case of autocorrelated service times in this paper. Accordingly, we will follow their notations and mathematical expressions.

This research paper is organized as follows. Section 2 presents the modeling of black box queueing systems, including some configurations and assumptions. Section 2.2 introduces a previous study that inferred a multi-server queue using external observations, and Sect. 3 relaxes the assumption used in the previous study. Section 4 describes some numerical results and Sect. 5 concludes this research study with a discussion about further work.

2 Research Background

2.1 Queueing Systems in the Black Box

The proposed systems in this paper are general multiple server queueing systems. As shown in Fig. 1, an arriving customer waits in the queue until a server is available. Having completed his service tasks, one then exits the system. All internal operations occur in a black box, from waiting in the queue to service completion, and thus cannot be observed externally. We know only the arrival and departure times that can be observed from outside the system. Not understanding the arrival process or the service time distribution will not hinder the applicability of our method to queue inference. In the case of autocorrelation among arrivals or services, we can still calculate exact solutions. Our method can calculate the solution even when the service time distributions differ between servers. The only assumption of our method is that the service discipline (or queueing discipline) is either 'first-come-first-served (FCFS, or FIFO, first in first out) or last-come-first served (LCFS, or LIFO, last in first out)'. Based on the service discipline and the external observations of the arrival and departure times, our method can infer the internal performance measures (queueing delays and service times) occurring within the black box.

Our proposed method is naturally applicable to any queueing system that has an unobservable interior. For example, we cannot follow the customers due to their simultaneity and privacy. The security policy also enhances the unobservable nature of the system. In these situations, using the solutions from our proposed method, we are able to infer the service distribution (or process) and identify the characteristics of the queueing system. In addition, we guarantee the applicability of our proposed method to real systems due to no assumption for the arrival process and relaxed assumption for the service time distribution.

2.2 The Earlier Study

In this part, we review a previous study that was introduced by Park *et al.* [22]. We assign the following notations. Let N, A_i and D_i, respectively, denote the number of

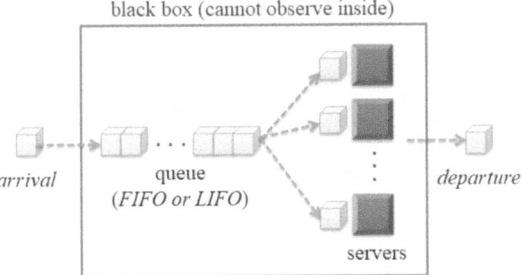

Fig. 1. Queuing system inside the black box.

observations, the arrival time and the departure time of customer i, the values of which are known from the observations. B_i is the time of service beginning for customer i, which is unknown. The inferred B_i will produce the queueing and the service time of customer i, which are denoted as Q_i $(= B_i - A_i)$ and S_i $(= D_i - B_i)$ respectively. The number of servers, c may be either known or unknown. The research goal is deducing the service time S_i, the queueing delay Q_i, and the number of servers c if it is unknown.

2.2.1 Known Number of Servers

In the circumstance of first come first out (FCFS) system, if the number of servers is known, the service beginning time is identified from the following equations presented by Hall [24]:

$$B_i = \begin{cases} A_i & (i \le c) \\ \max\{A_i, D_{(i-c)}\} & (o/w) \end{cases} \tag{1}$$

where $D_{(i)}$ is the ith sample order statistic.

To surmise the last come first served (LCFS) system, we employ three more notations. First, denotes N_i^A the number of customers seen by customer i upon arrival. Second, N_i^D denotes the number of customers that customer i leaves behind at departure. Third, $D^1(n, t)$ is the first departure time that leads to $N_i^D = n$, after time t. Using these three notations, the service beginning time B_i is obtained as follows:

$$B_i = \begin{cases} A_i & (i \le c) \\ D^1(N_i^A, A_i) & (o/w) \end{cases} \tag{2}$$

2.2.2 Unknown Number of Servers

In general, the number of servers may also be unknown. Therefore, the focus of the investigation will proceed to suppose the number of servers, c. Once c is determined, the queueing and service times can be calculated for each discipline; FCFS and LCFS. Accordingly, the unknown number of servers is the decision variable.

For the case of the FCFS system, the following optimization problem finds the unknown number of servers.

$$\underset{\hat{c}}{Minimize} \quad \frac{1}{N-1} \sum_{i=1}^{N} \left(\hat{S}_i - \overline{\hat{S}} \right)^2 \tag{3}$$

$$
\begin{aligned}
s.t. \ \hat{B}_i &= A_i & (1 < i \le \hat{c}) \ (4) \\
\hat{B}_i &= \max\{A_i, D_{(i-\hat{c})}\} & (\hat{c} < i \le N) \ (5) \\
\hat{S}_i &= D_i - \hat{B}_i & (1 < i \le N) \ (6) \\
\hat{S}_i &> 0 & (1 < i \le N) \ (7) \\
\hat{c} &: integer & (1 < \hat{c} \le N) \ (8)
\end{aligned}
$$

where hatted notations, $\hat{c}$, $\hat{B}$, and $\hat{S}_i$ are the estimated values, and $\overline{\hat{S}}$ is the sample mean of S_i values. For the LCFS case, (4) and (5) should be changed into the equations corresponding to (2). Park et al. [22] proved that the objective function (3) finds the exact solution if the service times are independent and the number of observations, N is large enough.

3   Relaxing the Assumption

3.1   For Positively Autocorrelated Service Times

By intuition, we can confirm that (1) and (2) are still applicable to a relaxed system having positively autocorrelated service times. Also, the optimization problem from (3) to (8) finds the unknown number of servers even though the assumption of independent service times is relaxed. *Theorem 1* is the key to verify that the optimization problem still works. *Lemma 1* supports *Theorem 1*.

Lemma 1: *Assume that we underestimate the number of servers, solving the optimization problem from (3) to (8). That implies $\hat{c} < c$, where c is the actual number of servers. In this case, as the number of arrivals increases; i.e., $N \to \infty$, there are always infeasible solutions if the service times have a variation; i.e., $Var(S) > 0$.*

Proof: If we underestimate the number of servers as $\hat{c} = c - k$ for a positive integer k, then the estimated service time of customer i, $\hat{s}_i^-$ is calculated as follows using (5) and (6):

$$\hat{s}_i^- = D_i - \max\{A_i, D_{(i-c+k)}\} \tag{9}$$

Equation (9) produces next two cases.

Case 1: $A_i < D_{(i-c+k)} \to \hat{s}_i^- = D_i - D_{(i-c+k)}$
Case 2: $D_{(i-c+k)} < A_i \to \hat{s}_i^- = D_i - A_i$

To produce feasible solutions, D_i should be greater than $D_{(i-c+k)}$ in the first case. However, as N increases, there must be the situation that $D_{(i-c+k)}$ is greater than or equal to D_i due to the variation among the service times; a counter-example for the constraint

(7). Therefore, if we underestimate the number of servers, there is always an infeasible solution. Note that *Lemma* 1 holds not only for the autocorrelated case but also for the independent case described in Sect. 3 if the service times have a positive variance.

Theorem 1: *The objective function* (3) *has an exceptional minimum at c, the actual number of servers in the FCFS queueing system as N increases if the service times have a positive autocorrelation.*

Proof: For the case of underestimation, in accordance with *Lemma* 1, there are always infeasible solutions as N increases. Therefore, we present only the overestimated cases using a reduction ad absurdum argument. We illustrate that overestimating the number of servers increases the sample variance of the estimated service times. Assume that the actual number of servers is c, but the estimated number of servers is $\hat{c} = c + k$. Analogous to (9), we get following $\hat{s}_i^+$ that denotes the estimated i service time of customer i.

$$\hat{s}_i^+ = D_i - max\{A_i, D_{(i-c-k)}\} \tag{10}$$

In this case, since $\hat{c}$ is always feasible, we should consider possible cases. To show the difference between the true and estimated values, we derive (11) from (10).

$$\hat{s}_i^+ = \left[D_i - max\{A_i, D_{(i-c)}\}\right] + \left[max\{A_i, D_{(i-c)}\} - max\{A_i, D_{(i-c-k)}\}\right] \tag{11}$$

Then there are three cases about (11) as follows:

Case 1: $A_i < D_{(i-c+k)} < D_{(i-c)} \rightarrow \hat{s}_i^+ = S_i + \left[D_{(i-c)} - D_{(i-c-k)}\right]$
Case 2: $D_{(i-c+k)} < A_i < D_{(i-c)} \rightarrow \hat{s}_i^+ = S_i + \left[D_{(i-c)} - A_i\right]$
Case 3: $D_{(i-c+k)} < D_{(i-c)} < A_i \rightarrow \hat{s}_i^+ = S_i$

The value of $\hat{s}_i^+$ in the first two cases is represented i as the sum of S_i and an extra term. Assuming the extra term in the third case to be zero, we can express the estimated service time as follows:

$$\hat{s}_i^+ = S_i + \Delta_i, (\Delta_i > 0) \tag{12}$$

where Si is the actual service time of customer i and Δi is the extra term. If the service times are positively autocorrelated and the sample size is large enough, we conclude that the sample variance of the estimated service times is greater than that of the actual service times via (12). As we increasingly overestimate the number of servers, the sample variance of the service times will increase.

Therefore, the decision variable $\hat{c}$ will accomplish a unique minimum when the estimated service capacity is equal to the actual value, c.

Remark 1: *Lemma* 1 also holds even when the service discipline is LCFS or random selection for service (RSS). We can simply prove these phenomena with similar arguments. Therefore, we can analogously extend *Theorem* 1 to the LCFS case.

Extending *Theorem* 1 to the LCFS case, we also ensure a correct solution for the decision variable $\hat{c}$ in the optimization problem. To solve these problems, the main characteristics of those optimization models are described in Park *et al.* [22]. The grid search over the range in (8) may be the simplest method.

Since there is still some uncertainty in the estimated number of servers, we present some remarks here.

Remark 2: The correctly estimated number of servers leads directly to the exact estimates of the queueing and service times via (1) or (2).

Remark 3: *Theorem* 1 assumes that the number of observations N is diverging to infinity. Therefore, if the number of observations is insufficient, the optimization problem from (3) to (8) cannot estimate the correct value of $\hat{c}$.

Remark 4: The optimization problem still fails if the workload (i.e., traffic intensity) is remarkably low, even though the number of observations N is sufficiently large. The low workload may produce a situation that the number of customers in the system is always smaller than the number of servers.

Remarks 2 and 4 are minutely described in Park *et al.* [22]. For *Remark* 4, the low workload system may have less interest in the queue analysis. *Remark* 3 explains the inapplicability of our proposed method when the data size is insufficient. Our numerical results in Sect. 4, however, show that our optimization problem still finds the exact solution under a small number of observations.

3.2 More Relaxation

Section 3.1 relaxes the assumption of independent service times to the positively autocorrelated service times. The inference is also analogous to the previous study of Park *et al.* [22]. For the case of negatively autocorrelated service times, *Theorem* 1 may not hold. Accordingly, in this section, we relax the assumption of both positive and negative cases using *Lemma* 1 and a simple objective function. Note that *Lemma* 1 holds for both cases of autocorrelated service times; i.e., positive and negative. Accordingly, we redefine the objective function for both cases. *Theorem* 2 is the key to relax the assumption.

Theorem 2: *As N increases, the sample mean of service times can be an objective function that finds the exact number of servers under the constraints (4)–(8) for an FCFS queue. We can determine the exact number of servers changing (3) into the following (13):*

$$\underset{\hat{c}}{Minimize} \quad \sum_{i=1}^{N} \hat{S}_i / N \tag{13}$$

Proof: For the case of underestimation, *Lemma* 1 still holds, and there is always no feasible solution as N increases. When we overestimate the number of servers, (12) naturally produces a greater sample mean than that of the actual service times S_i's. Therefore, the decision variable c will achieve an unique minimum when the estimated number of servers is equal to the true value, c.

Remark 5: In the case of independent service times, *Lemma* 1 is still holding, and the sample mean, the objective function in (13) of *Theorem* 2 is applicable to find the number of servers. However, *Lemma* 1 may not hold when the sample size is insufficient or the variance of service times is strictly small. In this situation, sample variance in (3) can still find the exact solution. Therefore, the objective function in (3) is valuable in itself.

Extending *Theorem* 2 to the LCFS case, we also ensure a correct solution for the decision variable c in the optimization problem with (4)–(8) and (13). The grid search over the range in (8) may also be the simplest method. In the real world, however, most of the queueing systems have independent or positively autocorrelated service times rather than negatively autocorrelated service times. Accordingly, we provide the results of experiments only for positively autocorrelated case.

4   Numerical Results

Theorem 1 guarantees that the optimization problem finds the exact solutions as the number of observations increases. In this section, we also verify that a small number of observations can find the exact solution. We perform some experiments with the simulation results of queueing systems to identify the number of data that finds exact solutions. Given only the arrival and departure times, we solve the optimization problems in Sect. 2. Finally, we compare our solutions with the values from the simulation results.

4.1   Experimental Design

We design the following queueing systems for realistic simulation. For arrivals, we employ MAP, modified autoregressive (AR) process, and autocorrelated exponential (AE) distribution which has the autocorrelation structures. We set the mean arrival rate to 1 and empirically discover the parameters for respective arrival processes. Following the notations of Lucantoni et al. (1990), the parameter matrices for the MAP used are given by:

$$C = \begin{pmatrix} -0.2 & 0.1 \\ 1.5 & -16.9 \end{pmatrix}, D = \begin{pmatrix} 0.1 & 0 \\ 0.1 & 15.3 \end{pmatrix}. \tag{14}$$

The modified AR arrival process is defined as follows:

$$X_t = |Z_t/1.82| \tag{15}$$

where $Z_t = 0.9Z_{t-1} + \varepsilon_t$, and $\varepsilon_t \sim N(0,1)$.

In the modified AR(1), X_t implies the time interval, which uses the absolute value of the AR(1) random variate due to the non-positivity of the AR(1).

For the AE distribution, we apply the transformexpand-sample (TES) process [25]. The TES process innovates the AE random variates with parameter λ. The parameter λ is the mean arrival rate in common with the original exponential distribution. By using TES process, we generate the autocorrelated random numbers with lag 1 correlation ρ_1 = 0.9, and variate the AE sample through the inverse transform method.

For the service time distributions, we use Pareto distributions with long-tailed probability, Markovian service process (MSP, identical to MAP), modified AR (MAR) process, and AE distribution from the TES process. We set the number of servers to 5 and 10. We fix the arrival rate and vary the service rate according to the number of servers to fit a server utilization rate of 0.9. The service time distributions used in the simulations are

Table 1. Service time distributions and their parameters used in simulation.

c	Distribution	Parameters
5	Pareto	$X = 2.25Y, f_Y(y) = 2y^{-3}$
	MSP	$C = \begin{pmatrix} -2/45 & 1/45 \\ 1/3 & -169/45 \end{pmatrix}, D = \begin{pmatrix} 1/45 & 0 \\ 1/45 & 17/5 \end{pmatrix}$
	MAR	$X_t = |Z_t/0.4062|, Z_t = 0.9Z_{t-1} + \varepsilon_t, \varepsilon_t \sim N(0, 1)$
	AE	$f(x) = 0.18e^{-0.18x}, \rho_1 = 0.677$
10	Pareto	$X = 4.5Y, f_Y(y) = 2y^{-3}$
	MSP	$C = \begin{pmatrix} -1/45 & 1/90 \\ 1/6 & -169/90 \end{pmatrix}, D = \begin{pmatrix} 1/90 & 0 \\ 1/90 & 17/10 \end{pmatrix}$
	MAR	$X_t = |Z_t/0.2031|, Z_t = 0.9Z_{t-1} + \varepsilon_t, \varepsilon_t \sim N(0, 1)$
	AE	$f(x) = 0.09e^{-0.09x}, \rho_1 = 0.677$

Table 2. Service time distributions of each server in the extra systems.

Server ID	Parameters
1	Pareto distribution and its parameters for $c = 5$ in Table 1
2	MSP and its parameter matrices for $c = 5$ in Table 1
3	MAR process and its parameters for $c = 5$ in Table 1
4	AE distribution and its parameters for $c = 5$ in Table 1
5	AE distribution and its parameters for $c = 5$ in Table 1

summarized in Table 1. In the case of AE, lag 1 correlation is reduced by the logarithm in the inverse form of the exponential distribution function.

Applying the FCFS and LCFS disciplines to those configurations, we first perform simulations of 48 queueing systems (3 arrival processes × 4 service time distributions × 2 cases of service capacities × 2 service disciplines). In addition, to verify that our method is successful even though the servers have different service time distributions, we simulate three extra systems for the respective discipline. In the extra systems, we reuse three arrival processes and set five servers combining the service distributions in Table 1 as listed in Table 2. In Tables 1 and 2, MAR implies the modified AR and c indicates the number of servers. X of the Pareto distribution denotes the service time with the long-tailed characteristic of Y, even though the service rate changes according to the server utilization. We heuristically find the values of parameters to satisfy the system condition as we fix.

4.2 Results and Analysis

Using the only pairs of arrival and departure times from the simulation results, we solve the optimization problems posed in Sect. 2. As a result, we confirm that the number of servers, the queueing delays, and the service times are all identical to the true values even when the number of data is relatively small if the workload is heavy.

Table 3. The minimum number of observations finding the exact solutions.

Arrival Process	Service Policy	Service Process								ES
		Pareto		MSP		MAR		AE		
		5	10	5	10	5	10	5	10	
MAR	FCFS	20	290	110	160	30	70	470	470	20
	LCFS	20	40	120	160	30	60	20	20	20
MAP	FCFS	30	180	100	170	50	50	70	500	30
	LCFS	20	30	100	140	20	50	20	70	20
AE	FCFS	30	40	110	160	20	40	180	180	20
	LCFS	20	50	110	160	40	40	20	170	20

Table 3 presents the minimum number of observations for each system that determines the exact solutions for c. We find those values by increasing the number of data in 10 counts. The term ES in Table 3 indicates the extra system. As mentioned previously, our proposed method finds the exact solutions even though the data size is relatively small.

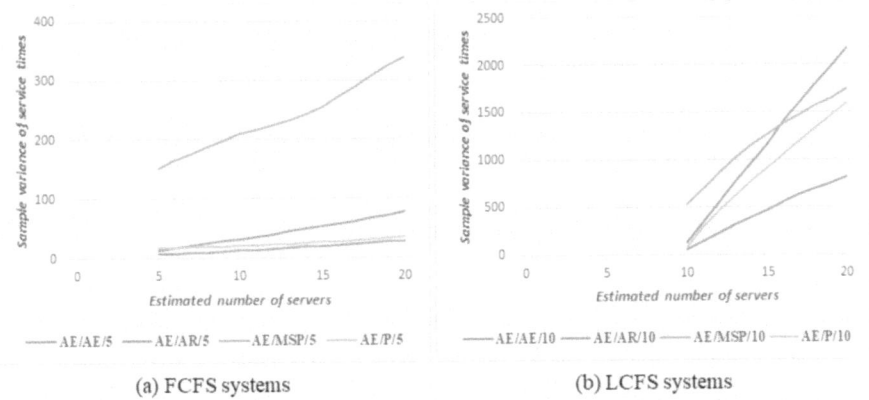

(a) FCFS systems   (b) LCFS systems

Fig. 2. Evaluations of objective function, sample variance of service times.

In a different respect, we show the evaluation plots of the objective functions in (3) to confirm the determination of the exact number of servers. Using 1,000 pairs of observations, we cite here some results of the inferences given by our method. In Fig. 2, (a) is the plot for FCFS systems and (b) for LCFS systems. In all of the graphs, the horizontal axis represents the number of servers, while the vertical axis shows the variance in the service time corresponding to the number of servers. In the label notations, AR stands for the modified AR process and P indicates the Pareto distribution. Areas with no data imply that there are infeasible solutions, which reconfirm the *Lemma* 1. In Fig. 2, the graphs show that the objective function values converge at the optimal number of servers. We recognize that these situations occur because the systems have the same distribution of service times.

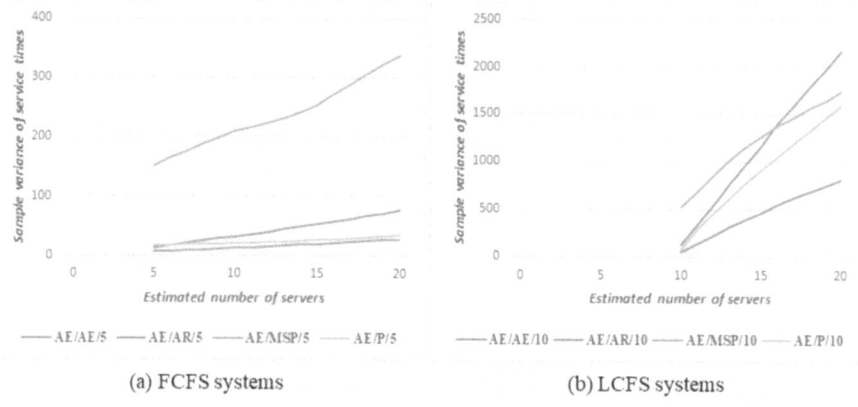

(a) FCFS systems   (b) LCFS systems

Fig. 3. Evaluations of objective function, sample mean of service times.

Figure 3 also shows the evaluation plots of the objective functions in (13) under the same condition as Fig. 2. We confirm that the objective function, the sample mean of the service times in (13) also obtain the exact number of servers.

The shapes of the objective functions explain that there are unique solutions in our optimization problems and that these solutions are exactly identical to the true values. Considering these results, we conclude that our method performs as intended. We do not list the queue statistics because the estimated queuing times and service times are identical to the outputs from the actual simulated systems.

5 Conclusions and Future Works

This paper introduces the inference method of Park *et al.* [22] to deduce internal operations based on external observations, arrival and departure times. We relax the assumption of independent service times. In the case of positively autocorrelated service times, the optimization problems still find the number of servers in a system to determine the exact queueing delays and service times, given sufficient data. Then we relax the assumption to the positively and negatively autocorrelated cases using a lemma and a simple objective function. This study also provides numerical results as verification. Our results consider reality and are applicable to FCFS or LCFS queueing systems. Since it is not necessary to assume the independent service times as well as the distribution of inter-arrival or service times, we expect that our method can be applied to any system that can be represented as a multi-server queueing system.

Our ongoing study aims to develop methods to describe the solutions starting with even more relaxed assumptions. The primary task is analyzing additional service disciplines, such as random selection for service. Another task is to eliminate the assumption that every customer is observed without error, by exploiting the simple order statistics of the arrival and departure times instead of matching them to customers.

References

1. Lucantoni, D.M., Meier-Hellstern, K.S., Neuts, M.F.: A single-server queue with server vacations and a class of non-renewal arrival processes. Adv. Appl. Probab. **22**(3), 676–705 (1990)
2. Lu, S., Molz, F.J., Liu, H.H.: An efficient, three-dimensional, anisotropic, fractional Brownian motion and truncated fractional Levy motion simulation algorithm based on successive random additions. Comput. Geosci. **29**(1), 15–25 (2003)
3. Taqqu, M.S.: Fractional Brownian motion and long-range dependence. In: Theory and Applications of Longrange Dependence, pp. 5–38 (2003)
4. Larson, R.C.: The queue inference engine: deducing queue statistics from transactional data. Manag. Sci. **36**(5), 586–601 (1990)
5. Larson, R.C.: The queue inference engine: addendum. Manag. Sci. **37**(8), 1062 (1991)
6. Larson, R.C.: Queue inference engine. In: Gass, S.I., Fu, M.C. (eds.) Encyclopedia of Operations Research and Management Science, pp. 1228–1234. Springer, Boston (2013). https://doi.org/10.1007/978-1-4419-1153-7_844
7. Bertsimas, D.J., Servi, L.D.: Deducing queueing from transactional data: the queue inference engine, revisited. Oper. Res. **40**(3–2), S217–S228 (1992)
8. Daley, D., Servi, L.D.: Exploiting Markov chains to infer queue length from transactional data. J. Appl. Probab. **29**, 713–732 (1992)
9. Dimitrijevic, D.D.: Inferring most likely queue length from transactional data. Oper. Res. Lett. **19**(4), 191–199 (1996)
10. Toyoizumi, H.: Sengupta's invariant relationship and its application to waiting time inference. J. Appl. Probab. **34**(3), 795–799 (1997)
11. Pickands, J., Stine, R.A.: Estimation for an M/G/∞ queue with incomplete information. Biometrika **84**(2), 295–308 (1997)
12. Mandelbaum, A., Zeltyn, S.: Estimating characteristics of queueing networks using transactional data. Queueing Syst. **29**(1), 75–127 (1998)
13. Jones, L.K.: Inferring balking behavior from transactional data. Oper. Res. **47**(5), 778–784 (1999)
14. Bingham, N., Pitts, S.M.: Non-parametric estimation for the M/G/∞ queue. Ann. Inst. Stat. Math. **51**(1), 71–97 (1999)
15. Jang, J., Suh, J., Liu, C.R.: A new procedure to estimate waiting time in GI/G/2 system by server observation. Comput. Oper. Res. **28**(6), 597–611 (2001)
16. Ross, J.V., Taimre, T., Pollett, P.K.: Estimation for queues from queue length data. Queueing Syst. **55**(2), 131–138 (2007)
17. Jones, L.K.: Remarks on queue inference from departure data alone and the importance of the queue inference engine. Oper. Res. Lett. **40**(6), 503–505 (2012)
18. Baccelli, F., Kauffmann, B., Veitch, D.: Inverse problems in queueing theory and Internet probing. Queueing Syst. **63**(1–4), 59–107 (2009)

19. Savariappan, P.R., Chandrasekhar, P., Jose, A.: Bayesian inference for an impatient M/M/1 queue with balking. J. Appl. Stat. Sci. **317**, 73–80 (2012)
20. Shestopaloff, A.Y., Neal, R.M.: On Bayesian inference for the M/G/1 queue with efficient MCMC sampling. arXiv preprint arXiv:1401.5548 (2014)
21. Acharya, S.K., Rodríguez-Sánchez, S.V., Villarreal-Rodríguez, C.E.: Maximum likelihood estimates in an M/M/c queue with heterogeneous servers. Int. J. Math. Oper. Res. **5**(4), 537–549 (2013)
22. Park, J., Kim, Y.B., Willemain, T.R.: Analysis of an unobservable queue using arrival and departure times. Comput. Ind. Eng. **61**(3), 842–847 (2011)
23. Keith, A., Ahner, D., Hill, R.: An order-based method for robust queue inference with stochastic arrival and departure times. Comput. Ind. Eng. **128**, 711–726 (2019)
24. Hall, R.W.: Queueing Methods for Services and Manufacturing. Prentice-Hall, Hoboken (1990)
25. Jagerman, D.L., Melamed, B.: The transition and autocorrelation structure of TES processes. Stoch. Models **8**(2), 193–219 (1992)

Interdisciplinary Simulation and Machine Learning

Development of Numerical Simulation Method for Analyzing the Refreezing Process of Ice Dust Irradiated by Cosmic Rays

Seiki Saito[1]([✉]) [ID], Hiroaki Nakamura[2,3] [ID], Gen Chiaki[4] [ID], Kensei Kobayashi[5,6] [ID], and Haruto Miura[1]

[1] Yamagata University, Yonezawa 992-8510, Japan
saitos@yz.yamagata-u.ac.jp
[2] National Institute for Fusion Science, Toki 509-5292, Japan
[3] Nagoya University, Nagoya 464-8603, Japan
[4] National Institute of Technology, Kochi College, Kochi 783-8508, Japan
[5] Yokohama National University, Yokohama 240-8501, Japan
[6] Tokyo Istitute of Technology, Tokyo 152-8550, Japan

Abstract. This paper introduces a numerical simulation approach to investigate the refreezing dynamics of the ice mantle of interstellar dust particle exposed to cosmic radiation. Using the binary collision approximation method, we estimate the energy transfer of cosmic rays in ice dust, revealing energy peaks at $100\,keV$ for protons and $1\,MeV$ for α particles. Subsequently, employing phase field modeling, we estimate the refreezing time scale, finding it to be approximately $1\,\mu s$. These findings enhance our understanding of the complex interactions between cosmic radiation and ice dust, shedding light on fundamental processes in the origin of amino acids related to the fabric of terrestrial life's beginnings.

Keywords: Ice Dust · Cosmic Rays · Phase Field Modeling

1 Introduction

The origin of amino acids is a topic intricately woven into the fabric of terrestrial life's beginnings. It is widely posited that the precursor molecules of amino acids might emerge from the irradiation of cosmic rays to the ice mantle of interstellar dust particle in molecular clouds.

Our understanding evolves through empirical and theoretical endeavors. In some experiments [1, 2], the synthesis of organic compounds is achieved by irradiating protons onto ice, mimicking the conditions of an ice mantle [3]. Furthermore, comprehensive research is underway to investigate the formation process of amino acids in water using molecular dynamics simulations [4]. The simulation recreates the scenario where ice dust melts as a result of cosmic ray radiation, subsequently interacting with solid ice within liquid water. Through this process, the simulation investigates the formation process while in a state of thermal non-equilibrium. However, in that scenario, despite the importance of understanding physical phenomena such as the amount of energy transferred to ice dust when irradiated with cosmic rays and the duration required for liquefied water to revert to solid ice, our comprehension remains inadequate. In this paper, therefore, we develop a combine simulation technique to investigate the amount of transferred energy by binary collision approximation (BCA) simulation, and time scale of refreezing process by the simulation of phase field modeling (Fig. 1).

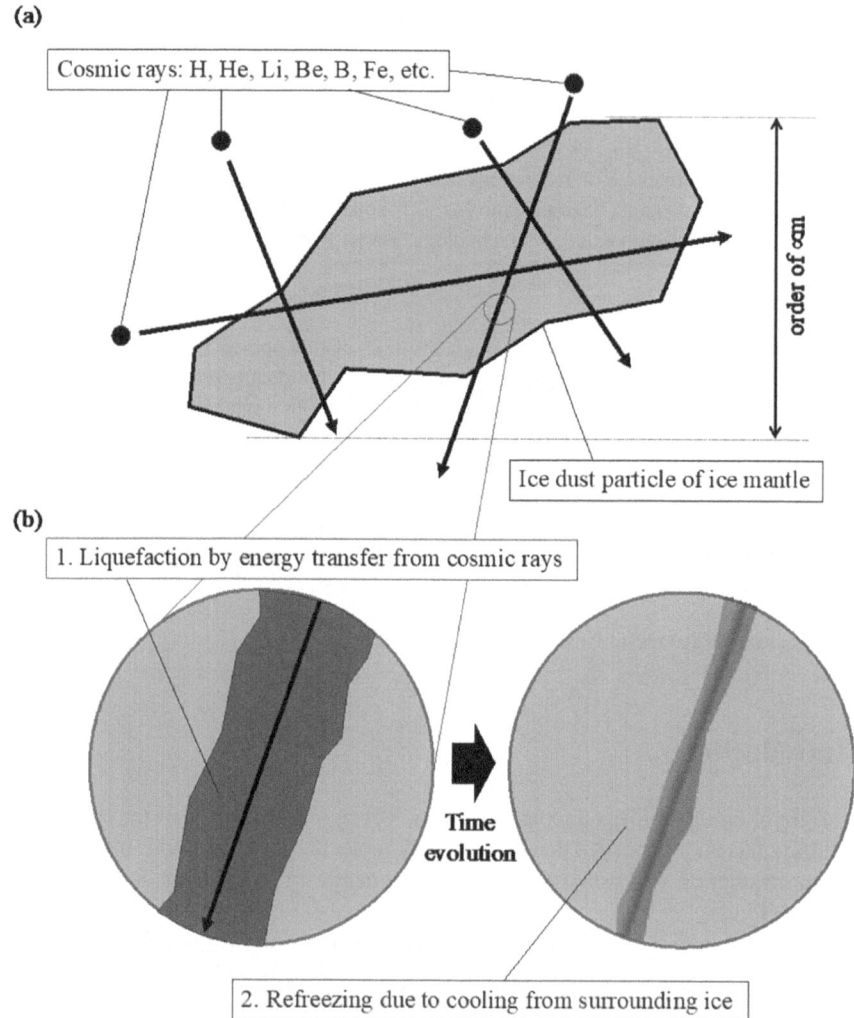

Fig. 1. Schematic diagram of (a) cosmic ray irradiation to ice dust particle of ice mantle and (b) process of liquefaction and refreezing as a result of the irradiation.

2 Estimation of Energy Transfer of Cosmic Rays in Ice Dust

2.1 Brief Introduction of Binary Collision Approximation Method

The binary collision approximation method [5–10] is a simulation technique employing the Monte Carlo method to track the collision process of ions and other particles entering a material, as well as the subsequent collision cascade process involving solid atoms energized by the incident particles. In this method, many-body interactions from surrounding atoms around the projectile are approximated as a series of binary collisions with the nearest atom in the direction of travel.

In the BCA simulation, the scattering angle Θ in center-of-mass system is calculated as:

$$\Theta = \pi - 2b \int_{r_0}^{\infty} \frac{1}{r^2 g(r)} dr,$$

where

$$g(r) \equiv \sqrt{1 - \frac{b^2}{r^2} - \frac{V(r)}{E_r}}.$$

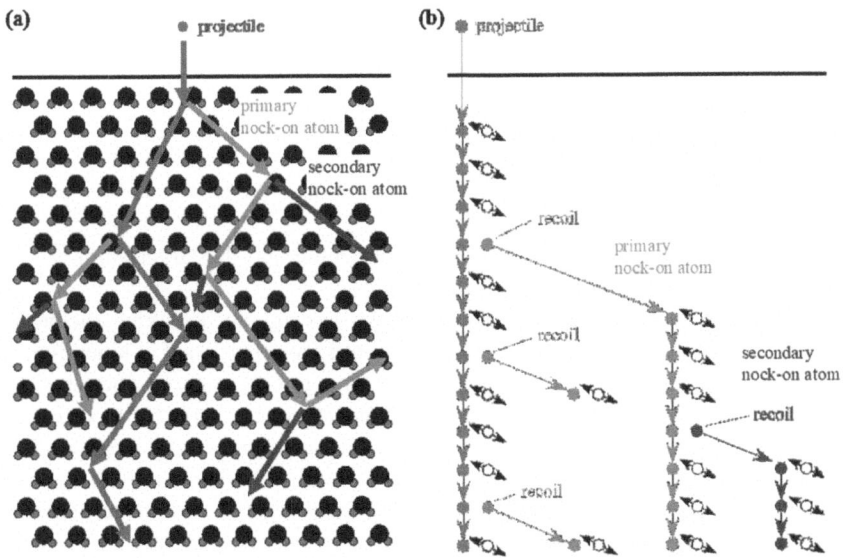

Fig. 2. (a) Schematic diagram depicting the cascade process of cosmic rays in the ice dust calculated by BCA simulation. The red, green, and blue lines represent the trajectories of the incident particle (projectile), atoms knocked by the incident particle (primary knock-on atom: PKA), and atoms knocked by the PKA (secondary knock-on atom), respectively. (b) Diagram explaining the cascade collision process and the generation of recoil atoms. (Color figure online)

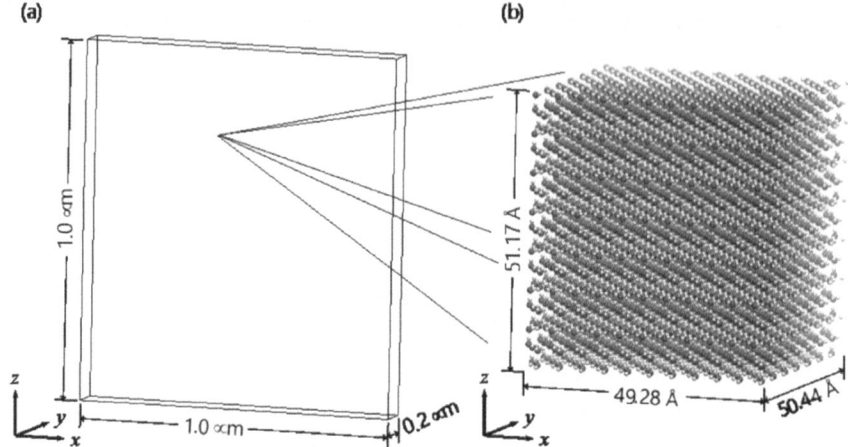

Fig. 3. (a) The ice plate serving as the target material, composed of arranged unit cells, and (b) the atomic structure of a unit cell filled with ice crystals.

E_r, b, r, and r_0 are the relative kinetic energy, the impact parameter of the binary collision, interatomic distance between the projectile and the target atom, the solution of $g(r_0) = 0$, respectively. $V(r)$ is the interatomic potential function. For our simulation, we employ the following Moliere approximation to the Thomas-Fermi potential [11] for $V(r)$:

$$V(r) \equiv \frac{Z_1 Z_2 e^2}{r} \Gamma \left(\frac{r}{a_s} \right)$$

where

$$\Gamma(x) \equiv 0.35e^{-0.3x} + 0.55e^{-1.2x} + 0.10e^{-6.0x}.$$

Z_1 and Z_2 are the atomic number of the projectile and the target atom, e is the electronic charge, and a_s is the screening length.

Figure 2(a) depicts a schematic diagram of the cascade process. When the original projectile with kinetic energy E collides with a target atom, a portion of its kinetic energy, denoted as ΔE, is transferred to the target atom, forming a collision pair. The target atom recoils when its kinetic energy ΔE exceeds the displacement energy E_d. In such cases, the target atom is also considered as a projectile with kinetic energy $\Delta E - E_B$, where E_B represents the bulk binding energy. The original projectile continues to move with kinetic energy $E' \equiv E - \Delta E$. When ΔE is less than E_d, the target atom remains stationary. In that case, the transferred energy ΔE is dissipated as heat through thermal processes. Figure 2(b) illustrates the cascade collision process and the generation of recoil atoms. The white circles represent target atoms that received less kinetic energy than E_d. Recoil atoms are depicted by circles filled with green or blue colors. The projectile is deemed to have stopped when its post-collision energy E' falls below a threshold value E_{stop}. At this point, the remaining energy E' is dissipated through energy dissipation mechanisms.

The BCA method also assesses the energy loss E_{loss} due to electron excitation [11, 12] for each collision, which is subsequently dissipated as heat. Consequently, the transferred energy ϵ from the projectile to the target material can be computed by aggregating the following lost energies:

$$\epsilon = \begin{cases} \Delta E + E_{loss} & \text{if } E' > E_{stop} \text{ and } \Delta E \leq E_d \\ E_{loss} & \text{if } E' > E_{stop} \text{ and } \Delta E > E_d \\ E' + E_{loss} & \text{if } E' \leq E_{stop} \end{cases} \tag{1}$$

2.2 Simulation Model

To calculate the energy transferred to the ice dust, we create an ice plate whose size is 1.0 μm × 0.2 μm × 1.0 μm, as illustrated in Fig. 3(a). The ice plate consists of arranging 203, 40, and 195 unit cells in x-, y-, and z-directions, respectively, as depicted in Fig. 3(b). Each unit cell comprises 4004 H_2O molecules in the structure of ice crystal (Ih).

30 protons or α particles with fixed incident energy are injected from the top surface in the x-y plane, with the incident angle parallel to the z-axis. The incident positions of y-coordinate are fixed at the center of the plate, while x-coordinates are randomly selected. The investigation examines the dependence of energy transfer from α particles to the ice plate by varying the incident energy from 100 eV to 1 GeV.

2.3 Results

Figure 4 illustrates the incident energy dependence of the trajectories of 30 protons and α particles injected into the ice plate. We only observed the generation of recoil atoms when the kinetic energy of projectiles becomes low. With increasing incident energy, the penetration depth also increases. For incident energies below 1 keV, most protons fail to penetrate the 1 μm-depth ice plate, while some succeed at 1 keV. At an incident energy of 100 keV, all 30 protons and α particles penetrate the plate, moving predominantly vertically with minimal motion in the x-y planes. Similar trajectories are observed for incident energies exceeding 100 keV. The simulations for protons and α particles yield comparable results despite slight differences in range.

Figure 5(a) presents the incident energy dependence of the transferred energy to the ice plate per injection, calculated by averaging over 30 trials. Figure 5(b) displays the mean penetration depth across 30 trials. Estimation of penetration depth is based on setting the ice plate depth to 25 mm, preventing penetration by incident protons and α particles. However, penetration depth calculation was infeasible for incident energies exceeding 10 MeV due to computational constraints. The mean penetration depth is less than 1 μm for incident energies up to 1 keV, so that, most of incident energy is transferred to the ice plate in that case. However, when the incident energy exceeds 1 keV, most of the incident particle penetrate the plate, resulting in less transferred energy than the incident energy. Notably, the transferred energy exhibits a peak at incident energies of 100 keV for protons and 1 MeV for α particles due to the decreasing scattering cross section with increasing incident energy.

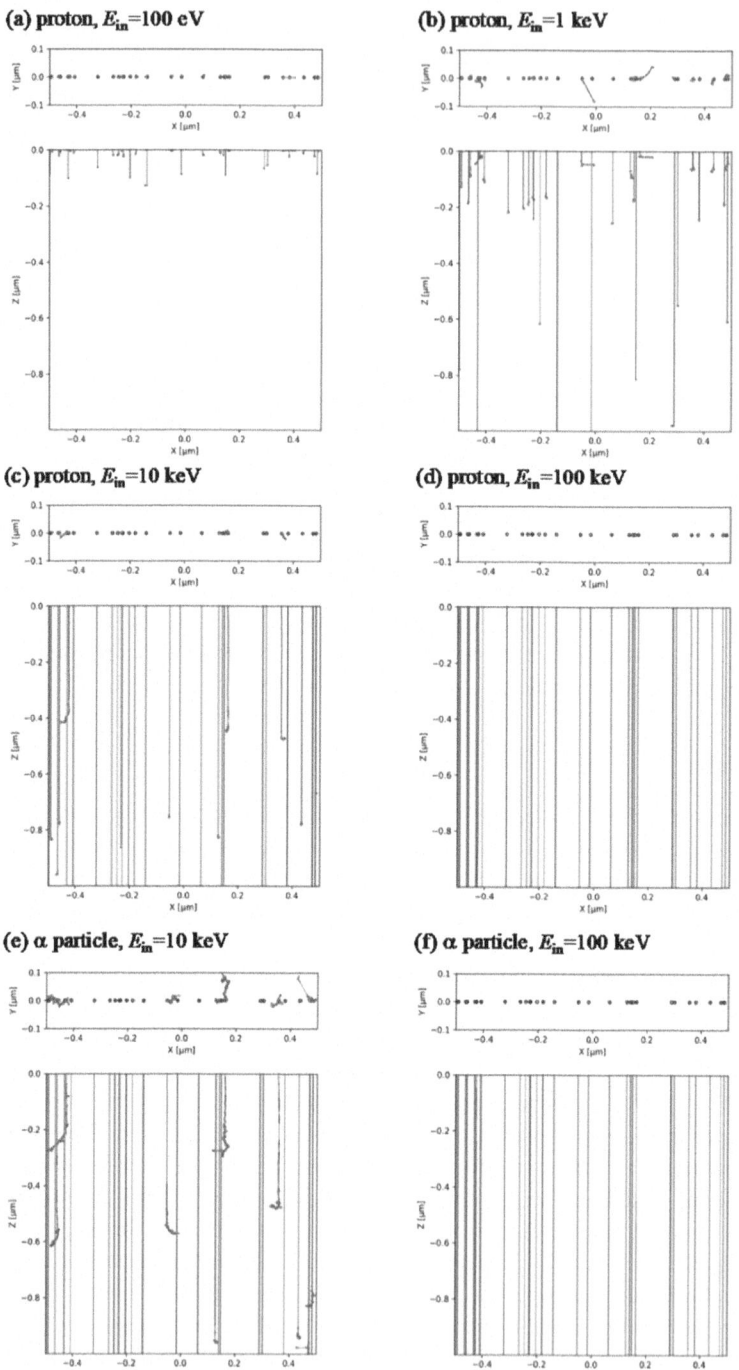

Fig. 4. Incident energy E_{in} dependence of trajectories of (a)–(d) 30 protons and (e)–(f) 30 α particles injected into an ice plate, with an incident angle parallel to the z-axis.

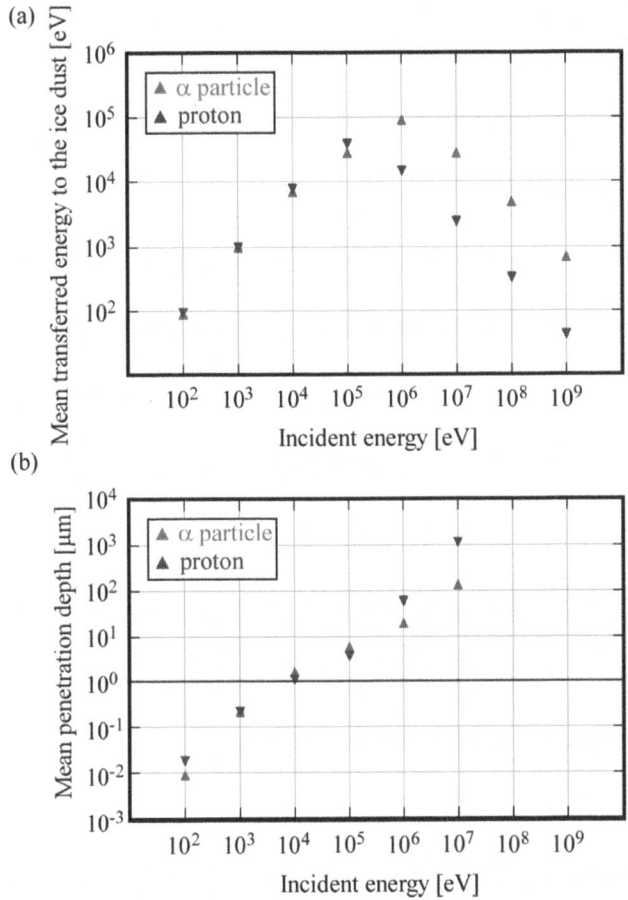

Fig. 5. (a) Incident energy dependence of the mean transferred energy to the ice plate per injection, calculated by averaging 30 trials. (b) Incident energy dependence of the mean penetration depth across 30 trials.

3   Estimation of Refreezing Time Scale in Ice Dust

3.1   Simulation Method: Phase Field Modeling

Due to energy transfer from proton or α particle irradiation, the region through which the projectile passes melts as water molecules receive thermal energy. Subsequently, after penetration, the melted region refreezes due to thermal transfer to the surrounding ice. To estimate the refreezing time scale, we conduct a simulation using phase field modeling [13]. In this simulation, we calculate the Gibbs free energy G of our system by integrating the local free energy density G_{chem}, G_{doub}, and G_{grad} over the volume V of the system as follows:

$$G = \int_V \left[G_{\text{chem}} + G_{\text{doub}} + G_{\text{grad}} \right] \mathrm{d}V.$$

G_{chem} represents the chemical free energy and is defined as follows:

$$G_{\text{chem}} \equiv h(\phi)G^{S}(T) + [1 - h(\phi)]G^{L}(T).$$

Here, $h(\phi)$ is the following function that smoothly transitions from 0 to 1 depending on the value of the phase field ϕ, which takes the value of 1 in the solid phase and 0 in the liquid phase.

$$h(\phi) \equiv \phi^{3}\left(10 - 15\phi + 6\phi^{2}\right)$$

G_{doub} represents the double well potential introduced to prevent the coexistence of the solid and liquid phases at the same position simultaneously. It is defined as:

$$G_{\text{doub}} \equiv W^{SL}\phi^{2}(1 - \phi)^{2}$$

The proportionality coefficient W^{SL} is defined as:

$$W^{SL} \equiv \frac{30\sigma}{d},$$

where σ and d are the gradient energy coefficent and diffuse interface width, respectively. We note that our model does not consider the anisotropy of σ, although some works related to phase field modeling consider the anisotropy of σ which comes from the atomic structure.

G_{grad} represents the gradient energy and is defined as:

$$G_{\text{grad}} \equiv \frac{a}{2}|\nabla\phi|^{2}.$$

We calculate the a as follows:

$$a \equiv \frac{3\sigma d}{5}.$$

Given the Gibbs free energy G, the time evolution of the phase filed ϕ can be calculated by the following equation:

$$\tau\frac{\partial\phi}{\partial t} = -\left(\frac{\delta G}{\delta t} + \xi\right).$$

Here, $\delta G/\delta t$ represents functional differentiation.

The constant τ, related to the diffusion time of ϕ, is calculated as follows:

$$\tau \equiv \frac{3dL}{5\mu T_{\text{m}}}.$$

ξ is a fluctuation term serving as the driving force for coagulation, defined by the following equation:

$$\xi \equiv 4W^{SL}(1 - \phi)\chi.$$

where χ is a uniform random number within the range $-0.05 \leq \chi \leq 0.05$.

The following time evolution equation is derived from functional differentiation.

$$\frac{\delta G}{\delta t} + \xi = 4W^{\mathrm{SL}}\phi(1-\phi)\left[\frac{1}{2} - \phi + \frac{15L(T-T_{\mathrm{m}})}{2W^{\mathrm{SL}}T_{\mathrm{m}}}\phi(1-\phi) + \chi\right] - \alpha\nabla^2\phi \quad (2)$$

Here, L represents the heat of solidification and T_{m} is the melting point. This equation is obtained by approximating the difference between $G^{\mathrm{S}}(T)$ and $G^{\mathrm{L}}(T)$ as:

$$G^{\mathrm{S}}(T) - G^{\mathrm{L}}(T) \sim \frac{-L(T_{\mathrm{m}} - T)}{T_{\mathrm{m}}}.$$

For the time evolution of the spatial distribution of temperature T, we simultaneously solve the following thermal diffusion equation:

$$\frac{\partial T}{\partial t} = \frac{K}{C_{\mathrm{p}}}\nabla^2 T + p(\phi)\frac{L}{C_{\mathrm{p}}}\frac{\partial\phi}{\partial t}. \quad (3)$$

Here, K is the thermal conductivity, C_{p} is the specific heat, and $p(\phi)$ is a function defined as:

$$p(\phi) \equiv \frac{\partial h}{\partial\phi} = 30\phi^2(1-\phi)^2.$$

The values for all parameters used in the simulation are shown in Table 1 as determined by reference to Ref. [14]. We note that these parameters are values at an atmospheric pressure of 1 atm. In the case of real ice dust placed in a vacuum, it sublimates at a lower temperature, and constants such as specific heat are thought to change. In this research, firstly, we aim to establish a simulation method and perform calculations under the condition of 1 atmosphere. We plan to perform calculations under vacuum conditions in the near future.

Table 1. Value of parameters used for the phase filed model.

Parameter	Variable	Value
Thermal conductivity	K [W $\cdot$ m^{-1} $\cdot$ K^{-1}]	0.569
Specific heat	C_{p} [MJ $\cdot$ K^{-1} $\cdot$ m^{-3}]	4.2
Heat of solidification	L [MJ $\cdot$ m^{-3}]	334.0
Melting point	T_{m} [K]	273.0
Kinetic growth coefficient	μ [m $\cdot$ K^{-1} $\cdot$ s^{-1}]	5.0
Gradient energy coefficient	σ [N $\cdot$ m]	0.022
Diffuse interface width	d [m]	8.0×10^{-8}

3.2   Simulation Model

First, we calculate the transferred energy to the ice plate, which has dimensions of $1.0\,\mu m$ $\times 0.2\,\mu m \times 1.0\,\mu m$ using the BCA simulation. In the simulation, we only calculate the case of α particle injection with incident energy of 1 MeV. Figure 6(a) illustrates 10 trajectories of the incident α particles. For the calculation, the incident particles are injected in the x-z plane at $y = 0.0\,\mu m$, as described in Sect. 2, with the incident polar angle set randomly. Periodic boundary conditions are applied in the x- and y-directions.

After completing the BCA calculation, we simulate the refreezing process using the introduced phase field model by solving the differentiated equation of Eq. 2 and 3 simultaneously in three dimensions with a time step of 1.0 ps. For this simulation, periodic boundary conditions are applied in the x- and y-directions, while Neumann boundary conditions are applied in the z-direction to maintain a zero gradient. The ice plate, with dimensions of $1.0\,\mu m \times 0.2\,\mu m \times 1.0\,\mu m$, is divided into $100 \times 20 \times 100$ cells, so that the volume of a cell is 10 nm $\times$ 10 nm$\times$ 10 nm. The initial temperature T_{ini} for all cells unaffected by collisions is set to 272 K, while for cells affected by collisions,

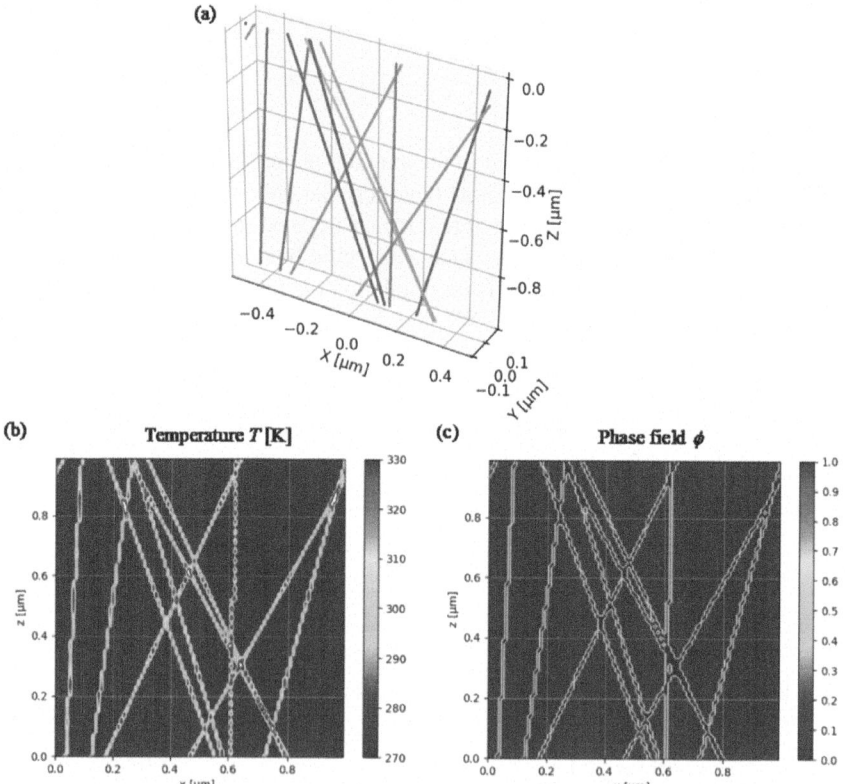

Fig. 6. (a) Trajectories of 10 incident α particles with incident energy of 1 MeV calculated by the BCA simulation. Spatial distribution of (b) initial temperature and (c) phase filed in the x-z plane at $y = 0$.

it is determined as follows:

$$T_{ini} = 272 + \frac{1}{C_p V_{cell}} \sum_{i \in A} \epsilon_i.$$

Here, V_{cell} represents the volume of the cell, A is the set of indices of collisions that occurred in the cell during 10 injections, and ϵ_i is the transferred energy of i-th collision calculated by Eq. 1.

Figure 6(b) depicts the spatial distribution of the initial temperature T_{ini} in the x-z plane at $y = 0.0$ μm. The maximum temperature of cells reaches 401.3 K. Figure 6(c) illustrates the spatial distribution of the initial phase field ϕ_{ini}. ϕ_{ini} is set to 0 or 1 for the cells with temperature higher than 272 or equal to 272, respectively.

3.3 Results

Figure 7 illustrates the time evolution of the spatial distribution of temperature T. It is evident that the temperature of the cells along the trajectories gradually decreases to match the temperature of the surrounding area within a time scale of 1 ns due to heat diffusion. Figure 8 depicts the time evolution of the spatial distribution of phase filed ϕ. Over time, the cells transition from the water phase to the liquid phase. It's notable that compared to the time scale of temperature T, the changes in ϕ occur a longer time scale of 1 μs.

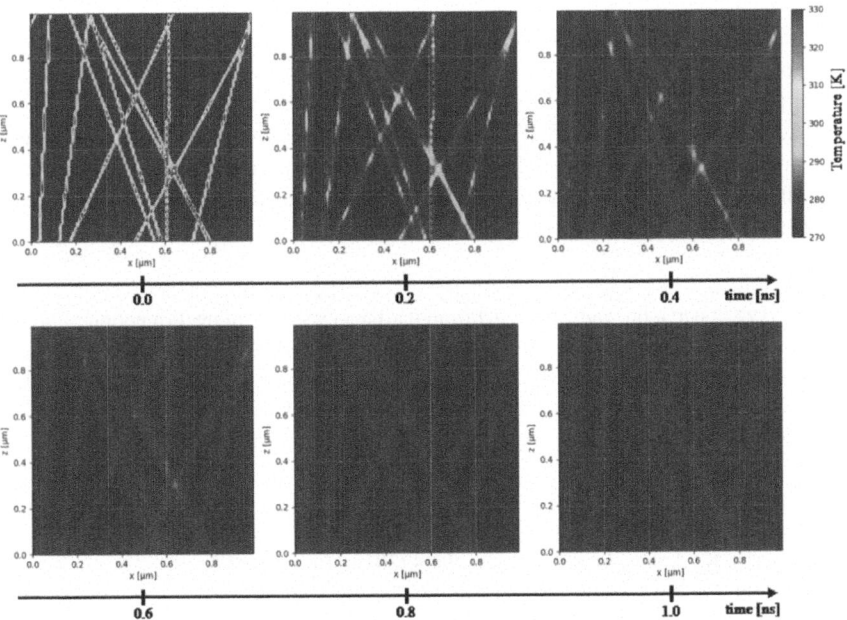

Fig. 7. Time evolution of spatial distribution of temperature T calculated by the phase field model.

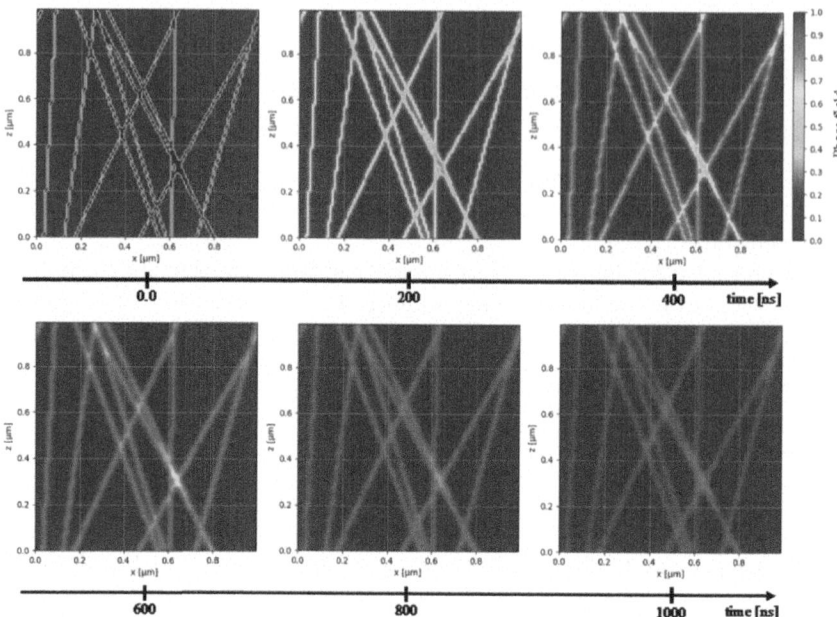

Fig. 8. Time evolution of spatial distribution of phase field ϕ calculated by the phase field model.

4 Summary

To explore the refreezing dynamics of ice dust irradiated by cosmic radiation, we initially examined the energy transfer from the radiation to the ice dust using the binary collision approximation method. Our findings revealed a peak energy transfer at incident energies of 100 keV for protons and 1 MeV for α particles. Subsequently, we conducted simulations employing phase field modeling to estimate the time scale for refreezing after ice was melted by the transferred energy from radiation. Our results indicate a refreezing time scale of approximately 1 μs.

Acknowledgments. The research was partially supported by Grant-in-Aid for Scientific Research No. 22K03572, 23H04609, and 23K03362 from the Japan Society for the Promotion of Science, and by the NIFS Collaborative Research Program NIFS22KISS021 and NIFS22KIGS002. The computations were performed using Research Center for Computational Science, Okazaki, Japan (Project: 24-IMS-C099) and Plasma Simulator of NIFS (Toki, Gifu, Japan).

References

1. Kobayashi, K.: Amino Acid Res. **12**(1), 7–14 (2019). (in Japanese)
2. Kobayashi, K.: Biotechnology **96**(11), 621–625 (2018). (in Japanese)
3. Takano, Y., et al.: Appl. Phys. Lett. **85**(9), 1633–1635 (2004)
4. Murai, M., Nakamura, H., Kobayashi, K., Masahiro, K., Usami, S., Goto, Y.: Molecular dynamics simulation for synthesis route of amino acid precursor in space. In: Proceedings of ISPlasma 2024, 05P-P1-50 (2024)

5. Robinson, M.T., Torrens, I.M.: Phys. Rev. B **9**, 5008 (1974)
6. Ohya, K., Kawakami, R.: Jpn. J. Appl. Phys. **40**, 5424 (2001)
7. Biersack, J.P., Eckstein, W.: Appl. Phys. A **34**, 73 (1984)
8. Yamamura, Y., Mizuno, Y.: Low-energy sputterings with the Monte Carlo program ACAT. Institute of Plasma Physics, Nagoya University, Japan, IPPJ-AM-40 (1985)
9. Saito, S., Ito, A.M., Takayama, A., Kenmotsu, T., Nakamura, H.: J. Nucl. Mater. (2011). https://doi.org/10.1016/j.jnucmat.2010.12.233
10. Saito, S., Ito, A.M., Takayama, A., Kenmotsu, T., Nakamura, H.: Prog. Nucl. Sci. Technol. **2**, 44–50 (2011)
11. Andersen, H.H., Ziegler, J.F.: The Stopping and Ranges of Ions in Matter, 3rd edn. Pergamon Press, Oxford (1977)
12. Bethe, H.A.: Zur Theorie des Durchgangs schneller Korpuskularstrahlen durch Materie. Ann. der Physik **5**, 325–400 (1930)
13. Koyama, T., Takaki, T.: Introduction to Phase Field Method. Maruzen Publissing (2013). (in Japanese)
14. Simulation of ice crystal growth on solid surface using phase field method. Trans. JSRAE **30**(3), 277–287 (2013). (in Japanese)

Model Calibration for Agent-Based Simulation Using a Pattern Clustering Network

Yuanjun Laili[1,2]($\boxtimes$), Jiabei Gong[1], Ke Hu[1], Lin Zhang[1], and Fei Wang[2]

[1] Beihang University, Beijing 100191, China
lailiyuanjun@buaa.edu.cn
[2] Zhongguancun Laboratory, Beijing 100094, China

Abstract. Agent-based simulation has become essential for simulating complex evolving systems, such as social systems and complex manufacturing systems. However, due to the uncertainty in agent behaviors, model calibration poses a significant challenge. Traditional methods like try-and-error and regression are inefficient, requiring numerous simulation runs under different parameter settings. To address this, we propose an online pattern clustering network-based calibration method, termed PCN-Calibration. This method establishes a pattern clustering network using reference data from the target system and compares simulation results under random parameter settings with this network. Ten weighting rules are introduced to estimate the best parameters based on the distance between the simulation results and the clustering network. Experimental results demonstrate that the proposed method can find feasible parameter settings within seconds. Furthermore, the experimental discussions provide guidelines for selecting suitable weighting rules for calibrating different models.

Keywords: Model calibration · Self-organizing incremental neural network · Pattern clustering and Distance measurement

1 Introduction

Agent-based simulation involves studying evolving systems by modeling agents with random states and autonomous behaviors [1]. It is particularly useful for System of Systems (SoS) scenarios [2], where mathematical modeling is impractical. However, one challenge is the lack of known statistics for simulation variables [3]. Model parameter calibration is crucial but often tackled using time-consuming trial and error methods [4]. Design of experiments (DoE) offers a more comprehensive exploration of parameter space, but it relies heavily on specific targets and models. Learning model-based approaches, such as Gaussian process regression [5] and Bayesian inference [6], predict suitable parameters using machine learning algorithms, but acquiring diverse simulation results for training is a major challenge [7–10].

© The Author(s), under exclusive license to Springer Nature Singapore Pte Ltd. 2024
S. Saito et al. (Eds.): AsiaSim 2024, CCIS 2170, pp. 152–164, 2024.
https://doi.org/10.1007/978-981-97-7225-4_12

A more versatile approach to model calibration involves framing it as an optimization problem [11,12]. If a mathematical model for simulation exists, derivative-based optimization methods like gradient descent, hill-climbing, and simplex methods can be employed [13,14]. Alternatively, derivative-free optimization methods such as least squares estimation, Expectation-Maximum (EM) algorithm, and evolutionary algorithms are utilized [15]. Various error, consistency, and divergence measurements serve as objective functions to guide optimization [16]. However, iterative optimization is time-consuming and may not adapt well to changing target characteristics. To automatically find a suitable model approximation, research also focuses on establishing models or posterior distributions to predict parameters yielding simulation results close to reference data. Regression methods like artificial neural networks, fuzzy regression, Gaussian mixture models, and Bayesian inference are employed for model calibration [8,9,17,18].

To efficiently calibrate agent-based simulation models with evolving states and high dynamics, this paper proposes a rapid pattern clustering network-based calibration method (PCN-Calibration). Utilizing a pattern clustering network established with reference data and the load balancing self-organizing incremental neural network (LB-SOINN) [19], the method evaluates simulation results' credibility online. It employs ten weighting rules to identify feasible parameter settings close to reference data, achieving rapid calibration within seconds, as demonstrated across two typical agent-based simulation models.

2   The PCN-Calibration Method

2.1   The Mainframe of the PCN-Calibration Method

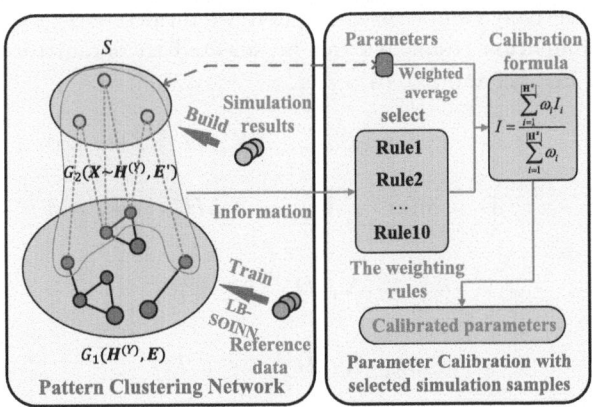

Fig. 1. Structure of PCN-Calibration Method.

As shown in Fig. 1, the PCN-Calibration method consists of two parts, the clustering network $G_1(\boldsymbol{H}^{(Y)}, \boldsymbol{E})$ and the association network $G_2(\hat{\boldsymbol{H}}^{(Y)}, \boldsymbol{E}')$. The clustering network is obtained by LB-SOINN using the reference data $\boldsymbol{Y}$ with a set of nodes $\boldsymbol{H}^{(Y)}$ and a set of edges $\boldsymbol{E}$. The association network $G_2(\hat{\boldsymbol{H}}^{(Y)}, \boldsymbol{E}')$ is established by connecting the most similar nodes of $\boldsymbol{X}$ to the compressed reference dataset $\boldsymbol{H}^{(Y)}$. As the parameters for generating $\boldsymbol{Y}$ is unknown, $\boldsymbol{X}$ is a group of simulation results generated under random parameter settings in specific ranges. One can deduce suitable parameters by finding the simulation results which are closest to the distribution region of the reference data. Therefore, suitable parameters are determined by the nodes that exist in G_1 and G_2 simultaneously. Finally, weighting rules are designed to determine the influence of the closest simulation samples $\boldsymbol{S}$ (each sample $\boldsymbol{S}_i = (\boldsymbol{I}_i^{(X)}, \boldsymbol{X}_i)$ represents a simulation parameter setting $\boldsymbol{I}_i^{(X)}$ with the corresponding simulation results $\boldsymbol{X}_i$) and deduce the most suitable parameter setting.

2.2   The Association Network

The association network G_2 designates the correlation between the reference data and the simulation results. It is determined in the following way.

For each piece of simulation result $\boldsymbol{X}_i$ in $\boldsymbol{X}$:

(1) Find the closest node $g_1 = \underset{j \in [1, |\boldsymbol{H}^{(Y)}|]}{\arg\min} (\left\| \boldsymbol{X}_i - \boldsymbol{H}_j^{(Y)} \right\|)$ of $\boldsymbol{H}^{(Y)}$ to $\boldsymbol{X}_i$.

(2) Find the second closest node $g_2 = \underset{j \in \boldsymbol{H}^{(Y)} \setminus \boldsymbol{H}_{g_1}^{(Y)}}{\arg\min} (\left\| \boldsymbol{X}_i - \boldsymbol{H}_j^{(Y)} \right\|)$ of $\boldsymbol{H}^{(Y)}$ to $\boldsymbol{X}_i$.

According to the PCN evaluation rules formulated in Eq. (1), the similarity thresholds that determine whether $\boldsymbol{X}_i$ belongs to the pattern clusters of $\boldsymbol{H}_{g_1}^{(Y)}$ and $\boldsymbol{H}_{g_2}^{(Y)}$, respectively can be obtained, here we termed as $\boldsymbol{T}_{g_1}^{(i)}$ and $\boldsymbol{T}_{g_2}^{(i)}$. Then, whether the simulation result $\boldsymbol{X}_i$ can be selected to parameter calibration is determined by Eq. (2) and Eq. (3).

$$T_g = \begin{cases} \underset{j \in \boldsymbol{\Psi}_g}{\max} \, d_{output}(\boldsymbol{X}_i, \boldsymbol{H}_j^{(Y)}) & \boldsymbol{\Psi}_g \neq \emptyset \\ \underset{j \in \boldsymbol{H}^{(Y)} \setminus \boldsymbol{H}_g^{(Y)}}{\min} \, d_{output}(\boldsymbol{X}_i, \boldsymbol{H}_j^{(Y)}) & \boldsymbol{\Psi}_g = \emptyset \end{cases} \tag{1}$$

$$d_{output}(\boldsymbol{X}_i, \boldsymbol{H}_{g_1}^{(Y)}) \leq \zeta_1 T_{g_1}^i \tag{2}$$

$$d_{output}(\boldsymbol{X}_i, \boldsymbol{H}_{g_2}^{(Y)}) \leq \zeta_2 T_{g_2}^i \tag{3}$$

where ζ_1 and ζ_2 are relaxation factors that satisfy $0.6 \leq \zeta_2 \leq \zeta_1 \leq 3$.

If Eq. (2) is satisfied, an edge connecting $\boldsymbol{X}_i$ and $\boldsymbol{H}_{g_1}^{(Y)}$ is added to $\boldsymbol{E}'$ in $G_2(\hat{\boldsymbol{H}}^{(Y)}, \boldsymbol{E}')$. Suppose the set of adjacent nodes that directly connected to the node $\boldsymbol{H}_{g_1}^{(Y)}$ is $\hat{\boldsymbol{H}}_{g_1}^{(Y)}$. To any $k \in [1, |\boldsymbol{H}^{(Y)}|]$, if $\langle \boldsymbol{H}_{g_1}^{(Y)}, \boldsymbol{H}_k^{(Y)} \rangle \in \boldsymbol{E}$, then $\hat{\boldsymbol{H}}_{g_1}^{(Y)} \leftarrow \boldsymbol{H}_k^{(Y)}$. $\hat{\boldsymbol{H}}^{(Y)} \leftarrow \hat{\boldsymbol{H}}_{g_1}^{(Y)} \cup \{\boldsymbol{H}_{g_1}^{(Y)}\}$, $\boldsymbol{E}'^{(1)} \leftarrow \langle \boldsymbol{H}_{g_1}^{(Y)}, \boldsymbol{H}_k^{(Y)} \rangle$. Similarly, if

Eq. (3) is satisfied, $\hat{\boldsymbol{H}}_{g_2}^{(Y)}$, G_2 and $\boldsymbol{E'}^{(2)}$ can be updated using the corresponding steps. Finally $\boldsymbol{E'} = \boldsymbol{E'}^{(1)} \cup \boldsymbol{E'}^{(2)}$.

2.3   The Weighting Rules

Based on the definition of the association network, the simulation samples that are close to the reference data are used to estimate the possible parameters. Using a group of weights, the suitable parameter can be determined by Eq. (4).

$$I = \frac{\sum\limits_{i=1}^{|H^{(X)}|} \omega_i I_i}{\sum\limits_{i=1}^{|H^{(X)}|} \omega_i} \tag{4}$$

where $\boldsymbol{I}_i$ is the parameter of $\boldsymbol{X}_i$.

Then, this paper proposes ten weighting rules to determine $\omega_i, i \in [1, |\boldsymbol{X}|]$, which are summarized in Table 1.

For $\boldsymbol{X}_i$ that connected to G_2, two labels L_{g_1} and L_{g_2} are introduced to demonstrate whether Eq. (2) and Eq. (3) are satisfied. If Eq. (2) is satisfied, $L_{g_1} = 1$. Otherwise, $L_{g_1} = 0$. If Eq. (3) is satisfied, $L_{g_2} = 1$. Otherwise, $L_{g_2} = 0$. For $\boldsymbol{X}_i$ that unconnected to G_2, $\omega_i = 0$. C_{g_1} and C_{g_2} are the number of nodes in the cluster associated with $\boldsymbol{H}_{g_1}^{(Y)}$ and $\boldsymbol{H}_{g_2}^{(Y)}$, respectively. To calculate the average density weight $\omega^{(2)}$, suppose the Euclidean distance and Cosine distance between two nodes $\boldsymbol{H}_p^{(Y)}$ and $\boldsymbol{H}_q^{(Y)}$ in the clustering network $G_1(\boldsymbol{H}^{(Y)}, \boldsymbol{E})$ is EU_{pq} and CO_{pq}, respectively. The maximum and minimum Euclidean distance and Cosine distance between any two nodes in G_1 is EU_{max}, EU_{min}, CO_{max} and CO_{min} separately. Hence, a combined distance $d_{com}(\boldsymbol{H}_p^{(Y)}, \boldsymbol{H}_q^{(Y)})$ as demonstrated in Eq. (5).

$$d_{com}(\boldsymbol{H}_p^{(Y)}, \boldsymbol{H}_q^{(Y)}) = \frac{1}{\eta^l} \frac{EU_{pq} - EU_{min}}{1 + EU_{max} - EU_{min}} + (1 - \frac{1}{\eta^l}) \frac{CO_{pq} - CO_{min}}{1 + CO_{max} - CO_{min}} \tag{5}$$

where η is the distance weight parameter, and l is the length of the reference data.

The average distance of a node $\boldsymbol{H}_j^{(Y)}$ from $G_1(\boldsymbol{H}^{(Y)}, \boldsymbol{E})$ is defined as Eq. (6). The density score p_j of it in each incremental learning iteration is defined in Eq. (7). Then the average cumulative density after N iterations can be calculated by Eq. (8).

$$\bar{d}(\boldsymbol{H}_j^{(Y)}) = \sum_{(\boldsymbol{H}_j^{(Y)}, \boldsymbol{H}_k^{(Y)}) \in \boldsymbol{E}} d_{dom}(\boldsymbol{H}_j^{(Y)}, \boldsymbol{H}_k^{(Y)}) \tag{6}$$

$$p_j = \begin{cases} 1 - \bar{d}(\boldsymbol{H}_j^{(Y)}) & \text{if } j \text{ is a winner} \\ 0 & \text{otherwise} \end{cases} \tag{7}$$

Table 1. Definition of Weighting Rules

Weighting rule	Expression
Cluster Proportion Weighting	$\omega_i^{(1)} = L_{g_1}|\boldsymbol{H}_{g_1}^{(Y)}|\dfrac{C_{g_1}^{(Y)}}{|\boldsymbol{H}^{(Y)}|} + L_{g_2}|\boldsymbol{H}_{g_2}^{(Y)}|\dfrac{C_{g_2}^{(Y)}}{|\boldsymbol{H}^{(Y)}|}$
Average Density Weighting	$\omega_i^{(2)} = L_{g_1}h_{g_1} + L_{g_2}h_{g_2}$
Euclidean Distance Weighting	$\omega_i^{(3)} = L_{g_1}\dfrac{1}{\left\|\boldsymbol{X}_i - \boldsymbol{H}_{g_1}^{(Y)}\right\|} + L_{g_2}\dfrac{1}{\left\|\boldsymbol{X}_i - \boldsymbol{H}_{g_2}^{(Y)}\right\|}$
COS Distance Weighting	$\omega_i^{(4)} = L_{g_1}\dfrac{\boldsymbol{X}_i \cdot \boldsymbol{H}_{g_1}^{(Y)}}{|\boldsymbol{X}_i||\boldsymbol{H}_{g_1}^{(Y)}|} + L_{g_2}\dfrac{\boldsymbol{X}_i \cdot \boldsymbol{H}_{g_1}^{(Y)}}{|\boldsymbol{X}_i||\boldsymbol{H}_{g_2}^{(Y)}|}$
Hybrid Distance Weighting	$\omega_i^{(5)} = L_{g_1}\left(\dfrac{\boldsymbol{X}_i \cdot \boldsymbol{H}_{g_1}^{(Y)}}{|\boldsymbol{H}_i^x||\boldsymbol{H}_{g_1}^{(Y)}|} + \dfrac{1}{\left\|\boldsymbol{X}_i - \boldsymbol{H}_{g_1}^{(Y)}\right\|}\right)$ $+ L_{g_2}\left(\dfrac{\boldsymbol{X}_i \cdot \boldsymbol{H}_{g_2}^{(Y)}}{|\boldsymbol{X}_i||\boldsymbol{H}_{g_2}^{(Y)}|} + \dfrac{1}{\left\|\boldsymbol{X}_i - \boldsymbol{H}_{g_2}^{(Y)}\right\|}\right)$
Average Connection Weighting	$\omega_i^{(6)} = \dfrac{|\boldsymbol{H}_{g_1}^{(Y)}| + |\boldsymbol{H}_{g_2}^{(Y)}|}{|\boldsymbol{H}^{(Y)}|}$
Association Weighting	$\omega_i^{(7)} = 1$
Maximum Cluster Weighting	$\omega_i^{(8)} = \begin{cases} 1 & \text{if } i = \underset{i\in[1,|\boldsymbol{H}^{(X)}|]}{\arg\max}\ \omega_i^{(1)} \\ 0 & \text{else} \end{cases}$
Maximum Density Weighting	$\omega_i^{(9)} = \begin{cases} 1 & \text{if } i = \underset{i\in[1,|\boldsymbol{H}^{(X)}|]}{\arg\max}\ \omega_i^{(2)} \\ 0 & \text{else} \end{cases}$
Combined Weighting	$\omega_i^{(10)} = \displaystyle\sum_{r\in[1,9]} \omega_i^{(r)}$

$$h_j = \sum_{r=1}^{N} p_j^{(r)} \tag{8}$$

Based on the definition in Table 1, the output parts of the samples that similar to the reference data (i.e., represented by the clustering network G_1) are selected and their weights are calculated. Then their input parts are used to calculate the possible parameter settings, according to Eq. (4).

3   Experiments and Discussions

This section evaluates the PCN-based calibration method with ten weighting rules through three main experiments using two representative Netlogo example models: the Wolf Sheep Predation model and the Traffic 2 Lanes model. A group of parameter settings generates reference data, while a wider range of settings generates simulation results (detailed in Tables 2 and 3).

Table 2. Parameter of the Wolf Sheep Predation Model

Parameter	Reference parameters	Simulation parameters	Target parameter range
Wolf-reproduce	2, 3.5, 5	2, 4, 10, 12	[2, 5]
Grass-regrowth-time	20, 25, 40	20, 40, 80, 100	[20, 40]
Sheep-reproduce	2, 3.5, 5	2, 4, 6, 8	[2, 5]

Table 3. Parameter of the Traffic 2 Lanes Model

Parameter	Reference parameters	Simulation parameters	Target parameter range
Acceleration	0.003, 0.005	0.0008, 0.0029, 0.005, 0.0071	[0.0029, 0.005]
Number-of-cars	40, 60	17, 39, 61, 83	[39, 61]
Deceleration	0.04, 0.06	0.016, 0.039, 0.062, 0.085	[0.039, 0.062]

3.1   Parameter Adjustment of the PCN-Calibration Method

The process begins by training the clustering network using reference data, followed by establishing the association network with simulation results. The corresponding confusion matrix $CM = \{\{TP, FN\}, \{FP, TN\}\}$ [20] is recorded.

FN: False Negative, a simulation result judged as a negative sample (i.e., inconsistent with the reference data), but in fact a positive sample (i.e., a sample that is credible).

FP: False Positive, a simulation result judged as a positive sample, but in fact a negative sample.

TN: True Negative, a simulation result judged as a negative sample, which in fact is a negative sample.

TP: True Positive, a simulation result judged as a positive sample, is actually a positive sample.

$$precision = \frac{TP}{(TP + FP)} \tag{9}$$

$$recall = \frac{TP}{(TP + FN)} \tag{10}$$

$$F_1 = \frac{2 \cdot precision \cdot recall}{precision + recall} \tag{11}$$

$$F_{0.5} = \frac{1.25 \cdot precision \cdot recall}{0.25 \cdot precision + recall} \tag{12}$$

By ensuring simulation results satisfy Eq. (1), they are deemed credible samples. The precision rate in Eq. (9) and recall rate in Eq. (10) are evaluated using the confusion matrix [21], with the F_1 score [22] in Eq. (11) providing a

Table 4. Threshold Optimization Experiments

ζ	0.5	0.7	0.9	1.1	1.3	1.5	1.7	1.9
Precision Rate	–	100.000%	100.000%	94.166%	86.631%	79.736%	76.400%	73.962%
Recall Rate	0	7.000%	29.000%	64.501%	81.000%	90.500%	95.500%	98.000%
F_1 Score	–	13.084%	44.961%	76.601%	83.721%	84.778%	84.889%	84.301%
$F_{0.5}$ Score	–	27.344%	67.130%	86.200%	85.443%	81.679%	79.583%	77.778%

combined accuracy assessment. Given the PCN-Calibration method's reliance on parameters from TP and FP samples, precision rate takes precedence over recall rate, with the $F_{0.5}$ score in Eq. (12) emphasizing precision. This paper takes the Wolf Sheep Predation model to discover suitable relaxation factor ζ. Suppose $\zeta_1 = \zeta_2$, the relaxation factors are set from 0.5 to 1.9. The results are shown in Table 4. It can be found that as ζ increases gradually, the accuracy rate degrades slightly, but the recall rate increases rapidly. When ζ increases to 1.1, the $F_{0.5}$ score is maximized. As ζ continues to increase, the $F_{0.5}$ score starts to decrease. Therefore, $\zeta = 1.1$ can be a good choice.

3.2   Comparison of Different Weighting Rules for PCN-Calibration

To compare the performance of the ten weighting rules, the normalized parameter values are introduced as demonstrated by Eq. (13).

$$I'_i = \frac{I_i - I_a^i}{I_b^i - I_a^i} \tag{13}$$

where $I = \{I_1, I_2, ..., I_n\}$ represents the parameter setting of a simulation result. n is number of parameters. I_b^i and I_a^i are the ith upper and lower bounds of target parameter range.

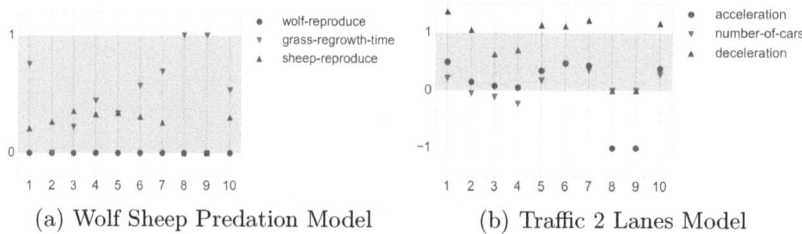

(a) Wolf Sheep Predation Model   (b) Traffic 2 Lanes Model

Fig. 2. Result of Weighting Rules on two models.

According to the above definition, Fig. 2 shows the normalized calibration results calculated by Eq. (13). The horizontal axis shows the serial number of the ten weighting rules, i.e. Cluster Proportion Weighting, Average Density Weighting, Euclidean Distance Weighting, COS Distance Weighting, Hybrid Distance

Weighting, Average Connection Weighting, Association Weighting, Maximum Cluster Weighting, Maximum Density Weighting, and Combined Weighting. The vertical axis demonstrate the normalized value of each parameter.

It can be observed from Fig. 2 that the result of Maximum Cluster Weighting and Maximum Density Weighting are always the largest or smallest, which have more risk in falling outside the target parameter range. While the results of Cluster Proportion Weighting, Average Connection Weighting, Association Weighting and Combined Weighting have smaller parameter errors and sometimes fall outside the target parameter range. The performance of Average Density Weighting and Hybrid Distance Weighting differs in the two models. Compared with other methods, Euclidean Distance Weighting and COS Distance Weighting got the best results.

Further, we define a comprehensive error to estimate the performance gap between different weighting rules, as shown by Eq. (14).

$$e_c = \sqrt{\sum_{i=1}^{n} k \min \left(\frac{I_i - I_{ij}}{I_b^i - I_a^i}\right)^2}, k = \begin{cases} 1 & \text{if } I_i \in [I_b^i, I_a^i] \\ 3 & \text{else} \end{cases} \tag{14}$$

The results are shown in Tables 5 and 6. It can be seen that the comprehensive error of Cluster Proportion Weighting, Average Connection Weighting, and Association Weighting are always high. The results of Maximum Cluster Weighting and Maximum Density Weighting are not stable. The performance of Combined Weighting performs medium and stable among these ten methods. Average Density Weighting, Euclidean Distance Weighting are always low and are recommended. In the next section, Euclidean Distance Weighting is adopted to the PCN-Calibration method to compare with other typical artificial intelligence-based calibration methods.

Table 5. Error of Weighting Rules by the Wolf Sheep Predation Model

Weighting rule	Red-reproduce	Resource-reproduce	Blue-reproduce	Comprehensive Error
Target Parameter Value	2, 3.5, 5	20, 25, 40	2, 3.5, 5	
Cluster Proportion Weighting	0	4.853	0.621	0.450
Average Density Weighting	0	0	0.709	**0.236**
Euclidean Distance Weighting	0	0.588	0.445	**0.178**
COS Distance Weighting	0	3.935	0.516	0.369
Hybrid Distance Weighting	0	1.673	0.475	0.242
Average Connection Weighting	0	6.455	0.573	0.514
Association Weighting	0	8.861	0.723	0.684
Maximum Cluster Weighting	0	0	0	**0.000**
Maximum Density Weighting	0	0	0	**0.000**
Combined Weighting	0	5.707	0.592	0.483

Table 6. Error of Weighting Rules by the Traffic 2 Lanes Model

Weighting rule	Acceleration	Number-of-cars	Deceleration	Comprehensive Error
Target Parameter Value	0.003, 0.005	40, 60	0.04, 0.06	
Cluster Proportion Weighting	0.00094	3.754	−0.0105	2.233
Average Density Weighting	0.00022	−2.121	−0.0033	**0.923**
Euclidean Distance Weighting	0.00007	−3.422	0.0065	**0.873**
COS Distance Weighting	0.00001	−5.889	0.0049	**1.133**
Hybrid Distance Weighting	0.00063	2.900	−0.0051	1.225
Average Connection Weighting	0.00090	9.267	−0.0047	1.618
Association Weighting	0.00082	6.812	−0.0071	1.816
Maximum Cluster Weighting	−0.00220	−1.000	−0.0010	3.600
Maximum Density Weighting	−0.00220	−1.000	−0.0010	3.600
Combined Weighting	0.00070	5.061	−0.0058	1.473

3.3 Comparison of the PCN-Calibration Method to Typical Artificial Intelligence-Based Calibration Method

In this section, a comparative experiment between the PCN-Calibration method and three other artificial intelligence-based calibration methods is conducted, i.e., Extreme Machine Learning (ELM) [23], Fast Incremental Gaussian Mixture Model (FIGMM) [24] and Metropolis-Hastings algorithm (MH) [25], which use simulation results as the training data while reference data as part of testing samples. To evaluate the precision and recall of the four methods, additional simulation results are generated under randomly selected parameters falling outside the target parameter range, forming the other part of the testing samples to test the ability of the algorithm to distinguish between credible samples and incredible samples.

$$F_2 = \frac{5 * precision * recall}{4 * precision + recall} \tag{15}$$

Table 7. Comparison of Algorithm Performance by the Wolf Sheep Predation Model

Calibration Method	Precision	Recall	F_1 Score	$F_{0.5} \backslash F_2$ Score	Calibration Result	Target Range
ELM	92.660%	20.741%	33.895%	24.552%	**[4.681, 28.347, 3.980]**	[2, 5]
FIGMM	75.012%	66.667%	70.594%	68.184%	**[4.622, 32.296, 3.970]**	[20, 40]
MH	71.836%	69.861%	70.835%	70.247%	**[3.900, 27.654, 3.145]**	[2, 5]
PCN-Calibration	94.166%	64.501%	**76.601%**	**86.200%**	[2.000, 24.412, 3.065]	

Table 8. Comparison of Algorithm Performance by the Traffic 2 Lanes Model

Calibration Method	Precision	Recall	F_1 Score	$F_{0.5} \backslash F_2$ Score	Calibration Result	Target Range
ELM	81.223%	20.221%	32.381%	23.795%	**[0.00375, 49.583, 0.0564]**	[0.0029, 0.005]
FIGMM	33.334%	27.567%	30.177%	28.555%	[0.00264, 50.000, 0.0637]	[39, 61]
MH	27.633%	52.533%	36.216%	**44.511%**	[0.00232, 50, 0.0591]	[0.039, 0.062]
PCN-Calibration	44.355%	31.336%	**36.726%**	40.952%	[0.00307, 36.578, 0.0535]	

For these three AI-based calibration methods, parameters are calculated by the parameters of TP and FN samples, so the recall is more important than precision. The F_2 score is used to represent the credibility of its calibration results in Eq. (15). The precision rate, recall rate, F_1 score and $F_{0.5}/F_2$ score of the four methods are shown in Tables 7 and 8. Under different scene, the performance of the three artificial intelligence methods fluctuates. It is mainly because the simulation models have different characteristics. As shown in Fig. 3, the simulation results of the Wolf Sheep Predation Model fluctuated obviously, and with the change of parameters, the amplitude and frequency of fluctuation change obviously, so every calibration method work well. However, the simulation results of the Traffic 2 Lanes Model has no big peaks and valleys, but remarkable high-frequency fluctuation. With the change of parameters, the trend change is unremarkable, which means that the model is relatively insensitive to these parameters, so the performance of the calibration methods degrade as well. Tables 7 and 8 show that the precision and recall of ELM vary greatly, indicating that its result is not credible. FIGMM performs stable and works better than ELM. MH has the highest $F_{0.5}/F_2$ score in the Traffic 2 Lanes Model. The results of the PCN-Calibration method are stable, with the best F_1 score in two models, while the $F_{0.5}/F_2$ score is also high, demonstrating its superior performance in the experiment.

Table 9. Comparison of Algorithm Running Time

Amount of Data (Thousand)	160	290	350	700	900
ELM	23.112 s	75.551 s	154.223 s	222.322 s	298.632 s
FIGMM	24.344 s	108.332 s	296.647 s	414.383 s	505.684 s
MH	1198.332 s	2356.321 s	5486.951 s	10221.336 s	19203.355 s
LB-SOINN	**11.332 s**	**26.332 s**	**45.832 s**	**94.378 s**	**160.062 s**

In Table 9, we test the performance of the four methods with increasing amount of simulation results for the model calibration. The calibration time of the PCN-Calibration method is the shortest. In addition, with increasing amount

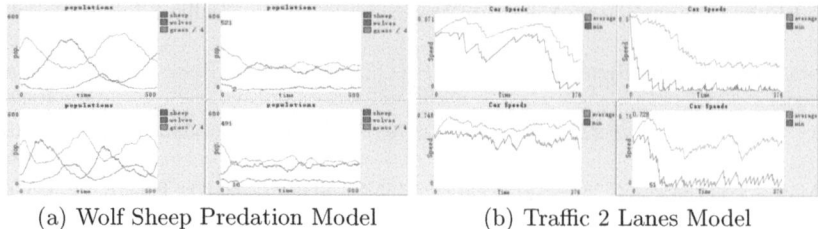

(a) Wolf Sheep Predation Model   (b) Traffic 2 Lanes Model

Fig. 3. Different Parameter Settings Result in Different Simulation Processes.

of data, the time of the PCN-Calibration method increases much slower due to its incremental learning ability. The PCN-Calibration runs at least twice faster than the other methods.

4 Conclusion and Future Works

This paper introduces a PCN-Calibration method for agent-based simulation model calibration, leveraging LB-SOINN to establish a pattern clustering network. By employing ten weighting rules, it efficiently adjusts simulation parameters based on nearest simulation results. Experiments were carried out to find the best configuration of the PCN-Calibration method and the best weighting rules. Comparative experiments with three artificial intelligence-based methods, including ELM, FIGMM, and MH, demonstrate PCN-Calibration's consistently superior $F_{0.5}/F_2$ score and significantly reduced runtime.

LB-SOINN is an incremental clustering network, which can be trained with data that does not have labels in an online manner. Compared with non-incremental methods, it can dynamically evaluate the agent-based simulation model and calibrate parameters during the simulation process. When it is necessary to obtain the possible parameter range which generates unknown data, the PCN-Calibration method can obtain very good results, especially for the simulation model that is sensitive to parameters.

We propose as future works studying whether the PCN-Calibration method can be combined with other methods to reduce its error in calibration of models which are insensitive to parameters, and tries to accelerate the calibration process within tens of seconds.

References

1. Kiesling, E., Günther, M., Stummer, C., Wakolbinger, L.M.: Agent-based simulation of innovation diffusion: a review. CEJOR **20**(2), 183–230 (2012)
2. Acheson, P., Dagli, C., Kilicay-Ergin, N.: Model based systems engineering for system of systems using agent-based modeling. Procedia Comput. Sci. **16**, 11–19 (2013)

3. Wen, G., Li, B.: Optimized leader-follower consensus control using reinforcement learning for a class of second-order nonlinear multiagent systems. IEEE Trans. Syst. Man Cybern. Syst. **52**, 5546–55555 (2021)
4. Darty, K., Saunier, J., Sabouret, N.: Calibration of multi-agent simulations through a participatory experiment. In: 14th International Conference on Autonomous Agents and Multiagent Systems, pp. 1683–1684. ACM (2015)
5. Guo, W., Pan, T., Li, Z., Chen, S.: Model calibration method for soft sensors using adaptive gaussian process regression. IEEE Access **7**, 168436–168443 (2019)
6. Xu, C., Wang, W., Liu, P., Li, Z.: Calibration of crash risk models on freeways with limited real-time traffic data using Bayesian meta-analysis and Bayesian inference approach. Accid. Anal. Prev. **85**, 207–218 (2015)
7. Hu, J., Zhou, Q., McKeand, A., Xie, T., Choi, S.K.: A model validation framework based on parameter calibration under aleatory and epistemic uncertainty. Struct. Multidiscip. Optim. **63**(2), 645–660 (2021)
8. Anirudh, R., Thiagarajan, J.J., Bremer, P.T., Germann, T.C., Del Valle, S.Y., Streitz, F.H.: Accurate calibration of agent-based epidemiological models with neural network surrogates. arXiv preprint arXiv:2010.06558 (2020)
9. Carrella, E., Bailey, R., Madsen, J.: Calibrating agent-based models with linear regressions. J. Artif. Soc. Soc. Simul. **23**(1), 1 (2020)
10. Chen, S., Desiderio, S.: A regression-based calibration method for agent-based models. Comput. Econ. **59**(2), 687–700 (2022)
11. Read, M.N., Alden, K., Rose, L.M., Timmis, J.: Automated multi-objective calibration of biological agent-based simulations. J. R. Soc. Interface **13**(122), 20160543 (2016)
12. Moya, I., Chica, M., Cordón, Ó.: A multicriteria integral framework for agent-based model calibration using evolutionary multiobjective optimization and network-based visualization. Decis. Support Syst. **124**, 113111 (2019)
13. Evangelista, S., Giovinco, G., Kocaman, S.: A multi-parameter calibration method for the numerical simulation of morphodynamic problems. J. Hydrol. Hydromechanics **65**(2), 175–182 (2017)
14. Singh, M., Marathe, A., Marathe, M.V., Swarup, S.: Behavior model calibration for epidemic simulations. In: Proceedings of the International Joint Conference on Autonomous Agents and Multiagent Systems: AAMAS. International Joint Conference on Autonomous Agents and Multiagent Systems. NIH Public Access, p. 1640 (2018)
15. Villanueva, R.J., Hidalgo, J.I., Cervigón, C., Villanueva-Oller, J., Cortés, J.C.: Calibration of an agent-based simulation model to the data of women infected by human papillomavirus with uncertainty. Appl. Soft Comput. **80**, 546–556 (2019)
16. Balakrishna, R., Antoniou, C., Ben-Akiva, M., Koutsopoulos, H.N., Wen, Y.: Calibration of microscopic traffic simulation models: methods and application. Transp. Res. Rec. **1999**(1), 198–207 (2007)
17. Platt, D.: A comparison of economic agent-based model calibration methods. J. Econ. Dyn. Control **113**, 103859 (2020)
18. Fadikar, A., Higdon, D., Chen, J., Lewis, B., Venkatramanan, S., Marathe, M.: Calibrating a stochastic, agent-based model using quantile-based emulation. SIAM/ASA J. Uncertainty Quantification **6**(4), 1685–1706 (2018)
19. Zhang, H., Xiao, X., Hasegawa, O.: A load-balancing self-organizing incremental neural network. IEEE Trans. Neural Netw. Learn. Syst. **25**(6), 1096–1105 (2013)
20. Zhou, X., Valle, A.D.: Range based confusion matrix for imbalanced time series classification. In: 6th Conference on Data Science and Machine Learning Applications (CDMA), pp. 1–6. IEEE (2020)

21. Wong, A., Kamel, M.S.: Classification of imbalanced data: a review. Int. J. Pattern Recogn. Artif. Intell. **23**(4), 687–719 (2009)
22. Foresti, G.L., Visentini, I., Snidaro, L.: Diversity-aware classifier ensemble selection via f-score. Inf. Fusion **28**, 24–43 (2016)
23. Deng, C., Huang, G., Xu, J., Tang, J.: Extreme learning machines: new trends and applications. Sci. China Inf. Sci. **58**(2), 1–16 (2015)
24. Pinto, R.C., Engel, P.M.: A fast incremental Gaussian mixture model. PLoS ONE **10**(10), e0139931 (2015)
25. Griffin, J.E.: On adaptive metropolis-hastings methods. Stat. Comput. **23**(1), 123–134 (2013)

An Intelligent Simulation Result Validation Method Based on Variational Autoencoder

Fan Yang[1,2] (ID), Ping Ma[1,2], Wei Li[1,2(✉)], Chao Tao[1], Ming Yang[1,2], Jianchao Zhang[3], and Huichuan Cheng[3]

[1] Control and Simulation Center, Harbin Institute of Technology, Harbin 150080, China
fleehit@163.com
[2] National Key Laboratory of Complex System Modeling and Simulation, Harbin 150080, China
[3] Chinese Aeroengine Research Institute, Beijing 101304, China

Abstract. Validation for complex system simulation faces the challenges of a large amount of validation data, which comes in the form of extensive samples and prolonged time series. To address this issue, this paper proposes an intelligent simulation result validation method based on Variational Autoencoder (VAE), achieving model validation through intelligent feature extraction and feature consistency metrics. Initially, we present the overall validation framework, which consists of feature extraction and simulation result validation, and detail the steps of the validation process. Subsequently, the principles and procedures of the feature extraction and simulation result validation methods are thoroughly discussed. The feature extraction is achieved by learning the distribution parameters of the latent space using a VAE model. Feature differences across various aspects are measured by three distance metrics and are further converted into simulation credibility. Finally, the effectiveness and superiority of this method are demonstrated through a case study on validating a drone delivery model.

Keywords: Model Validation · Data Feature · Variational Autoencoder (VAE) · Distance Metrics

1 Introduction

With the widespread application of modeling and simulation technology in complex system research, simulation model validation has emerged as a hotspot in the field of simulation science [1, 2]. It serves as an important means to analyze and assess the credibility of simulation models. Simulation result validation, one of the most direct and effective methods for validating simulation models, measures the consistency between simulation outputs and reference data [3, 4]. Considering that the intricate evaluation objects and large data volumes in complex system simulation, validating simulation results in the form of extensive samples and prolonged time series presents a significant challenge.

Most dynamic output consistency metrics rely directly on the original data. However, these methods not only fail to capture the intrinsic characteristics of the validation data but also exhibit high computational complexity. Another type of method validates simulation results based on data features, comparing the differences in features between the simulation outputs and the reference data [5–7]. These features are generally designed and selected manually according to different tasks and data types. However, manual feature extraction often suffers from problems such as one-sided characterization and difficulty in feature selection, which not only rely heavily on experience and luck but are also very time-consuming and laborious when dealing with large amounts of data.

Deep learning methods can autonomously and rapidly learn effective features, which are more representative and distinguishable than hand-designed features [8, 9]. Hence, considering that validation data are massive and unlabeled, we seek to implement feature extraction of validation data through unsupervised learning methods. The Variational Autoencoder (VAE) is adept at capturing the intrinsic properties of the data and extracting interpretable features, [10] making it a good choice for intelligent extraction of features from validation data.

In this paper, we propose an intelligent validation framework for validating simulation results in the form of extensive samples and prolonged time series. There are two main parts in the validation process: feature extraction and simulation result validation. A VAE model is first constructed and trained to perform intelligent feature extraction on the validation data. Compared to traditional feature-based model validation methods, it enables the extracted features to characterize the data more comprehensively and significantly improves the efficiency of feature extraction. Then, the differences in mean features and feature distributions are measured and converted into the simulation credibility. This approach not only represents the consistency of the intrinsic characteristics between the simulation outputs and the reference data but also reflects the consistency of the uncertainty between them.

This paper is organized as follows. An intelligent validation framework is primarily exhibited in Sect. 2. Next, a feature extraction method for validation data based on VAE is proposed in Sect. 3. Then, a simulation result validation method based on feature distance is presented in Sect. 4. A case study on validating a drone delivery model is provided to verify the effectiveness of the proposed method in Sect. 5. Finally, the conclusions and future work are presented in Sect. 6.

2   Intelligent Validation Framework

In this section, for the validation data in the form of extensive samples and prolonged time series, we propose a framework for intelligently validating simulation results. Firstly, we extract the features of the validation data by a VAE model, enabling similar inputs to be represented similarly in the latent space. Then, the similarity of features corresponding to two sets of validation data is calculated based on feature distances and used to represent the credibility of simulation results. The framework is illustrated in Fig. 1.

The specific steps are as follows:

Step 1: According to the characteristics of the validation data, design the structure and parameters of the network to construct a VAE model, mainly including the number of hidden layers, the dimensions of hidden layers, the dimension of the latent space and learning rate.

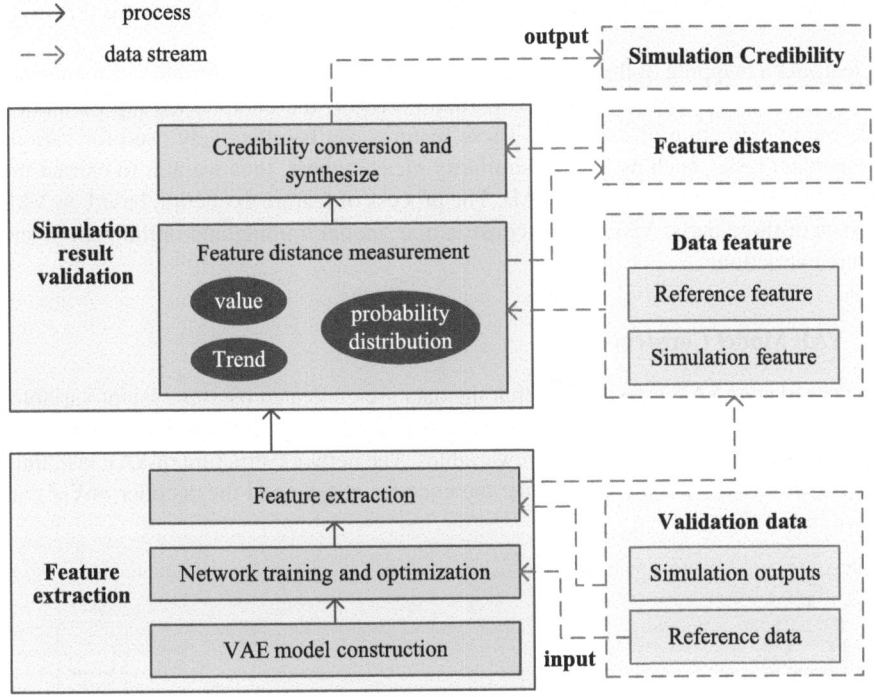

Fig. 1. Intelligent validation framework

Step 2: Normalize two sets of validation data including the reference data X_R and the simulation outputs X_S.

Step 3: Input the normalized reference data X_{Rnorm} into the constructed VAE model for training.

Step 4: Adjust the model structure, training parameters and iteration termination condition to optimize the model.

Step 5: After training and optimization, input the normalized reference data X_{Rnorm} and the simulation outputs X_{Snorm} to the VAE model to obtain two sets of distribution parameters of the latent spaces, and take the mean matrices μ_R and μ_S as the corresponding features of two sets of validation data.

Step 6: Measure the feature distances of two sets of features in terms of value, trend, and probability distribution by using Euclidean distance, cosine distance and Jensen-Shannon (JS) divergence.

Step 7: Convert and synthesize the feature distance metrics to obtain the similarity of mean features and the similarity of feature distributions through the corresponding conversion methods.

Step 8: Configure appropriate weights for the two similarity metrics and synthesize them to obtain the overall credibility of the simulation model.

3   Feature Extraction Method for Validation Data Based on VAE

By learning a mapping of the input data to a structured and meaningful low-dimensional latent space, VAE provides a powerful framework for the compressed representation and feature extraction of data [11]. These features can be effectively used for various downstream tasks, such as feature similarity measurement, thus we aim to extract the features of the validation data by VAE. The process of feature extraction based on VAE consists of three parts: VAE model construction, model training and optimization, and feature extraction.

3.1   VAE Model Construction

The core idea of VAE is to assume that the data are generated by some latent variables through a certain probability distribution and to realize feature extraction by learning the distribution parameters of these latent variables. The network structure of VAE is mainly composed of two critical components: the encoder $q(Z|X)$ and the decoder $p(X|Z)$, as shown in Fig. 2.

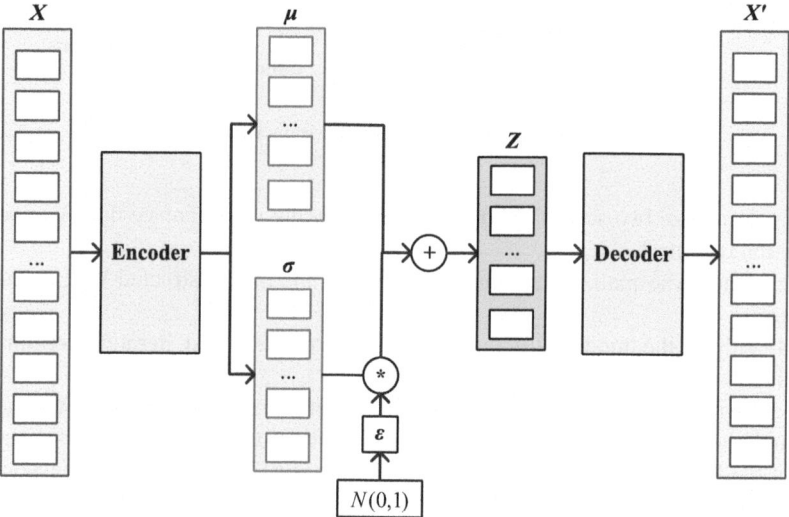

Fig. 2. Structure of Variational Auto-Encoder

The encoder processes the original data, transforming it through a sequence of layers, such as fully connected or convolutional layers, ultimately yielding parameters—typically the mean and variance—that characterize the distribution within the latent space. Then the latent representation can be obtained by sampling from the distribution through the Reparameterization Trick [12, 13] as follows.

$$Z = \mu + \varepsilon \times \sigma \qquad (1)$$

where μ is the mean and σ is the standard deviation, ε is sampled from a standard Gaussian distribution $N(0, 1)$.

The decoder, receiving the latent representation, reconstructs the original data through a sequence of reverse transformations, including layers such as transposed convolutional layers and fully connected layers [14].

Applying VAE to realize feature extraction from the validation data requires designing the structure and parameters of the network according to the characteristics of the validation data. The hidden layers can consist of fully connected layers, convolutional layers, or other layers. The configuration of the hidden layers, specifically their depth, denoting the layer count, and their width, indicating the neuron count per layer, is determined by the length and complexity of the time series. For computational simplicity, the VAE's network architecture, as illustrated in Fig. 3, is designed to include a hidden layer within both the encoder and decoder, each configured as a fully connected layer. In practice, if the input data are long time series with complex characteristics, additional hidden layers can be added as needed.

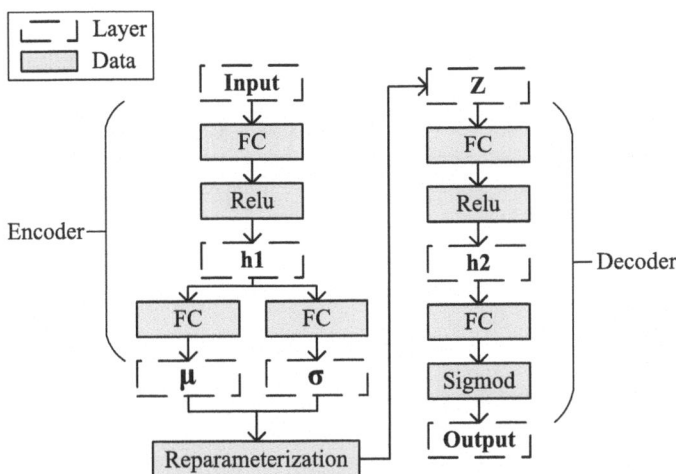

Fig. 3. Design of the VAE's network architecture

Here are the methods for setting the dimensions of the hidden layer and the latent space for the network above. The hidden layer is used to capture important features of the input data. When the input dimension is relatively small (e.g., within a few hundred dimensions), the dimension of the hidden layer can be set to 2 to 10 times the input dimension. When the input dimension is large (e.g., several thousand dimensions), the dimension of the hidden layer is usually set as a fraction of the input dimension (e.g., 1/2, 1/4, 1/10, etc.).

The latent space dimension should be small enough to ensure the model learns a compact and meaningful representation of the input data, but also large enough to store important information. It is commonly set based on the complexity of the input data. For simpler low-dimensional data, the latent space dimension can be set to a relatively small value (between 2 and 20); for more complex high-dimensional data, it can be set between 20 and 100.

When constructing the VAE model, the appropriate initial dimensions for the hidden layer and potential space are first set according to the above rules, and the specific dimension size is adjusted through the subsequent model training process.

3.2 Model Training and Optimization

In this paper, the univariate output data is used as the validation target to guide the validation method. The validation data include the reference data $X_R = \left[X_R^{im} \right]_{N_R \times M}$ and the simulation outputs $X_S = \left[X_S^{jm} \right]_{N_S \times M}$, which are respectively derived from the simulation model and the real system in repeated experiments under the same input conditions. X_R is an $N_R \times M$ matrix representing N_R samples of reference data across M timestamps, while X_S is an $N_S \times M$ matrix representing N_S samples of simulation outputs across M timestamps.

In order to improve the training efficiency and model performance, it is desirable that the input of the model be normalized. Hence, the reference data X_R and the simulation outputs X_S are normalized separately to obtain X_{Rnorm} and X_{Snorm} as follows:

$$X_{Rnorm}^{im} = \frac{X_R^{im} - \min\left\{ X_R^{im}, X_S^{jm} \right\}}{\max\left\{ X_R^{im}, X_S^{jm} \right\} - \min\left\{ X_R^{im}, X_S^{jm} \right\}} \tag{2}$$

$$X_{Snorm}^{jm} = \frac{X_S^{jm} - \min\{ X_R^{im}, X_S^{jm} \}}{\max\{ X_R^{im}, X_S^{jm} \} - \min\{ X_R^{im}, X_S^{jm} \}} \tag{3}$$

where $m = 1, 2, ...M$; $i = 1, 2, ...N_R$; $j = 1, 2, ...N_S$; $\min\{ X_R^{im}, X_S^{jm} \}$ denotes the minimum element in X_R and X_S, and $\max\{ X_R^{im}, X_S^{jm} \}$ denotes the maximum element in X_R and X_S.

The normalized reference data X_{Rnorm} are then input into the VAE model for training. The input data are mapped to the latent space by the encoder and then reconstructed by the decoder. The reconstruction loss and Kullback-Leibler (KL) divergence can then be computed to further obtain the total loss, and the model parameters are updated by backpropagation.

The loss function of the VAE, comprising two components, reconstruction loss and KL divergence, is designed to learn effective latent representations while maintaining the continuity of the latent space [15]. Due to space constraints, the detailed principles will not be further elaborated here; see reference [10] for more information. The loss function is as follows:

$$L\left(\varphi, \theta; X^{(i)} \right) = \mathbb{E}_{q_\varphi(Z|X^{(i)})} \left[\log p_\theta \left(X^{(i)} | Z \right) \right] - \mathrm{KL}\left(q_\varphi \left(Z | X^{(i)} \right) \| p_\theta(Z) \right) \tag{4}$$

Typically, the reconstruction loss functions employed are mean squared error (MSE) and binary cross-entropy (BCE), with the selection determined by the type of data. Given that the VAE's input and output in the model validation context are both continuous numerical data, MSE is adopted to compute the reconstruction loss.

Modifications to the VAE model's architecture, including the number of hidden layers, the dimensions of the hidden layers, and the dimension of the latent space, along with adjustments to training parameters such as the learning rate, are strategically made to optimize the VAE model's performance. This optimization is guided by monitoring changes in the training loss and observing the distribution parameters of the latent space obtained at the end of training.

3.3 Feature Extraction

After the training and optimization, the feature extraction from the validation data can be accomplished using the trained VAE model. The normalized reference data X_{Rnorm} and simulation outputs X_{Snorm} are fed into the trained encoder, which outputs the distribution parameters of the latent space corresponding to the two sets of validation data, including the mean and variance.

Each data sample input into the encoder is mapped to a pair of vectors in the latent space, representing the mean and variance, respectively. The mean vector is typically regarded as a compressed representation of the input data, capturing the main features of the input data. The variance indicates the range of possible variations of the latent representation around a given mean while does not directly reflect the content features of the input data.

Hence, to simplify the calculation, two mean matrices $\mu_R = \left[\mu_R^{id}\right]_{N_R \times D}$ and $\mu_S = \left[\mu_S^{jd}\right]_{N_S \times D}$ are respectively considered as the features of the simulation outputs and the reference data, where D is the dimension of the latent space.

4 Simulation Result Validation Based on Feature Distance

Simulation result validation is a common way of assessing simulation credibility, measuring the consistency between the simulation outputs and the reference data. Since the features extracted by VAE define the location in the latent space, thus directly relating to the content and structure of the input data, we analyze the consistency of the two sets of validation data by measuring the difference of corresponding features.

4.1 Feature Distance Measurement

Since the features consist of multiple samples, we compare the feature differences in two aspects. One is to compare the differences of the mean features of multiple samples in values and trends. The other is to compare the distribution differences of multiple samples for each dimensional feature.

To compare the differences of mean features, the two feature matrices μ_R and μ_S are respectively averaged by column to obtain the mean feature vectors $F_{1R} = \left[F_{1R}^d\right]_{1 \times D}$ and $F_{1S} = \left[F_{1S}^d\right]_{1 \times D}$. Then, we measure the differences in values and trends between the two mean feature vectors F_{1R} and F_{1S} by distance metrics.

As we know, Euclidean distance can measure the absolute difference in position, while cosine distance can measure the relative difference in direction. Therefore, we measure the difference in values between the two vectors F_{1R} and F_{1S} by Euclidean distance, and the difference in trends by cosine distance, which are shown in Eq. (5) and Eq. (6).

$$D_{eu} = \left\| F_{1S} - F_{1R} \right\|_2 \tag{5}$$

$$D_{cos} = \frac{F_{1S} \cdot F_{1R}}{\| F_{1S} \| \| F_{1R} \|} \tag{6}$$

where D_{eu} and D_{cos} denote the results measured by Euclidean distance and cosine distance, respectively.

Due to the aleatory uncertainty in modeling and simulation, it can be assumed that the multi-sample validation data at each timestamp follow a normal distribution [16, 17]. Since the feature matrix is a compressed representation of the validation data, it can be assumed that multiple samples for each dimensional feature follow a normal distribution. Then we measure the distribution differences between two sets of features column by column.

JS divergence is a method that can effectively measure the difference between two probability distributions. Assum that the d_{th} column data of the two feature matrices be μ_R^d and μ_S^d, we first calculate the mean and standard deviation of μ_R^d and μ_S^d, denoted as $F_{2R} = \begin{bmatrix} F_{\mu R}^d \\ F_{\sigma R}^d \end{bmatrix}_{2 \times D}$ and $F_{2S} = \begin{bmatrix} F_{\mu S}^d \\ F_{\sigma S}^d \end{bmatrix}_{2 \times D}$, and then use JS divergence to measure the probability distribution difference as follows:

$$D_{JS}^d = JS(A||B) = \frac{1}{2} KL\left(A || \frac{A+B}{2} \right) + \frac{1}{2} KL\left(B || \frac{A+B}{2} \right) \tag{7}$$

where D_{JS}^d denotes the JS divergence between the d_{th} column feature vectors μ_R^d and μ_S^d; $d = 1, 2, ...D$. $A \sim N\left(F_{\mu R}^d, \left(F_{\sigma R}^d \right)^2 \right)$ and $B \sim N\left(F_{\mu S}^d, \left(F_{\sigma S}^d \right)^2 \right)$ denote the two normal distributions estimated from μ_R^d and μ_S^d.

The JS dispersion is calculated based on the KL dispersion. Given that two normal distributions $P(x) \sim N\left(\mu_1, \sigma_1^2 \right)$ and $Q(x) \sim N\left(\mu_2, \sigma_2^2 \right)$, the KL divergence between them can be calculated as follows:

$$KL(P||Q) = \log \frac{\sigma_2}{\sigma_1} + \frac{(\sigma_1)^2 + (\mu_1 - \mu_2)^2}{2(\sigma_2)^2} - \frac{1}{2} \tag{8}$$

4.2 Credibility Conversion and Synthesize

The more similar the features corresponding to the two sets of validation data are, the more consistent the two sets of validation data are, and consequently, the more credible the simulation model is. Hence, we convert feature distance into feature similarity and use the similarity metrics to represent the simulation credibility. Additionally, we aim to map the range of feature similarity to [0, 1].

The differences of mean features in values and trends are respectively measured by D_{eu} and D_{cos}. The range of D_{eu} is $[0, +\infty]$, and the larger the D_{eu}, the greater the feature difference. The range of D_{cos} is $[-1, 1]$, and the smaller the D_{cos}, the greater the feature difference. Hence, D_{eu} and D_{cos} are converted and synthesized as follows:

$$C_1 = \frac{1}{2}\left(\frac{1}{1 + D_{eu}} + \frac{1 + D_{cos}}{2}\right) \tag{9}$$

where $C_1 \in (0, 1]$ denotes the similarity of mean features, representing the consistency of intrinsic characteristics between two sets of validation data.

The distribution difference of the d_{th} dimensional feature is measured by D_{JS}^d. The range of D_{JS}^d is $[0, \log(2)]$, and the larger the difference between the two distributions, the larger the D_{JS}^d. Hence, $D_{JS}^1, D_{JS}^2, ...D_{JS}^D$ are converted and synthesized as follows:

$$C_2 = \frac{1}{D}\sum_{d=1}^{D}\left(1 - \frac{D_{JS}^d}{\log(2)}\right) \tag{10}$$

where $C_2 \in [0, 1]$ denotes the similarity of feature distributions, representing the consistency of uncertainty between two sets of validation data.

The results of the two feature similarity metrics are both restricted to the range of [0, 1], and the closer the results are to 1, the higher the consistency of the two groups of features. Then we synthesize them to get the total feature similarity, further representing the credibility of the simulation results.

$$C(X_S, X_R) = \omega_1 C_1 + \omega_2 C_2 \tag{11}$$

where, $C(X_S, X_R)$ denotes the credibility of the simulation results; ω_1, ω_2 respectively denote the contribution of the two feature similarity to the total simulation credibility, and satisfy $0 < \omega_1, \omega_2 < 1$; $\omega_1 + \omega_2 = 1$. In general, the value of ω_1 is set to be larger than the value of ω_2. The higher the consistency requirement for uncertainty, the larger ω_2. In practical application, ω_1, ω_2 can be adjusted according to the simulation demand.

According to Eq. (11), we can know that the range of $C(X_S, X_R)$ is (0, 1]. More specifically, $C(X_S, X_R) = 1$ when the simulation outputs X_S and the reference data X_R are completely consistent, and $C(X_S, X_R) \to 0$ as the consistency between X_S and X_R decreases.

5 Experimental Studies

In this section, we evaluate the effectiveness and superiority of the proposed validation method by an application example of a drone delivery simulation model, as shown in Fig. 4.

There are many parameters in the model, including initial flight path angle θ_0, atmosphere density C_ρ, disturbance of lift coefficient D_{CL} and disturbance of drag coefficient

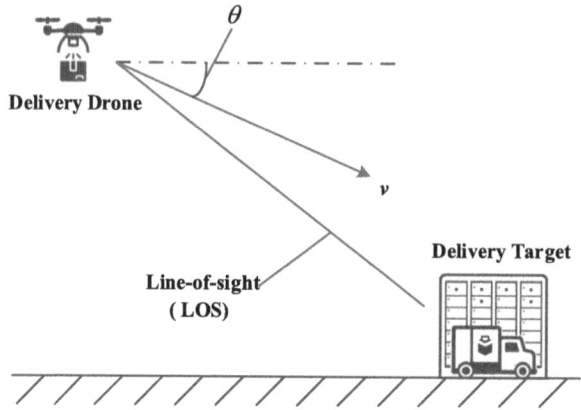

Fig. 4. Drone delivery simulation model

D_{CD}. We design two sets of experiments to verify the effectiveness of the proposed method,using a simulation model under certain parameters as a reference model.

The atmosphere density C_ρ is set as a uncertain parameter so that the outputs are uncertain for simplicity, keeping consistent with $N(0, 0.033)$ in each model. We adjust the initial flight path angle θ_0 in experiment I while keeping the values of other parameters the same. Similarly, the disturbance of drag coefficient D_{CD} is adjust in experiment II. The parameters and sample size are shown in Table 1.

Table 1. Information of experiment I and experiment II.

Model	Output type	Sample size	Experiment I ($\theta_0/°$)	Experiment II (D_{CD})
R	reference	500	30	0.033
S1	simulation	1000	10	0.033
S2	simulation	1000	15	0.05
S3	simulation	1000	20	0.1
S4	simulation	1000	25	0.2
S5	simulation	1000	30	0.3

We taking the flight path angle θ as the model output to perform simulation model validation. The validation process is firstly explained. Since the dynamic outputs are across 1000 timestamps, we adopted the network structure shown in Fig. 3 and set the dimensions of the hidden layer and latent space to 200 and 50, respectively. After training and optimization, features of 6 sets of validation data can be extracted by VAE and the mean feature vectors are shown in Fig. 5 for a more intuitive comparison.

It can be seen that the smaller the parameter difference between the simulation model and the reference model, the closer the corresponding feature distributions are. Then 3 types of similarity metrics can be calculated. Furthermore, the final validation results in

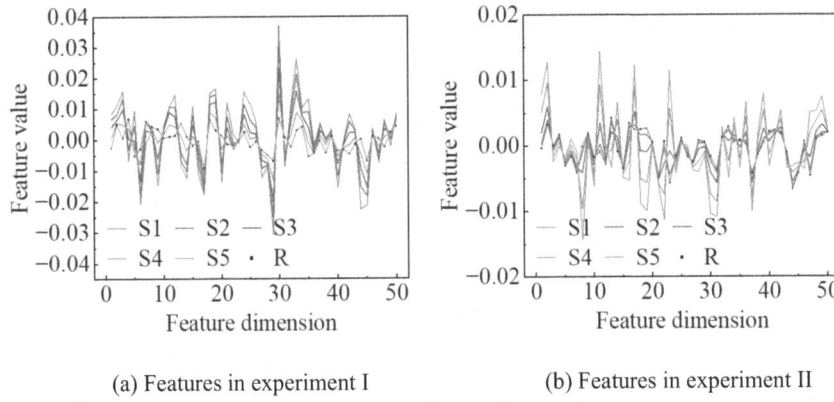

(a) Features in experiment I   (b) Features in experiment II

Fig. 5. Features of validation data.

two experiments could be obtained with $\omega_1 = 0.7, \omega_2 = 0.3$, as seen in Table 2 and Table 3.

Table 2. Validation results in experiment I.

Model	C_1	C_2	C_3	$C(X_S, X_R)$
S1	0.6909	0.9196	0.5565	0.7307
S2	0.7294	0.9367	0.6825	0.7879
S3	0.7918	0.9552	0.8212	0.8578
S4	0.9028	0.9780	0.9632	0.9472
S5	1	0.9997	1	0.9999

Table 3. Validation results in experiment II.

Model	C_1	C_2	C_3	$C(X_S, X_R)$
S1	1	1	1	1
S2	0.9908	0.9965	0.9996	0.9954
S3	0.8729	0.9866	0.9985	0.9504
S4	0.6521	0.9691	0.8993	0.8372
S5	0.5596	0.9542	0.7388	0.7515

In addition, we compared the proposed method in this paper with another simulation dynamic output validation method based on the data feature proposed in reference [5], which validates the simulation results using the improved Bayesian factor (BF). The validation method based on BF is used to validate the simulation results in experiment

II. The comparison of the validation results and the time consumption of the two methods are presented in Table 4.

Table 4. Comparison of validation results and time consumption.

Method	Validation results					Time/s
	S1	S2	S3	S4	S5	
VAE	1	0.9954	0.9504	0.8372	0.7515	14.62
BF	0.5839	0.5077	0.4615	0	0	297.17

When using the validation method based on BF, it is assumed that the simulation output is consistent with the reference data, and the null hypothesis is accepted if the validation result exceeds 0.5. From the table above, it can be observed that when the parameters of the simulation model and the reference model are consistent, both methods can provide appropriate metrics and draw correct validation conclusions. However, the validation method based on VAE significantly reduces the validation time and enhances the efficiency of model validation.

6   Conclusions

A large amount of validation data presents a significant challenge in model validation for complex system simulation. Traditional validation methods often suffer from difficulties in feature selection and extensive time consumption. In this work, an intelligent validation method based on VAE has been proposed. A VAE model is designed to capture the intrinsic properties of the data, enabling the intelligent extraction of features from two sets of validation data, including the simulation outputs and the reference data. Then, feature differences in values, trends and distributions are measured by Euclidean distance, cosine distance and JS divergence and are further converted and synthesized into the final validation result.

The proposed method can be applied in validating dynamic single outputs, while multivariate and correlated outputs are more common in the field of complex system simulation. Future work will focus on addressing the model validation challenges with the multivariate and correlated outputs.

Acknowledgments. This study was funded by National Science and Technology Major Project (J2019-I-0004-0005).

References

1. Durst, P.J., Anderson, D.T., Bethel, C.L.: A historical review of the development of verification and validation theories for simulation models. Int. J. Model. Simul. Sci. Comput. **8**(2), 1730001 (2017)

2. Wang, Y.N., Li, J.Q., Sun, H.B.: A survey on VV&A of large-scale simulations. Int. J. Crowd Sci. **3**(1), 63–86 (2019)
3. Zhou, Y.C.: Research on verification method of complex simulation model. Harbin Institute of Technology (2019)
4. Li, W., Zhou, Y.C., Lin, S.L.: Review of simulation model validation methods. J. Syst. Simul. **31**(7), 1249–1256 (2019)
5. Yang, M., Qian, X.C., Li, W.: A dynamic output verification method for simulation based on data characterization. Syst. Eng. Electron. **38**(2), 457–463 (2016)
6. Li, Z., Zhang, H., Wu, S., Zhao, Y.: Similarity measure of time series based on feature extraction. In: 2020 IEEE 5th International Conference on Cloud Computing and Big Data Analytics, ICCCBDA 2020, pp. 13–16 (2020). Art. no. 9095654
7. Li, H.L., Liang, Y.: Similarity measure based on numerical symbolic and shape feature for time series. Kongzhi yu Juece/Control Decis. **32**(3), 451–458 (2017)
8. Arora, H., Bansal, M.: Feature extraction through sentiment analysis of tourist sentiments using deep learning techniques like CNN, RNN and LSTM. Int. J. Recent Technol. Eng. (IJRTE) **9**(1), 2254–2261 (2020)
9. Iqbal, A., Amin, R.: Time series forecasting and anomaly detection using deep learning. Comput. Chem. Eng. **182**, 108560 (2024)
10. Kingma, D.P., Welling, M.: Auto-encoding variational bayes. CoRR abs/1312.6114 (2013)
11. Ai, Q.Z.: Research on learning and inference methods in deep generative models. University of Electronic Science and Technology (2023)
12. Yao, R., Liu, C., Zhang, L.: Unsupervised anomaly detection using variational auto-encoder based feature extraction. In: 2019 IEEE International Conference on Prognostics and Health Management, ICPHM 2019 (2019). Art. no. 8819434
13. Mahmud, M.S., Fu, X.: Unsupervised classification of high-dimension and low-sample data with variational autoencoder based dimensionality reductionMahmud. In: 2019 4th IEEE International Conference on Advanced Robotics and Mechatronics, ICARM 2019, pp. 498–503 (2019). Art. no. 8834333
14. Zhang, N.: Unsupervised remote sensing image super-resolution based on deep learning. University of Chinese Academy of Sciences (2022)
15. Ma, X.R., Lin, Y.Z., Nie, Z.H.: Structural damage identification based on unsupervised feature-extraction via variational auto-encoder. Measurement **160**, 107811 (2020)
16. Qian, X.C.: Research on simulation model validation and calibration method considering the effect of uncertainty. Harbin Institute of Technology (2016)
17. Li, W., Lin, S.L., Qian, X.C.: An evidence theory-based validation method for models with multivari-ate outputs and uncertainty. SIMULATION **97**(12), 821–834 (2021)

Improved Priority-Based Hindsight Experience Replay in Reinforcement Learning

Hongwei Han, Guanghong Gong$^{(\boxtimes)}$, and Ni Li

School of Automation Science and Electrical Engineering, Beihang University, Beijing 100191, China
ggh@buaa.edu.cn

Abstract. Reinforcement Learning (RL) algorithms, crucial for robotic tasks in sparse reward settings, often require extensive interactions with the environment to learn effectively. Hindsight Experience Replay (HER) addresses this challenge by introducing virtual goals, thereby enhancing sample efficiency by learning from transitions in which the original goal was not achieved. However, randomly sampled additional goals in HER may hinder learning due to their irrelevance. To mitigate this, we propose the improved Priority-based Hindsight experience replay method to prioritize additional goals, improving learning speed and robustness across various robotic tasks. Our improved Priority-based Hindsight Experience Replay method samples with higher probability on the trajectories with temporal difference error and reducible loss, prioritizing those with greater learning potential. Through empirical validation, strategies taken by improved Priority-based Hindsight experience replay method have been demonstrated to enhance training performance and sample efficiency, offering promising avenues for advancing RL in robotic control tasks.

Keywords: Reinforcement Learning · Priority Experience Replay · Hindsight Experience Replay

1 Introduction

Reinforcement Learning has been applied to a variety of domains, including Atari games [1], Poker [2], Go [3], robotics [4] and complex continuous control tasks [5].

In general, the learning problem in reinforcement learning is concerned with controlling a system to maximize a numerical value that represents a long-term objective. In this context, the learner is referred to as the agent, and it operates within an environment. The standard formalism of reinforcement learning [6] corresponds to a decision-making framework, where an agent interacts with an environment and enhances its performance based on feedback. The agent receives a state and chooses an action, after which the environment provides a reward and a new state. Overall, the objective is to maximize the cumulative reward.

Although RL holds promise, its implementation in real-world scenarios often entails high costs due to sampling inefficiency [7]. This implies that the algorithm requires

S. Saito et al. (Eds.): AsiaSim 2024, CCIS 2170, pp. 178–190, 2024.
https://doi.org/10.1007/978-981-97-7225-4_14

numerous iterations to achieve satisfactory results. One approach to mitigate this challenge is through the adoption of experience replay [8], which involves reusing previous experiences. Additionally, there are other methods available to tackle this problem. Recent alternatives include utilizing Gaussian processes [9] and employing babbling to expedite the learning process [10]. However, this paper specifically focuses on experience replay and its various forms.

One of the primary obstacles hindering the realization of RL's full potential in robotics applications is the challenge of efficiently learning from sparse and binary rewards, which are prevalent in various robotic tasks [11]. This challenge arises because the robot is likely to encounter zero rewards during exploration, leaving them clueless about areas for improvement. To address this issue, one of the most effective approaches thus far involves the utilization of HER [11] in conjunction with an off-policy RL algorithm like Deep Deterministic Policy Gradient (DDPG) [12]. By altering the original goal, HER enables the robot to extract valuable insights even from failed experiences.

One limitation of HER is that it employs a predefined strategy to randomly select additional goals from each episode of transitions for replay [13]. This lack of an evaluation mechanism introduces the possibility of choosing suboptimal transitions, which can potentially hinder the learning process.

The current challenge lies in establishing a criterion to assess the importance of transitions for replaying. In this work, we propose an improved priority-based hindsight experience replay method to enhance training performance. We utilize the Temporal Difference error and Reducible loss to calculate the priority of a transition, which serves as an implicit measure of learning progress. Specifically, within a learning process, a transition with a high priority is more likely to be sampled. And we empirically show that our method is more robust than random sampling and also better than prioritizing only with respect to TD error.

2 Related Work

In this section, we will introduce the preliminaries, including the RL approaches employed and the experience replay methods.

2.1 Reinforcement Learning

The standard RL formalism consists of an agent interacting with an environment. The interaction between the agent and the environment in reinforcement learning can be modelled by Markov Decision Process (MDP). MDP refers to the fact that the system obeys the Markov property: Given the current state and all previous states, the probability of a future state depends solely on the most recent state and action instead of the previous state or action in a sequential decision-making problem. Generally, the MDPs are defined as a 5-tuple (S, A, R, P, ρ_0). Where S represents the collection of all valid states, $s_t \in S$ represents the state of the agent at time t; A represents all valid action, $a_t \in A$ is the action taken by the agent at time t; R represents the reward function, $r_t \in R$ is the reward obtained by the agent taking action a_t at state s_t; P is the transition function with $P(s'|s, a)$, which means the probability of transitioning to the next state s' after performing action a_t at state s_t. ρ_0 represents the starting state distribution. In RL, the action performed by the agent is selected by the strategy π, which maps state s_t to action a_t. The goal of RL is to discover the optimal policy π^*, specifically,

$$\pi^* = \underset{\pi}{\mathrm{argmax}}\, E_\pi \left[\sum_{t=0}^{\infty} \gamma^t r_t | S_t = s, A_t = a \right] \tag{1}$$

where $\gamma \in [0, 1]$ is the discount factor.

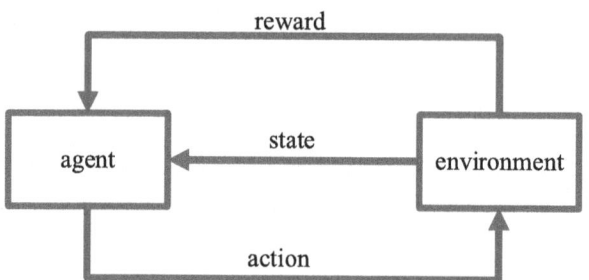

Fig. 1. The interaction of RL

Figure 1 is the interaction process of RL. First, the agent take an action a_t to the environment according to the current state s_t and strategy π. After the environment receives the action, the environment returns the reward of the current action and the next state s_{t+1}, and finally the agent updates the strategy π according to the reward.

In order to evaluate the value of each action in a certain state, the action with the greatest value can be selected for execution, an action-value Q function is introduced. The training process of RL is essentially the process of continuously iterating the value of Q until Q converges to its maximum value.

$$Q^\pi (s_t, a_t) = E[R_t | s_t, a_t] \tag{2}$$

The π in the above formula represents the action selection strategy. Using the Bellman equation, Q-value function can be defined as follows:

$$Q^\pi (s_t, a_t) = r_t + \gamma \max_a Q^\pi (s_{t+1}, a) \tag{3}$$

Let Q^* denote an optimal action value Q-function, and it can be obtained as:

$$Q^*(s, a) = E_{s' \sim p(\cdot | s, a)} \left[r(s, a) + \gamma \max_{a' \in A} Q^* (s', a') \right] \tag{4}$$

This is the basic value-based RL algorithm, which is to calculate the value of each action based on the value. If we skip the intermediate steps and directly choose action based on the current state instead of using the value as the evaluation data, we arrive at another significant algorithm in RL, which is, the policy gradient.

Temporal difference error (TD error) is the difference between $Q^*(s, a)$ and $r(s, a) + \gamma \max\limits_{a' \in A} Q^*(s', a')$. Methods that utilize Q values aim to reduce the TD error of the Q function, Q^θ, where θ represents the parameters of a neural network. Therefore, the loss function for the Q network is expressed as:

$$L_\theta = \left(Q^\theta(s_t, a_t) - \left(r_t + \gamma \max\limits_a Q^\theta(s_{t+1}, a) \right) \right)^2 \tag{5}$$

2.2 Deep Deterministic Policy Gradient

DDPG combines the learning of an approximation to $Q(s, a)$ with the learning of an approximation to $\mu(s)$, in a manner specifically designed for environments with continuous action spaces.

The Actor uses the strategy function θ, responsible for generating actions and interacting with the environment. Meanwhile, the critic utilizes the value function Q to assess the actor's performance and inform its actions in subsequent stages. The Actor-Critic algorithm schematic diagram is shown in the Fig. 2.

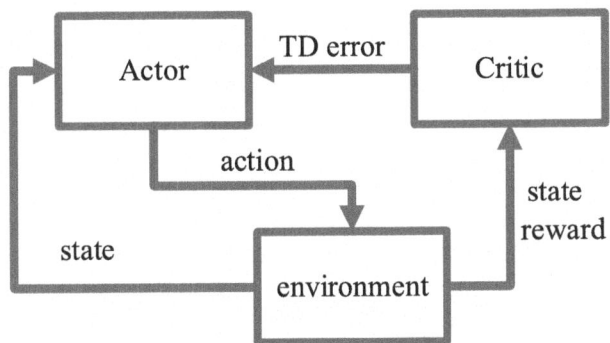

Fig. 2. Actor-Critic Algorithm

The actor adjusts the parameters θ of the stochastic policy $\pi_\theta(s)$ by stochastic gradient ascent of policy gradient, which is shown below:

$$g = E_{s \sim \rho^\pi, a \sim \pi_\theta} \left[\nabla_\theta log \pi_\theta(a|s) Q^\pi(s, a) \right] \tag{6}$$

The expression of TD error in the figure is shown in the following formula, serves the purpose of maximizing the utilization of training data through the utilization of the time-series difference method.

$$\delta(t) = R_{t+1} + gV(S_{t+1}) - V(S_t) \tag{7}$$

Actor policy function parameter update formula is:

$$\theta = \theta + \alpha \nabla_\theta log \pi_\theta(s_t, a_t)\delta(t) \tag{8}$$

Using an appropriate policy evaluation algorithm, the Critic utilizes the action-value function $Q^\omega(s, a)$ with parameter vector ω to estimate $Q^\pi(s, a)$ in Eq. 6.

For the Critic model parameter ω, the update formula of the model parameter ω of Critic itself can be expressed as:

$$\delta = R_{t+1} + gQ(S_{t+1}, Q_{t+1}) - Q(S_t, A_t) \tag{9}$$

Critic Q function parameter update formula is:

$$\omega = \omega + \beta\delta\nabla_\omega Q^\omega(s_t, a_t) \tag{10}$$

The Actor-Critic algorithm raises the upper limit of reinforcement learning algorithms and enables them to be applied to more complex scenarios.

Mathematically, DDPG utilizes a deterministic policy $\mu_\theta(s)$ parameterized by θ that determines the action taken in a given state, and a critic $Q_\omega(s, a)$ parameterized by ω that predicts the action-value function. The DDPG algorithm alternates between collecting experience by running the policy and updating the parameters. Experience is collected by a behavior policy, which is a variant of the deterministic policy incorporating noise. The behavior policy is defined as $\pi_b(s) = \mu_\theta(s) + N$, where N represents noise, specifically mean-zero Gaussian noise. The critic network is trained by minimizing the following loss function:

$$L(\omega) = \mathbb{E}_{(s,a,r,s')\sim D}\left[y - Q_\omega(s, a)\right]^2 \tag{11}$$

where, $y = r + \gamma Q_{\omega^-}\left(s', \mu_{\theta^-}(s')\right)$. Q_{ω^-} is the target Q-value function, μ_θ represents the target policy, and D is the experience replay buffer containing a set of experience $\{(s, a, r, s')\}$.

To ensure stable training, a separate network often maintains the target Q-value function Q_{ω^-}, with its weights periodically averaged over the weights of the main network. In DDPG, the actor is then learned by maximizing the following objective function:

$$J(\theta) = \mathbb{E}_{s\sim D}\left[Q_\omega(s, \mu_\theta(s))\right] \tag{12}$$

The derivative of the objective function is calculated utilizing the deterministic policy gradient theorem:

$$\nabla_\theta J(\theta) = \mathbb{E}_{s\sim D}\left[\nabla_\theta \mu_\theta(s)\nabla_a Q_\omega(s, a)|_{a=\mu_\theta(s)}\right] \tag{13}$$

Throughout this step, the parameter ω of the Q-value function remains constant. To enhance training stability, a separate network is employed to maintain the target policy, with its weights periodically adjusted through moving averages. In practice, the Eq. (5) and (7) are approximated as follows:

$$L(\phi) \approx \frac{1}{|B|}\sum_{i=1}^{|B|}\left[r_i + \gamma Q_{\phi^-}\left(s_i', \mu_{\theta^-}\left(s_i'\right)\right) - Q_\phi(s_i, a_i)\right]^2 \tag{14}$$

And the gradient is approximated by:

$$\nabla_\theta J(\theta) \approx \frac{1}{|B|}\sum_{i=1}^{|B|}\left[\nabla_\theta \mu_\theta(s_i)\nabla_a Q_\phi(s_i, a)|_{a=\mu_\theta(s_i)}\right] \tag{15}$$

2.3 Prioritized Experience Replay

Online RL algorithms execute updates as soon as a transition is obtained. However, this approach not only diminishes learning efficiency but also induces high correlation among recent transitions. To mitigate this issue, Lin [8] introduces experience replay, which stores obtained transitions and offers a means for sampling batches of transitions. The method of experience replay is when a single agent obtains a set of interactive data, such as (S, A, R, S'). This set of interactive data will not be used directly, but stored in the experience buffer, and then specific samples will be drawn from the experience buffer for policy training.

In the process of sample extraction, sample data is randomly selected from the experience buffer to make the extracted data independent of each other. Although this satisfies the requirements of neural network training data and can achieve better training effects, this approach remains inefficient since not all data carries equal significance.

Schaul [14] proposes Prioritized Experience Replay (PER), which samples transitions with probabilities based on the transitions' priority. And the priority p_i is calculated proportional to their TD error. So the probability P_i is calculated as follows:

$$P_i = \frac{p_i^{\alpha^p}}{\sum_j p_j^{\alpha^p}} \tag{16}$$

where $\alpha^p \in [0, 1)$ is a hyperparameter. PER uses importance sampling (IS) weights ω^p to ease the non-uniform:

$$\omega_i^p = \left(\frac{P_{uniform}}{P_i}\right)^{\beta^p} \tag{17}$$

where $\beta^p \in [0, 1]$ controls the degree of correction for gradient changes.

2.4 Hindsight Experience Replay

Despite numerous advancements in Deep RL, training agents with sparse rewards continues to be a significant challenge. In such tasks, a commonly utilized reward function is as follows:

$$r_t = \begin{cases} 0, & \text{if } \|s_t - g\|_2 < \rho \\ -1, & \text{otherwise} \end{cases} \tag{18}$$

In this reward function, s_t represents the current state, g represents the desired goal and ρ is a tunable hyperparameter. So, this provides a reward to the agent if the current state is within a certain threshold of the desired goal g. This rewards the agent for being close to the goal.

In a challenging environment with sparse rewards, the agent might encounter prolonged periods without receiving any positive feedback, making learning difficult and leading to sample inefficiency. To address this issue, Andrychowicz [11] propose a simple yet effective approach called Hindsight Experience Replay (HER), which involves relabeling the goals of existing experiences.

HER leverages the concept of hindsight goals by duplicating episodes in the replay buffer but with a different set of goals [15]. It should be noted that HER can be integrated with any off-policy RL algorithm. This assumption holds true in our experiments.

3 Improved Priority-Based Hindsight Experience Replay

In this section, we present a formal description of our approach, which includes the motivation behind our method, the derivation of the priority method, and the improved Priority-based Hindsight Experience Replay framework.

In a nutshell, we propose an improved priority method and then sample the trajectories with higher priority for replay.

Nowadays, the majority of mainstream reinforcement learning algorithms utilize the experience replay method of random sampling. This method involves sequentially storing interactive data in an experience buffer, from which the training agent selects a batch size of data. During training, new interaction transitions replace old transitions in a first-in-first-out manner when the experience buffer becomes full. However, this approach can lead to inadequate training because it overlooks rare valuable and learnable transitions.

To address the problem of selecting replay transitions from the replay buffer, we introduces an improved priority combined with TD error and Reducible Loss. The TD error enables the agent to prioritize learning from transitions with high TD error, emphasizing the significance of information that has not yet been incorporated into the model. However, some transitions may have high TD error due to inherent noise or inherent difficulty in learning for the model. In order to overcome this issue, the priority of a transition should consider the potential decrease in TD error, which can be defined as the learnability of the data. Shivakanth [16] defines the ReLo for reinforcement learning as the difference between the loss of a data point with respect to the online network parameterized by θ and with respect to the target network parameterized by $\overline{\theta}$, which can be computed as:

$$ReLo = L_\theta - L_{\overline{\theta}} \tag{19}$$

There are analogous traits between ReLo and the sampling behavior with low priority in comparison to the PER framework. Specifically, transitions that are deemed inconsequential under the PER framework, as indicated by their low L_θ values, similarly retain their marginal significance within the ReLo paradigm. This alignment arises from the inherent relationship between low L_θ values and correspondingly diminished ReLo scores, as expressed by Eq. (19).

However, transitions that are noisy or unlearnable will receive low ReLo priority because both L_θ and $L_{\overline{\theta}}$ will be high. This prioritization ensures that the agent focuses more on transitions that are worth sampling and useful, resulting in more stable training.

In order to prioritize transitions with high learnability, we introduce an improved priority parameter, which combines TD error with ReLo score. This priority serves as the basis for selecting transitions from the replay buffer, taking into account both high TD error transitions and low-priority transitions that may be noisy or unlearnable. This improves the efficiency of training in reinforcement learning.

The TD error δ_i is calculated using Eq. (9). Based on this, the priority of a transition is represented by p_{T_i}, given as:

$$p_{TD_i} = |\delta_i| + \epsilon \tag{20}$$

In practice, a small ϵ is added to ensure that every transition has a minimum probability of being sampled. Thus, the probability of sampling a transition based on TD error can be represented as:

$$P_{TD}(i) = \frac{p_{TD_i}^{\alpha}}{\sum_k p_{TD_k}^{\alpha}} \tag{21}$$

Here, α is the priority adjustment factor. When $\alpha = 0$, it degenerates to uniform sampling.

However, the probability P_{T_i} only considers TD error and does not take into account transitions with high learnability. Therefore, the ReLo score is introduced to modify the selection probability based on the learnability of transitions. The ReLo parameter is given by Eq. (14). As ReLo does not have a non-negative property, negative values are clipped to zero and a small ϵ is added in practice. Thus, the priority for sampling a transition based on ReLo becomes:

$$p_{RL_i} = \max(ReLo, 0) + \epsilon \tag{22}$$

The probability of sampling a transition based on ReLo priority can be represented as:

$$P_{RL}(i) = \frac{p_{RL_i}^{\alpha}}{\sum_k p_{RL_k}^{\alpha}} \tag{23}$$

Here, k is the mini-batch size. The transition probability of the TDRl experience replay method is:

$$P(i) = \beta P_{TD}(i) + (1 - \beta)P_{RL}(i) \tag{24}$$

where β is the regulatory factor and $0 < \beta < 1$.

One transition is sampled from each group as the training data, based on the probability $P(i)$. This approach not only avoids the problem of training overfitting but also enhances the stability of the training process and improves exploration ability simultaneously.

Complete Algorithm: We summarize the complete training algorithm in Algorithm 1.

Algorithm 1 Improved Priority-based Hindsight Experience Replay algorithm

---

Given the online Q network parameters θ, target Q network parameters $\bar{\theta}$, replay buffer B, the priority adjustment factors α, β, epsilon priority ϵ, training episode numbers N, target network update frequent T_{target}, batch size k.

Initialize the network.

for episode $= 1, ..., N$ do

    Sample a goal g and an initial state s_0.

    for t $= 0, ..., T - 1$ do

        Sample an action a_t using the behavior policy:

$$a_t \rightarrow \pi_b(s_t||g)$$

        Execute the action a_t and observe a new state s_{t+1}

    end for

    Calculate TD error from Eq.9

    Calculate TD priority p_{TD_i} from Eq.20, and then calculate TD probability $P_{TD}(i)$ from Eq.21

    Calculate ReLo priority p_{RL_i} from Eq.22, and then calculate TD probability $P_{RL}(i)$ from Eq.23

    Calculate the transition probability of the improved priority $P(i)$ from Eq.24

    for t $= 0, ..., $ T-1 do

        $r_t := r(s_t, a_t, g)$

        Store the transition $(s_t||g, a_t, r_t, s_{t+1}||g, p)$ in R

        Sample transition T for replay based on priority $P(i)$

        Sample transitions (s_t, a_t, s_{t+1}) from T

        Sample virtual goals $g' \in \{s_{t+1}, ..., s_{T-1}\}$ at a future timestep in T

        $r_t' := r(s_t, a_t, g')$

        Store the transition $(s_t||g', a_t, r_t', s_{t+1}||g', p)$ in R

    end for

    for $t = 1, ..., M$ do

        Sample a minibatch k from the replay buffer R

        Update policy network $\theta \leftarrow \theta + \alpha\nabla_\theta log\pi_\theta(s_t, a_t)\delta(t)$

        if $t \% T_{target} == 0$:

            Soft update target network $\bar{\theta}$ with online policy network θ

    end for

end for

---

4   Experiments and Results

Experiments were conducted within a multi-goal reinforcement learning environment designed for continuous control tasks, as described in [17]. The experimental environment was developed by integrating the OpenAI Gym interface [18] with the MuJoCo physics engine [19]. Specifically, the research focused on the Fetch environment, where the primary objective was to control a 7-degree-of-freedom (DoF) robot arm. The efficacy of the proposed method was evaluated through two distinct tasks outlined.

1) FetchPush: The robot shifts a box situated on a table to a specified location by applying pushing and rolling maneuvers. A successful episode is determined when the box is within 7 cm of the target location.
2) FetchPickAndPlace: Here, the robot secures a box and transports it to a designated spot, which may be aerial or situated on the table's surface. Achievement is recorded if the box is positioned less than 7 cm from the target upon episode completion.

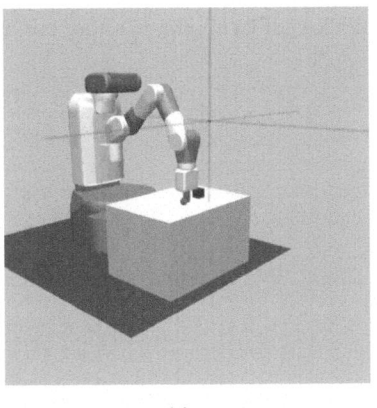
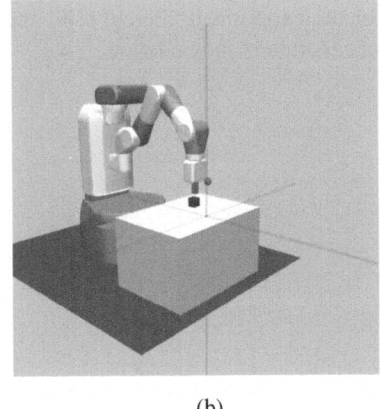

(a)   (b)

Fig. 3. (a) FetchPush (b) FetchPickAndPlace

Figure 3 presents a visual depiction of the experimental environment during an episode for each of the described tasks. The multi-goal reinforcement learning framework for the aforementioned tasks can be characterized as follows.

1) States: The joint angles and velocities for the robot, as well as the object's position, rotation, and both linear and angular velocities.
2) Goals: The object's target position within a three-dimensional space, surrounded by a tolerance sphere with radius ξ. This radius is set to 7 cm which is the same as the original paper [11]. Hence, the goal fulfilment function $f_g(s)$ is defined as:

$$f_g(s) = \begin{cases} 1, & |g - s_{\text{object}}| \leq \xi \\ 0, & |g - s_{\text{object}}| \geq \xi \end{cases} \tag{25}$$

where g represents the target position and s_{object} represents the position of the object.
3) Rewards: The experimental setup utilizes binary sparse rewards as follows:

$$r(s_t, g, a_t) = \begin{cases} 0, & f_g(s_{t+1}) = 1 \\ -1, & f_g(s_{t+1}) = 0. \end{cases} \tag{26}$$

4) Actions: The action space includes the distance between the gripper's fingers and three additional dimensions relating to gripper movements.

Both the actor and critic networks employ a multilayer perceptron (MLP) architecture, utilizing rectified linear units (ReLUs) as non-linear activation functions for each layer. Training is accomplished via backpropagation, with the ADAM optimizer [30] facilitating the process.

We choose the base algorithm to be Deep Reinforcement Learning with DDPG. And we implement the DDPG algorithm using the pytorch framework [20]. The hyperparameter of the actor and critic network and the network update weight are the same in the DDPG paper [21]. The experience buffer is 100,000. The batch size of transitions is 32, and α, ϵ are the same in the PER paper. And we set β to be 0.5.

For the experiment "FetchPush" and "FetchPickAndPlace", the learning curve of the success rate is shown below.

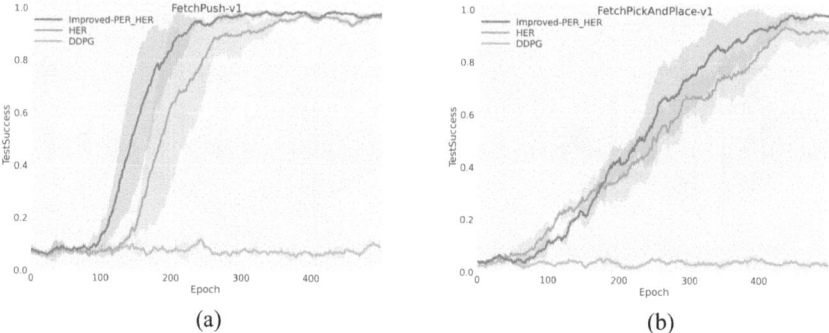

(a)   (b)

Fig. 4. (a) Learning curve of Push (b) Learning curve of PickAndPlace

We run five random seeds in every experiment to evaluate the stability of training. Figure 4 illustrates that the improved Priority-based HER achieves faster convergence than both standard DDPG and vanilla HER across both tasks. As we can see in the success rate of "FetchPush" in Fig. 4(a), the improved Priority-based Hindsight Experience Replay significantly improved the training speed. Notably, the "FetchPickAndPlace" experiment in Fig. 4(b) the improved Priority-based HER agent not only converges more quickly but also demonstrates superior performance by the end of the training period. It's particularly appealing that the improved Priority-based HER results in improved performance in both tasks, especially in "FetchPickAndPlace" experiment which gains about seven percentage points over HER alone. These results underscore the efficiency of our method.

5 Conclusion

This study explores a replay prioritization technique for multi-goal scenarios within sparse-reward environments. Drawing inspiration from the prioritization approach utilized in PER, this work proposes an improved prioritization scheme for experience replay.

The improved prioritization scheme considers the TD error and the ReLo loss together. TD error confirms the transitions with high TD error will get high priority during learning. And ReLo loss make sure that the transitions with high learnability will have higher priority while the transitions with low learnability will have low priority. By utilizing the improved prioritization with HER, we have demonstrated encouraging outcomes across complex robotic manipulation tasks and outperforming vanilla HER and achieving significantly superior results compared to DDPG algorithms. The implementation of the improved Priority-based HER method has expedited progress in reinforcement learning, and we anticipate that its application across a broader range of scenarios will unveil a plethora of exhilarating opportunities for RL agents.

Disclosure of Interests. The authors have no competing interests to declare that are relevant to the content of this article.

References

1. Mnih, V., et al.: Playing atari with deep reinforcement learning, p. 9
2. Zha, D., et al.: DouZero: mastering DouDizhu with self-play deep reinforcement learning. arXiv (2021). http://arxiv.org/abs/2106.06135. Accessed 15 Mar 2024
3. Silver, D.: Mastering the game of go with deep neural networks and tree search
4. Luo, Y., et al.: D2SR: transferring dense reward function to sparse by network resetting. In: 2023 IEEE International Conference on Real-Time Computing and Robotics (RCAR), Datong, China, pp. 906–911. IEEE (2023). https://doi.org/10.1109/RCAR58764.2023.102 49999
5. Seraj, E.: Embodied team intelligence in multi-robot systems (2022)
6. Sutton, R.S., Barto, A.G.: Reinforcement Learning: An Introduction
7. Zhang, S., Sutton, R.S.: A deeper look at experience replay. arXiv (2018). http://arxiv.org/ abs/1712.01275. Accessed 24 Nov 2023
8. Lin, L.-J.: Self-improving reactive agents based on reinforcement learning, planning and teaching. Mach. Learn. **8**(3–4), 293–321 (1992). https://doi.org/10.1007/BF00992699
9. Grande, R., Walsh, T., How, J.: Sample efficient reinforcement learning with gaussian processes. In: International Conference on Machine Learning, PMLR, pp. 1332–1340 (2014)
10. Kwiatkowski, R., Lipson, H.: Task-agnostic self-modeling machines. Sci. Robot. **4**(26), eaau9354 (2019)
11. Andrychowicz, M., et al.: Hindsight experience replay. arXiv (2018). http://arxiv.org/abs/ 1707.01495. Accessed 24 Nov 2023
12. Lillicrap, T.P., et al.: Continuous control with deep reinforcement learning, p. 14 (2016)
13. Zhao, R., Tresp, V.: Energy-based hindsight experience prioritization. In: Conference on Robot Learning, PMLR, pp. 113–122 (2018)
14. Schaul, T., Quan, J., Antonoglou, I., Silver, D.: Prioritized experience replay. arXiv (2016). http://arxiv.org/abs/1511.05952. Accessed 25 Nov 2023
15. Luu, T.M., Yoo, C.D.: Hindsight goal ranking on replay buffer for sparse reward environment. IEEE Access **9**, 51996–52007 (2021). https://doi.org/10.1109/ACCESS.2021.3069975
16. Sujit, S., Nath, S., Braga, P.H.M., Kahou, S.E.: Prioritizing samples in reinforcement learning with reducible loss. arXiv (2023). http://arxiv.org/abs/2208.10483. Accessed 28 Feb 2024
17. Plappert, M., et al.: Multi-goal reinforcement learning: challenging robotics environments and request for research. arXiv, abs/1802.09464 (2018). https://api.semanticscholar.org/Cor pusID:3619030

18. Brockman, G., et al.: OpenAI gym. arXiv (2016). http://arxiv.org/abs/1606.01540. Accessed 15 Mar 2024
19. Todorov, E., Erez, T., Tassa, Y.: MuJoCo: a physics engine for model-based control. In: 2012 IEEE/RSJ International Conference on Intelligent Robots and Systems, pp. 5026–5033 (2012)
20. Paszke, A., et al.: PyTorch: an imperative style, high-performance deep learning library. arXiv (2019). http://arxiv.org/abs/1912.01703. Accessed 15 Mar 2024
21. van Hasselt, H., Guez, A., Silver, D.: Deep reinforcement learning with double Q-learning. arXiv (2015). http://arxiv.org/abs/1509.06461. Accessed 15 Mar 2024

Deep Learning for Ultrafiltration Membrane Prediction in Drinking Water Treatment

Nur Sakinah Ahmad Yasmin, Norhaliza Abdul Wahab[✉], Mashitah Che Razali, and Nurul Adilla Mohd Subha

Faculty of Electrical Engineering, Universiti Teknologi Malaysia, Johor Bahru, Johor, Malaysia
aliza@fke.utm.my

Abstract. Ultrafiltration membrane (UFM) technology offers an efficient filtration solution for purifying underground water sources into potable drinking water. However, membrane fouling is one of main problems in this technology. Developing membrane filtration models is imperative for predicting and managing fouling occurrences during the filtration process. In this study, two deep learning models, namely long-short term memory (LSTM) and a hybrid GRU-LSTM, were employed to forecast transmembrane pressure (TMP) in UFM systems. Leveraging the capacity of deep learning LSTM to manage extensive dependencies inherent in long-range data, a dataset of 6686 observations was utilized. The results revealed that the hybrid gated recurrent unit long-short term memory (GRU-LSTM) model outperformed the LSTM model, achieving an R^2 value of 97% compared to LSTM's 92.5%. This underscores the significance of integrating multiple architectural components to enhance the learning capability of neural networks for time-series forecasting tasks, as demonstrated by the hybrid GRU-LSTM model in comparison to LSTM alone.

Keywords: Deep Learning · GRU-LSTM · Ultrafiltration Membrane

1 Introduction

Ultrafiltration membranes (UFMs) have been employed in water treatment industry for many years [1, 2]. Despite their longstanding use, fouling remains a significant constraint to the membrane, imposing limitations on performance and durability. This process is exhibited when an UFM operates at a constant flowrate and has an increase in transmembrane pressure (TMP) or decrease in flowrate when it is operated at a constant TMP. The performance of membrane fouling can be recover by reversible fouling, however irreversible fouling poses a challenge reducing membrane lifespan and defying traditional cleaning methods [3, 4]. Conventional techniques for monitoring and mitigating fouling have not been effective in offering up-to-date information or the ability to forecast fouling occurrences accurately. Conventional methods may rely on periodic sampling or manual observations, which may not capture real-time changes in fouling conditions or provide early warnings of potential fouling events. As a result, these methods may not be able to proactively manage fouling issues or prevent them from causing significant disruptions to membrane performance.

S. Saito et al. (Eds.): AsiaSim 2024, CCIS 2170, pp. 191–201, 2024.
https://doi.org/10.1007/978-981-97-7225-4_15

Predicting a TMP in an UFM has been demonstrated to be closely related to the mechanisms of membrane fouling [5–7]. A data-driven modeling approach has attracted many researchers to use it as a predicting tool in membrane fouling. D. J. Kovacs et al. [8], used a machine learning which including random forest, artificial neural network, and long-short term memory (LSTM) in predicting TMP at various stages of the membrane bioreactor production cycles. O. Kisi *et al.* [9], support vector machine learning with firefly algorithm as optimization in predicting water level fluctuation has shown an improvement in the predictive accuracy and capability of generalization. Additionally, W. Li [10], applied machine learning techniques and compared artificial neural networks (ANN) and support vector machines (SVM) for predicting membrane flux. Their study revealed that random forest (RF) achieved the highest prediction accuracy among the models tested. Apart from that, Mirbagheri et al. [11], conducted research where they constructed multi-layer perceptron (MLP) and radial basis function (RBF) models to predict transmembrane pressure (TMP) and permeability. These models were trained using a dataset spanning 60 days from a pilot-scale membrane bioreactor (MBR), with input variables including time, total suspended solids (TSS), chemical oxygen demand (COD), solids retention time (SRT), and mixed liquor suspended solids (MLSS).

Despite the promising outcomes demonstrated by these data-driven models, they are subject to various limitations. One such limitation is the small size of the datasets, often characterized by low temporal resolutions. Building a reliable membrane fouling model becomes challenging when working with data collected over short periods, as the severity of fouling tends to escalate over time. Larger datasets are essential for extracting more comprehensive information regarding the mechanisms and patterns associated with fouling. LSTM networks were developed with a specific focus on mitigating the vanishing gradient issue, a challenge frequently encountered in conventional Recurrent Neural Networks (RNNs). The vanishing gradient problem arises in RNNs due to the tendency of gradients to diminish significantly during backpropagation through successive time steps. This difficulty hampers the ability of RNNs to effectively capture long-term dependencies [12]. LSTMs tackle this challenge by incorporating gating mechanisms which are forget gate, input gate and output gate. These gates are designed to retain and preserve information over long sequences such as time series data, rendering them suitable for tasks where temporal dependencies are crucial. Unlike conventional RNNs, which face challenges in retaining information over time due to the vanishing gradient problem, LSTMs excel at retaining and utilizing past information effectively. To address this issue, more advanced architecture like LSTM were introduced by Hochreiter, S. and J. Schmidhuber [13]. Another gated architecture which is gated recurrent unit (GRU) were introduced by Chung, J., et al. [14]. Both methods has been proven their ability to solved vanishing gradient problem during training, thereby enabling them to handle input and output data with long-range dependencies more effectively [15–17].

Therefore, in this paper, we delve into the application of GRU-LSTM deep learning architecture to investigate fouling phenomena in UFMs. By leveraging the inherent capabilities of recurrent neural networks (RNNs), particularly in capturing temporal dependencies and long-range dependencies within data sequences, we aim to develop a predictive model capable of discerning fouling patterns and forecasting membrane performance degradation.

2   Materials and Methods

A pilot plant employing ultrafiltration (UF) membrane technology was designed and established for the treatment of drinking water. The plant utilized river water sourced from Universiti Teknologi Malaysia (UTM). The schematic diagram of the drinking water treatment process is depicted in Fig. 1. Initially, $0.1 \, m^3$ of river water was collected and stored in an influent tank. The feed water was then pumped through a proportional valve (PV) set at 50% opening before passing through a 0.5 mm pre-filter. Subsequently, the feed water from the influent tank was directed into UF modules at a constant flow rate of $0.5 \, m^3/h$. A $1 \, m^3$ filtration tank was employed to collect the treated water post-filtration. During backwashing cycles, these collected waters were utilized to backwash the membranes, eliminating any accumulated particles from the membrane surface and pores.

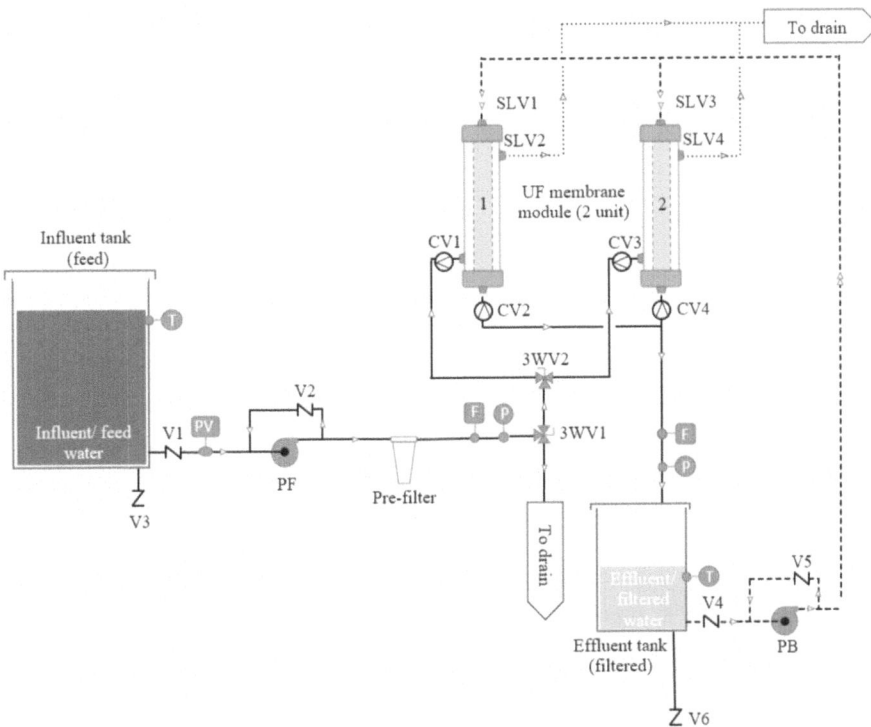

Fig. 1. Schematic diagram hollow fibre of UFM system.

2.1   UFMs Data Collection

In this study, the time series data were collected through automated system through comprehensive Laboratory Virtual Instrument Engineering Workbench (LabVIEW) system for real-time monitoring and control. Membrane cleaning was performed solely through physical backwashing, omitting the use of chemicals, to examine the impact of fouling on TMP and membrane resistance over a prolonged duration. As for our preliminary experiment work, filtration mode was on for 80 s and relation mode for 20 s for 2 h to observe the formation of fouling on the UFM. The input of this UFM process is the flowrate and voltage while the output is TMP. Figure 2 displays the UFM dataset, which comprises 6866 data points. At intervals of 1 s, measurements of three parameters—flow rate, voltage, and TMP—were continuously recorded throughout the duration of the experiment.

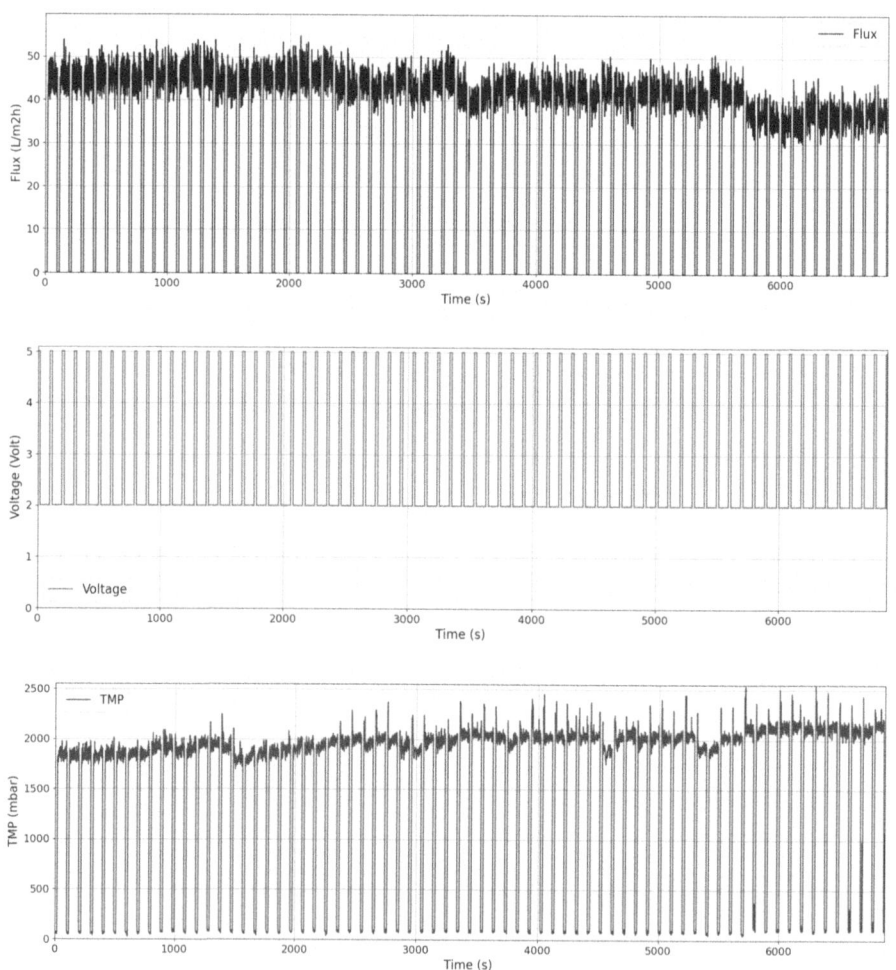

Fig. 2. UFM datasets.

2.2 Data Pre-processing

Data preprocessing involves two key steps: normalization and data partitioning. It is necessary to preprocess data to convert raw data into a format that is interpretable. Real-world data often lacks certain trends or behaviors and may introduce errors. In this study, the data was adjusted to an appropriate range to facilitate optimal learning during model development.

Normalization ensures that attributes with large numeric scales do not dominate those with smaller scales. This simplifies calculations and reduces the risk of data overflow or underflow. Typically, data is normalized to fit within the range of (-1, + 1) or (0, 1). It is recommended to normalize data before training and testing. Equation (1), sourced from [18], is employed for data normalization as described below:

$$x = \frac{0.8}{d_{max} - d_{min}}(d_i - d_{min}) + 0.1 \tag{1}$$

where d_i represents the ith input/output data, d_{max} signifies the maximum value of input/output data, and d_{min} denotes the minimum value of the input/output data.

The dataset undergoes scaling within the range of (0, 1). This normalization process offers a significant advantage by preventing the influence of large original input data on the solution. Additionally, it mitigates potential numerical issues during calculations. Then the dataset is partitioned into two sets: a training dataset and a testing dataset, utilizing the hold-out technique. This method is preferred, particularly when dealing with large experimental datasets, to mitigate the risk of overfitting [19]. The hold-out strategy allocates 60% of the data for training (4035 data) and reserves the remaining 40% for testing (2831). Maintaining the model's accuracy necessitates that the training dataset represents more than half of the entire experimental dataset [20].

2.3 Deep Learning of GRU-LSTM

The GRU-LSTM hybrid model was utilized, aiming to overcome the limitations observed in the standalone of LSTM model in predicting UFM datasets. A GRU neural network is a recently developed RNN-optimized model that features a less pronounced forget gate compared to the well-known LSTM network, making it easier to converge. It regulates the flow of information through a unique gate structure. GRU alters the computation mode of a hidden layer state variable in a recurrent neural network by employing resetting and updating gates, thereby optimizing the traditional RNN network [21].

The purpose of combines two different types of network which are GRU and LSTM is to employ the GRU to extract the local features in time series data which might be overlooked by the LSTM [22]. Our proposed hybrid model is built using 3 layers, where the first layer contains LSTM with 100 hidden neurons followed by second layer contains GRU with 100 hidden neurons and then the third layer contains the dense layer with 1 hidden neuron. Figure 3 shows the hybrid architecture of GRU-LSTM.

2.4 Hyperparameter Optimization

Within the realm of deep learning, hyperparameters—such as the learning rate and the amount of historical data utilized for predicting future values—are crucial components,

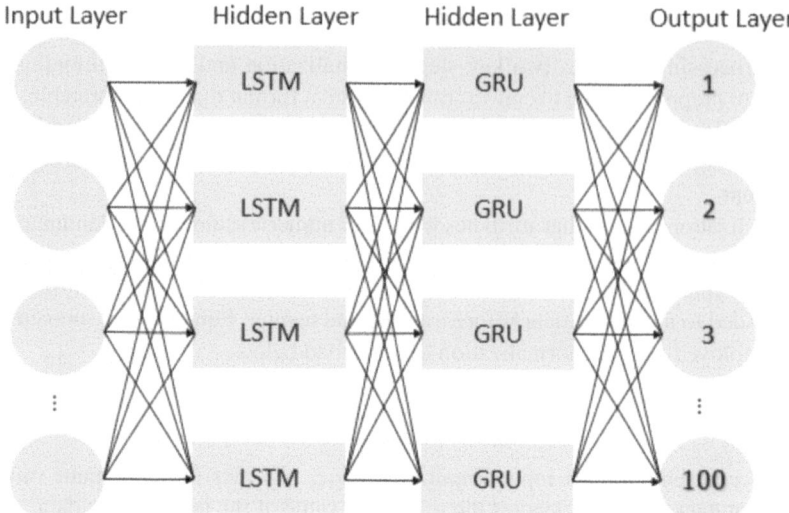

Fig. 3. Structure of GRU-LSTM.

as they directly impact model performance. Consequently, optimizing these hyperparameters becomes paramount. In our study, we employed the Adam optimization method. Adam aims to discover a set of weights, denoted as w, that minimizes the mean squared error. Unlike some optimization methods, Adam can be used with LSTM neural networks to minimize the mean squared error, which is the goal when training such networks. Adam is a versatile optimization method that adapts well to changing data distributions and objectives over time, unlike some other methods that might perform better with static data. It is particularly effective at handling sparse gradients and automatically adjusts the learning rate during training. Empirical evidence suggests that Adam performs well in practice, offering significant advantages over other stochastic optimization techniques like AdaGrad and RMSProp, especially in non-stationary settings [23].

The validation mean square error (MSE) and mean absolute error (MAE) was used as the objective function during hyperparameter optimization, (Eqs. (2) and (3)) as presented below. The optimized hyperparameters were chosen when the minimum validation MSE was observed during iterations [24].

$$MAE = \frac{\sum_{i=1}^{n} \left| Y_i - \hat{Y}_i \right|}{n} \tag{2}$$

$$MSE = \frac{\left[\sum_{i=1}^{n} \left(Y_i - \hat{Y}_i \right)^2 \right]}{n} \tag{3}$$

2.5 Model Evaluation

The UFM prediction results were evaluated using MSE, MAE, correlation coefficient (R^2) and root mean square error (RMSE) (Eqs. (4) and (5)), as presented below:

$$R^2 = 1 - \frac{\sum_{i=1}^{n}\left(\widehat{Y}_i - Y_i\right)^2}{\sum_{i=1}^{n}\left(Y_i - \overline{Y}\right)^2} \tag{4}$$

$$RMSE = \sqrt{\frac{\left[\sum_{i=1}^{n}\left(Y_i - \widehat{Y}_i\right)^2\right]}{n}} \tag{5}$$

where observed data is denoted as Y_i, predicted data denoted as $\widehat{Y}_i$, mean of the test data is denoted as $\overline{Y}$ and the size of the test set is denoted as n.

3 Results and Discussion

3.1 Deep Learning Model Optimization

Figure 4 shows the training loss and validation loss for two deep learning models which are hybrid LSTM and hybrid GRU-LSTM. Both models have an epoch size of 300, batch size 64 and previous time step 3. The x-axis indicates the number of epochs, while the y-axis displays both training and validation loss, measured in MAE and MSE accuracy. The orange trend line illustrates validation loss, while the blue trend line depicts training loss. From the plot, Fig. 4(a) and Fig. 4(b) demonstrate that the training and validation loss steadily followed the line up to 300 epoch. A lower loss value of <0.003, MAE were maintained from the 100th epoch while MSE from the 40th epoch. A similar trend is also shown in Fig. 4(c) and Fig. 4(d) where the MSE and MAE value. These indicate that both models LSTM and LSTM-GRU were effectively trained before being utilized for prediction purposes. The hyperparameters during training for both models (convergence speed, the number of data processed simultaneously, etc.) were optimized using Adam optimization, which designed for optimizing stochastic objective functions using first-order gradients, employing adaptive estimates of lower order moments [23].

3.2 TMP Prediction Results

Figure 5 shows the comparison plotted graph between LSTM and hybrid LSTM-GRU model. From the plotted graph, the developed LSTM-GRU model exhibited superior performance accuracy model of TMP during training results for UFM with R^2 of 97.7% with a lower MSE of 6.150 and RMSE of 2.480. Meanwhile, the LSTM model only achieved 88.5% of R^2 with MSE of 13.111 and RMSE of 5.880. These results proved that the LSTM-GRU model outperforms the LSTM model in terms of accuracy and efficiency, indicating the effectiveness of the combined LSTM and GRU architecture in capturing and predicting complex temporal dependencies within the dataset.

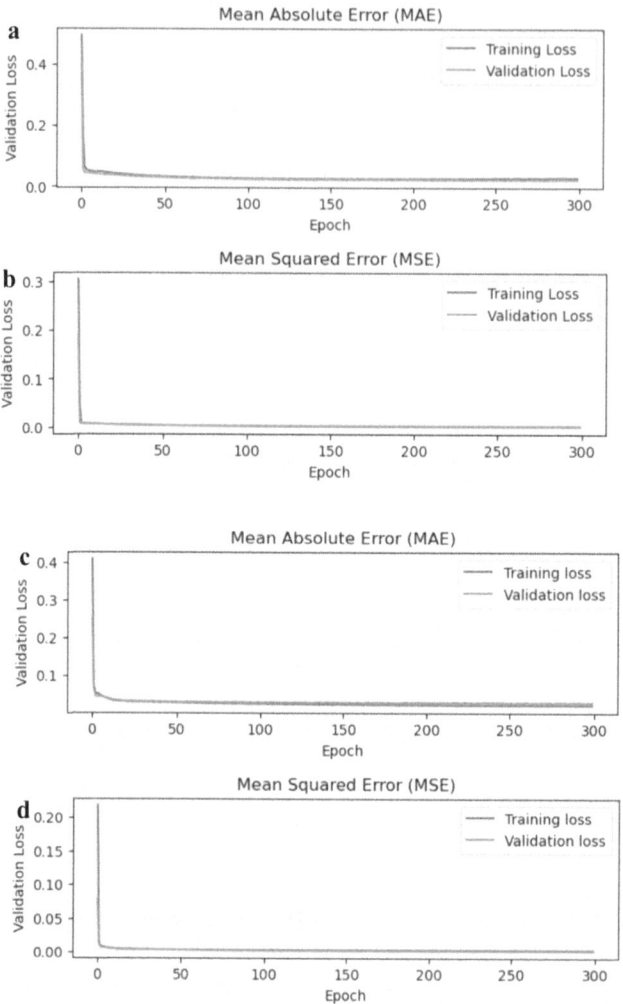

Fig. 4. Validation loss during hyperparameters optimization using MAE and MSE (a) MAE accuracy for LSTM (b) MSE accuracy for LSTM (c) MAE accuracy for LSTM-GRU (d) MSE accuracy for LSTM-GRU

Comparing the forecast results of LSTM and hybrid GRU-LSTM as shown in Fig. 6, it was observed that GRU-LSTM performed better performance with 97% of R^2 with lower value of MSE of 8.866 and RMSE of 2.978. Meanwhile, LSTM achieved 92.5% of R^2 with MSE of 10.866 and RMSE of 4.984. In addition to the forecasting outcomes, the GRU-LSTM models demonstrate promising predictive performance, exhibiting strong efficacy in both training and testing datasets. However, in the case of the LSTM model, the forecasting results indicate signs of overfitting, with the training performance notably surpassing that of the testing performance.

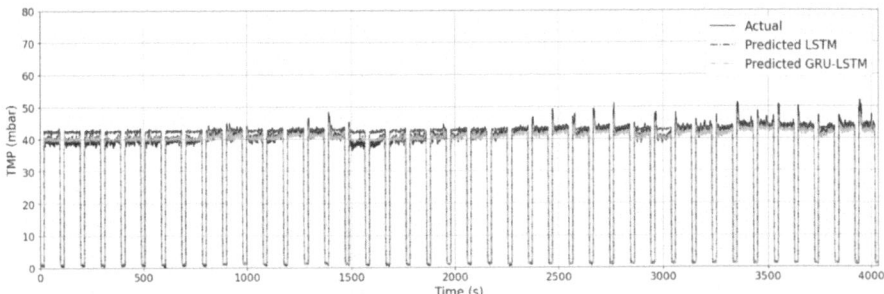

Fig. 5. TMP prediction plotted graph during training for LSTM and hybrid GRU-LSTM.

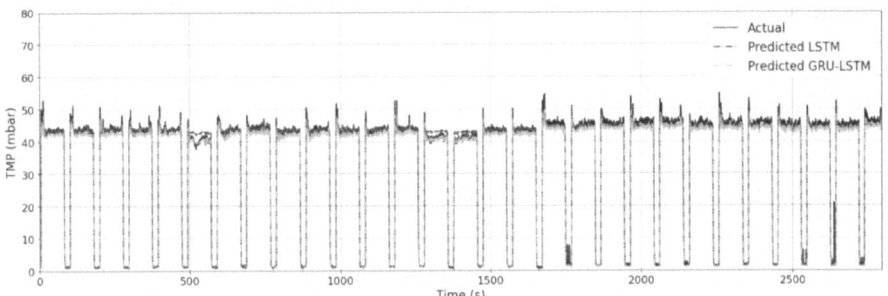

Fig. 6. TMP prediction plotted graph during testing for LSTM and hybrid GRU-LSTM.

4 Conclusion

This study employed two distinct machine learning architectures: the LSTM model and a hybrid GRU-LSTM model, to forecast TMP in ultrafiltration membranes utilized in drinking water treatment processes over time. These models underwent training and evaluation using a dataset consisting of 6866 observations gathered from a pilot plant. With two input variables used to predict a single output variable, namely TMP, the findings of this investigation lead to several conclusions. Firstly, the GRU-LSTM model demonstrated generally strong performance in predicting TMP in the UFM membrane pilot plant. Additionally, both the training and testing accuracy of the GRU-LSTM model surpassed that of the standalone LSTM model.

The superior performance of the GRU-LSTM model compared to the LSTM model can be attributed to its ability to effectively capture and retain relevant temporal dependencies while mitigating the vanishing gradient problem. The GRU-LSTM model achieves this by leveraging the gating mechanisms of both GRU and LSTM architectures, allowing it to maintain a balance between remembering and forgetting past information.

As for the observed overfitting in the LSTM model, where the testing performance is lower than the training performance, it suggests that the LSTM model may have excessively memorized the training data, leading to reduced generalization ability when presented with unseen data. This discrepancy between training and testing performance highlights the importance of regularization techniques and model architecture optimization to prevent overfitting and improve the model's ability to generalize to new data.

In conclusion, the utilization of a hybrid GRU-LSTM model demonstrates the significance of leveraging multiple architectural components to enhance the learning capacity of neural networks for time-series forecasting tasks. Additionally, the observed overfitting in the LSTM model underscores the necessity of careful model selection and regularization strategies to ensure optimal performance and generalization capability in predictive modeling applications.

Acknowledgments. This work was supported in part by the Universiti Teknologi Malaysia High Impact University Grant (UTMHI) vote Q.J130000.2451.08G74 and the Ministry of Higher Education under Prototype Research Grant Scheme (PRGS/1/2019/TK04/UTM/02/3).

Disclosure of Interests. The authors have no competing interests to declare that are relevant to the content of this article.

References

1. Ultrafiltration Membranes: Technologies and Global Markets. BCC Publishing (2020)
2. Nakatsuka, S., Nakate, I., Miyano, T.: Drinking water treatment by using ultrafiltration hollow fiber membranes. Desalination **106**(1–3), 55–61 (1996)
3. Yamamura, H., Kimura, K., Watanabe, Y.: Mechanism involved in the evolution of physically irreversible fouling in microfiltration and ultrafiltration membranes used for drinking water treatment. Environ. Sci. Technol. **41**(19), 6789–6794 (2007)
4. Gao, W., et al.: Membrane fouling control in ultrafiltration technology for drinking water production: a review. Desalination **272**(1–3), 1–8 (2011)
5. Shi, Y., et al.: Recent advances in the prediction of fouling in membrane bioreactors. Membranes **11**(6), 381 (2021)
6. Hai, F.I., Yamamoto, K.: Membrane biological reactors (2011)
7. Iorhemen, O.T., Hamza, R.A., Tay, J.H.: Membrane bioreactor (MBR) technology for wastewater treatment and reclamation: membrane fouling. Membranes **6**(2), 33 (2016)
8. Kovacs, D.J., et al.: Membrane fouling prediction and uncertainty analysis using machine learning: a wastewater treatment plant case study. J. Membr. Sci. **660**, 120817 (2022)
9. Kisi, O., et al.: A survey of water level fluctuation predicting in Urmia Lake using support vector machine with firefly algorithm. Appl. Math. Comput. **270**, 731–743 (2015)
10. Li, W., Li, C., Wang, T.: Application of machine learning algorithms in MBR simulation under big data platform. Water Pract. Technol. **15**(4), 1238–1247 (2020)
11. Mirbagheri, S.A., et al.: Evaluation and prediction of membrane fouling in a submerged membrane bioreactor with simultaneous upward and downward aeration using artificial neural network-genetic algorithm. Process. Saf. Environ. Prot. **96**, 111–124 (2015)
12. Noh, S.-H.: Analysis of gradient vanishing of RNNs and performance comparison. Information **12**(11), 442 (2021)
13. Hochreiter, S., Schmidhuber, J.: Long short-term memory. Neural Comput. **9**(8), 1735–1780 (1997)
14. Chung, J., et al.: Empirical evaluation of gated recurrent neural networks on sequence modeling. arXiv preprint arXiv:1412.3555 (2014)

15. Tran, Q.-K., Song, S.-K.: Water level forecasting based on deep learning: a use case of Trinity River-Texas-The United States. J. KIISE **44**(6), 607–612 (2017)
16. Son, H., Kim, S., Jang, Y.: LSTM-based 24-h solar power forecasting model using weather forecast data. KIISE Trans. Comput. Pract. **26**, 435–441 (2020)
17. Yi, H., Bui, K.-H.N., Seon, C.-N.: A deep learning LSTM framework for urban traffic flow and fine dust prediction. J. KIISE **47**(3), 292–297 (2020)
18. Sanjay, C., Jyothi, C.: A study of surface roughness in drilling using mathematical analysis and neural networks. Int. J. Adv. Manuf. Technol. **29**(9–10), 846–852 (2006)
19. Yusuf, Z., Wahab, N.A., Sahlan, S.: Modeling of submerged membrane bioreactor filtration process using NARX-ANFIS model. In: 2015 10th Asian Control Conference (ASCC). IEEE (2015)
20. Dahmani, K., et al.: Estimation of 5-min time-step data of tilted solar global irradiation using ANN (Artificial Neural Network) model. Energy **70**, 374–381 (2014)
21. Gao, S., et al.: Short-term runoff prediction with GRU and LSTM networks without requiring time step optimization during sample generation. J. Hydrol. **589**, 125188 (2020)
22. Son, M., et al.: Deep learning for pH prediction in water desalination using membrane capacitive deionization. Desalination **516**, 115233 (2021)
23. Chang, Z., Zhang, Y., Chen, W.: Effective adam-optimized LSTM neural network for electricity price forecasting. In: 2018 IEEE 9th International Conference on Software Engineering and Service Science (ICSESS). IEEE (2018)
24. Bergstra, J., et al.: Algorithms for hyper-parameter optimization. In: Advances in Neural Information Processing Systems, vol. 24 (2011)

Autonomous Navigation with Feature Points Map for Self-position Estimation for High-Accuracy Landing on Unknown Celestial Bodies

Hiroaki Miura$^{(\boxtimes)}$ ⓘ and Hiroyuki Kamata ⓘ

Meiji University, 1–1–1 Higashi-mita, Tama-ku, Kawasaki 214-8571, Kanagawa, Japan
{ce231070,kamata}@meiji.ac.jp

Abstract. This paper presents a novel method for the highly accurate landing of small unmanned space probes on unknown celestial bodies. In this method, a map of feature points is generated, and the self-position is estimated precisely. Creating high-precision maps is a challenge owing to numerous space-specific disturbances, limited computational resources, and the need for high reliability. Here, we demonstrate highly reliable map generation and self-position estimation using only generic feature points (AKAZE and SURF) detected in images captured by a space probe for high-precision landing on unknown celestial bodies. This study focuses on visual simultaneous localization and mapping (V-SLAM) in space. We prepared four image datasets consisting of 80 images with randomly added space-specific disturbances. We tested the reconstruction using the dataset in a low computational resource environment, such as a Raspberry Pi 4B and Mac Studio, for repetitive simulations. The simulations indicated that a balance between computation time and accuracy can be achieved, even with limited computational resources, by adjusting the number of feature points. We also demonstrated that our proposed method can operate on a resource-limited Raspberry Pi 4B, meeting both accuracy and time requirements. Our experiments highlight that the feature point map is capable of guiding space probes to achieve highly accurate landings, and this algorithm raises the possibility of practical use for space probe operations.

Keywords: Map Generation · Self-Position Estimation · Feature Points

1 Introduction

In the field of space exploration, our understanding of celestial bodies is expanding through numerous observations and explorations. Consequently, the target points for landing explorations are becoming more pinpointed. Efficient space exploration and effective mission outcomes in the future are expected to involve landing with high precision toward target points. Landing on small celestial bodies (i.e., those with low gravity) is relatively easier, as the process can be controlled remotely from Earth. Missions to more distant objects, such as Hayabusa (JAXA, 2003) [1] and Hayabusa2 (JAXA, 2014) [2], require significant round-trip communication times—approximately 20 min each way. As the size of celestial bodies increases, space probes experience stronger gravitational forces. While awaiting commands from Earth, the space probe consumes more fuel for hovering. Probes cannot feasibly carry excessive fuel for this purpose, necessitating minimal fuel usage for landing. This requires a fast descent, making it difficult to rely solely on earth-based guidance. Consequently, space probes must autonomously and in real-time navigate their descent. To make significant advances in space development and increase exploration frequency, it is ideal for space probes to be small and lightweight.

In collaboration with JAXA, we conducted a study on self-positioning methods for lunar exploration using a small, lightweight unmanned space probe as part of the SLIM project [3]. The aim of this project is to achieve lunar landings with accuracy within 100 m. On January 20, 2024, the probe successfully landed with an error of only 55 m [4]. This achievement utilized a terrain-relative navigation technique called image-matching navigation. Image-matching navigation involves estimating the probe's position by precisely matching two images: (1) a prepared map image created through careful exploration and preparation, and (2) an image captured by the space probe during its descent. Furthermore, the Moon exerts gravitational forces, requiring space probes to autonomously navigate to their destinations.

Previous studies have utilized a lightweight crater-based position estimation method on a space probe [5], using a space-grade FPGA through principal component analysis and ETSM [6, 7]. These efforts drew upon detailed observations by SELENE (JAXA, 2007) and LRO (NASA, 2009). Additionally, backup processing for feature point-based position estimation [8–10] was performed on Earth due to the ease of real-time communication between the Moon and Earth. However, crater-based position estimation methods face limitations in regions with few craters and require advanced definitions of the expected solar altitude and flight direction. Similar to ground-based backup processing, careful exploration and preparation of crater maps and images are necessary, reducing the method's applicability, especially for other distant celestial bodies, due to associated financial and time costs.

Thus, we proposed a method to generate a map image by stitching together partial images captured by a space probe [11]. However, this approach incurs a high processing load, making real-time wide-area map generation extremely difficult. Subsequently, we proposed a method to detect feature points from partial images captured by a space probe and generate a map based on these feature points [12]. By avoiding the use of image information in the map, we reduced the processing load and improved accuracy, enhancing the method's potential for future production applications. Nevertheless, given the anticipated execution of these processes in low-computing-resource environments such as on space probes, further weight reduction and higher reliability are required.

In this study, we identified the requirements necessary for the algorithm to function effectively in space and conducted repeated simulations in various environments, including the Raspberry Pi 4B, to simulate low-computational-resource environment assuming a space probe. This helped ascertain the current capabilities and potential of our proposed algorithm, providing a guideline for its actual implementation on a space probe, which will form the basis for future research. Unlike on Earth, images captured in space can be subject to numerous space-specific disturbances. Failure in estimation could lead to the loss of space probes. Specifically, our study assumes a scenario where the space probe autonomously creates a map to approach an unknown celestial body and subsequently lands using this map. We employed a general-purpose feature-point detection method to continuously detect feature points from surface images captured by the space probe during navigation. These points are then precisely aligned on a map in real time, allowing for the expansion of the map while simultaneously estimating the probe's position. These processes are performed on the space probe, enabling fully autonomous navigation without the need for communication with the ground.

2 Principles

2.1 Self-position Estimation Using Feature Points

The self-position of a space probe is estimated by matching two sets of data: (1) feature points detected in an image captured by the space probe, and (2) stored feature points on the map. The method for accurately matching these feature points to estimate the self-position is described as follows:

1. Feature Detection: AKAZE [13] and SURF [14] features are detected and described in images captured by the space probe. AKAZE features are particularly effective for lunar images [15], and SURF is also considered due to its active use by JAXA. Here, we consider the case in which the information describing the AKAZE and SURF features is already included in the map generation process, as described in Sect. 2.2.
2. Corresponding Point Search: A correspondence point search is performed to associate the features detected in the captured image with the feature points from the map. This is achieved using the k-nearest neighbor (k-NN) method [16] along with a ratio test [17]. Note that this process may result in some incorrect pairs of correspondence points.
3. Removing Incorrect Correspondence Pairs: Incorrect correspondence pairs are removed using random sample consensus (RANSAC) [18]. However, RANSAC may not be effective if incorrect pairs make up more than half of all corresponding pairs. Therefore, it is important to remove incorrect pairs effectively in a single step.
4. Estimating Transformation Matrix: A transformation matrix is estimated to project the captured image onto the feature-point map. This transformation matrix is calculated using singular value decomposition (SVD) [19] with the remaining pairs of correspondence points. This step accurately determines the location of the captured image on the feature-point map.

The flowchart below shows the sequence of steps involved in self-position estimation by matching feature points (Fig. 1).

2.2 Map Generation

For the simulations, it is assumed that (1) the images captured by the space probe follow a continuous sequence, with each image following directly from the previous one. This continuity helps in generating a coherent map of the terrain; (2) the attitude control system of the space probe is correct, ensuring that the probe consistently maintains a

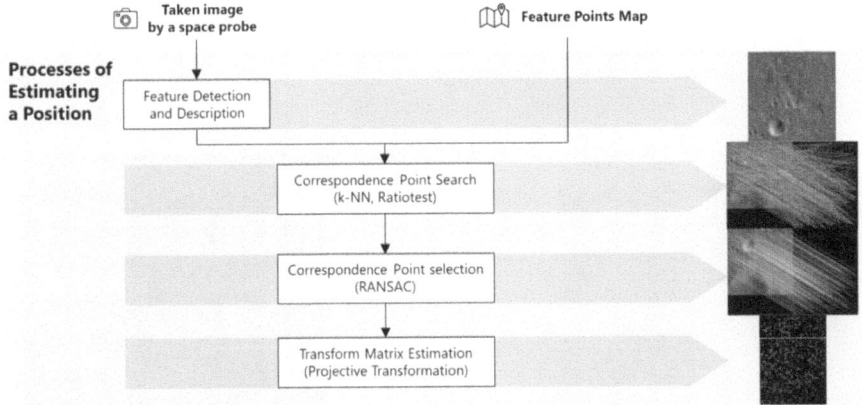

Fig. 1. Flowchart of the matching method using feature points and its sample images.

straight-down orientation towards the ground surface while capturing images; and (3) the space probe creates several orbits, including the landing site, to generate a more extensive map. The conceptual diagram is shown below (Fig. 2).

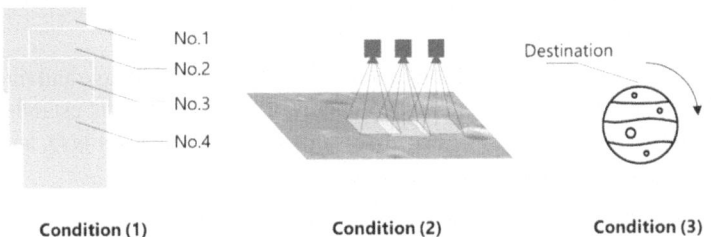

Fig. 2. Conceptual diagram of conditions for simulations.

We generated the map using the result of the position estimation in Sect. 2.1 because it provides information on where to add new features to the feature point map.

When extending the map using the detected feature points, it is necessary to determine whether the feature points already exist on the map. It also aims to reduce the processing load by reducing the number of feature points.

The proposed method uses the matching results between the feature point map and the captured image to identify new feature points. Correctly matched points are treated as matched (existing) points, and unmatched points are treated as new feature points. This allows only unknown feature points to be extracted and the map area to be extended with a lower processing load. A diagram of the map generation and self-position estimation methods is presented in Sect. 3.1.

2.3 Accuracy Evaluation

Figure 3 shows the original map image provided by JAXA (left) and the feature point map generated using the procedure described in Sects. 2.1 and 2.2 (right): The feature points detected in the left image and the generated feature-point map on the right should be placed at the same location with the same feature values.

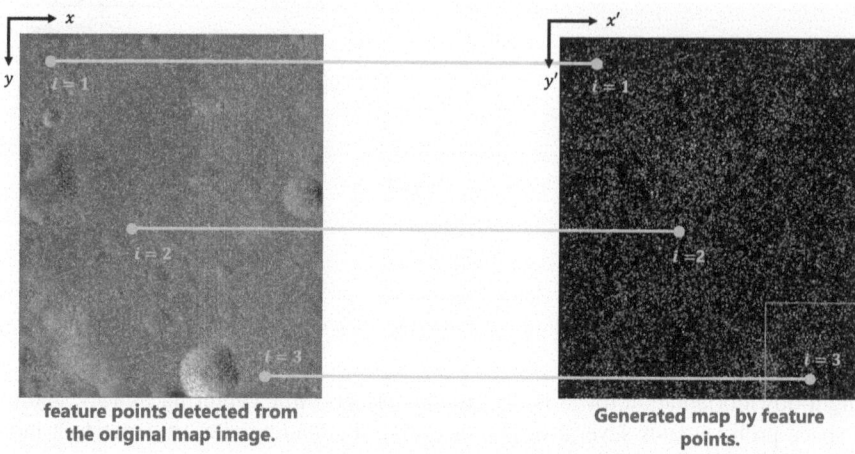

feature points detected from
the original map image.

Generated map by feature
points.

Fig. 3. Comparison of positions between two feature point groups. The coordinates of $i = 1, 2, 3$ should be identical to each other with reference to the top left corner.

Therefore, the accuracy can be compared by matching the left image and right feature point map and calculating the difference in coordinate values for the established pairs of corresponding points. If the average error value is A, the total error is D, and the total number of pairs of corresponding points is N, then

$$A = \tfrac{D}{N} = \tfrac{1}{N} \cdot \sum_{i=1}^{N} \sqrt{(x_{i\prime} - x_i)^2 + (y_{i\prime} - y_i)^2} \qquad (1)$$

is the average (A) and cumulative (D) errors, respectively. A small cumulative error indicates that the map was generated with high accuracy and also indicates that the self-position was estimated with high accuracy.

The average error for each feature point was calculated by dividing the number by the number of feature points. This value is used as the average error.

For a more detailed analysis, the positions of the feature points on the map and the actual positions of the feature points were visualized using vectors connected by arrows. This clarifies where the errors were generated on the map (self-position estimation). It also reveals how accuracy deteriorates as the map is generated and where errors may have increased rapidly as the map was generated (Fig. 4).

2.4 Space-Specific Disturbances

Unlike Earth, outer space is not surrounded by an atmosphere, making it an extremely harsh environment. Temperatures can drop below -100 °C or rise above 100 °C, and there can be strong radiation. This poses a significant challenge for current electronic

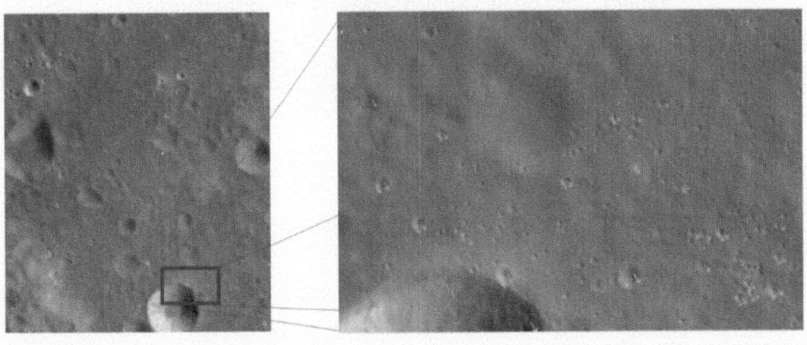

Fig. 4. Image of arrows on the map indicates a significant deterioration in accuracy due to a very strong disturbance. The arrow points from where the feature point should be (true point) to where it is actually located.

equipment, which operates on weak currents. An important instrument used in this study is a navigation camera made of a semiconductor (CMOS). When CMOS cameras are used in space, they experience space-specific disturbances that are superimposed onto images due to the harsh environment. Therefore, we prepared an image dataset (consisting of images taken by a space probe) with various disturbances that could occur in space and performed simulations.

Next, we discuss the degree of disturbance that should be added to the image datasets. Adding a strong disturbance may cause image breaks, whereas adding a small disturbance renders the simulation ineffective. We used the disturbance parameter values considered by JAXA for the SLIM project for the upper and lower limits.

Here, the disturbance limit parameters were set to values closer to those of the actual space environment. The assumed disturbances and their parameters, along with examples of actual images, are presented in the following Table 1 and Fig. 5. The brightness fluctuation is generated by

$$dst(x, y) = src(x, y) + n \tag{2}$$

and the contrast fluctuation is generated by

$$dst(x, y) = ((src(x, y) - 0.5) \times 10^n) + 0.5 \tag{3}$$

where $src(x, y)$ means nominal image and $dst(x, y)$ means damaged image.

The radiation noise is created by introducing data loss through randomly selected pixels, up to a total of $n \times 100\%$ of all the pixels in total (Table 2).

3 Verification Method

3.1 Overview of Our Processing Method

The experimental procedure is divided into four steps. In this study, we used the map images provided by JAXA for the SLIM project.

Table 1. List of all assumed space-specific disturbances.

Dataset No	Disturbance type name	Lower limit	Upper limit
1	Nominal (No Disturbance)	-	-
2	Brightness fluctuation	–0.3	0.3
3	Contrast fluctuation	–0.4	0.4
4	Radiation noise	0	5e–6

Table 2. Parameter values (fixed) and the range of values (varied).

Type	Parameter name	Lower limit	Step	Upper limit
AKAZE	akaze_descriptor_size	1e-4	0	9e−6
	akaze_descriptor_channels		3	
	akaze_threshold		[0.5]e−[1]	
	akaze_nOctaves		4	
	akaze_nOctaveLayers		4	
SURF	Surf_extended	0	False	100
	Surf_threshold		1	
	surf_nOctaveLayers		3	
	surf_nOctaves		4	
	Surf_Upright		False	
k-NN	k-NN k		2	
Ratio test	Matchratio		0.84	

Step 1: The map image is cut into 80 pieces. These images are designed to overlap in specific areas, as described in Sect. 2.2.

Step 2: Disturbances are introduced to the images cropped in Step 1 as necessary, resulting in a dataset consisting of 80 pieces.

Step 3: The process of self-position estimation and map generation is repeated.

Step 4: The accuracy of the final map is assessed by comparing it with the correct map image provided by JAXA.

The following diagram shows the experimental procedure and workflow, encompassing all previous explanations (Fig. 6).

3.2   Step 1: Test Image Cropping

We cropped the map images of the Moon provided by JAXA for simulations within the SLIM project, as shown in the following Fig. 7. The test image dataset was created using images, assuming the following four events: (1) the space probe is navigating over a celestial body with controlled direction; (2) there is a potential for navigation error

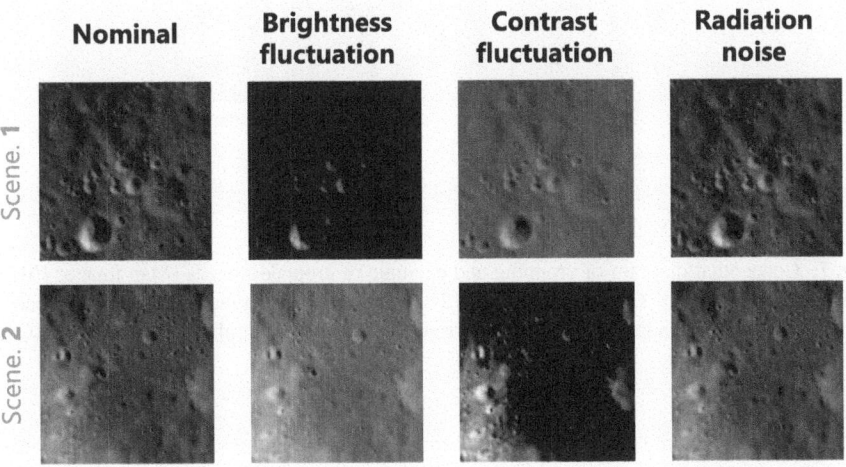

Fig. 5. Images showing actual specific examples of disturbances.

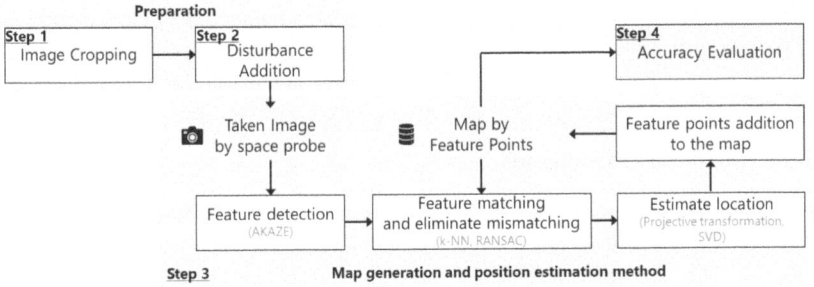

Fig. 6. Diagram covering all processes of the verification method described in this paper.

(<50 px); (3) space-specific disturbance may occur; and (4) the space probe completes eight orbits around the celestial body while moving the flying position slightly along with the red arrows in Fig. 7. The entire map was generated gradually through successive simulations and self-position estimations.

3.3 Step 3: Map Generation and Self-position Estimation

The processes of map generation and self-location estimation are illustrated in Sect. 3.1.

We hypothesized that there is a relationship between the number of feature points and the accuracy of location estimation. This hypothesis was based on the prediction that the impact of inaccurate feature points can be reduced by increasing the number of feature points incorporated into a map. Increasing estimation accuracy is a key objective of this study. However, simply increasing the number of detected feature points is insufficient. The resources required to run the program and the execution time were limited. To achieve this goal, striking a balance between quantity and accuracy is important.

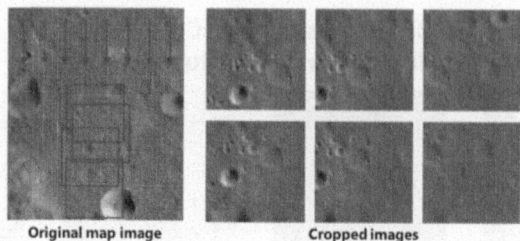

Original map image   Cropped images

Fig. 7. Conceptual diagram of cropping and example of cropped images (Map image: 1615 ×
1883 px, BMP format; Cropped images: 512 × 512 px, BMP format). 1 px is equivalent to
approximately 30 m on the Moon. The red arrows indicate the flight plan for eight orbits. (Color
figure online)

In the experiment, the parameters determining the degrees of feature detection for
AKAZE and SURF were set within the ranges listed in the following table. The number
of feature points detected was controlled by varying the values within the specified range
at a specific rate, and the dependence of accuracy on these parameters was investigated.
Three runs were performed for each value during the adjustment process. Based on the
results, the optimal parameters were selected to strike a balance between accuracy and
speed. We tested the proposed system by running it on a Raspberry Pi 4B system. The
process was repeated ten times to check for variations in calculation times.

3.4   Experimental Environments

We performed all the experiments described in this paper using a Mac Studio for repetitive simulations and a Raspberry Pi 4B for emulating a low computational resource
environment. The programs were developed using Visual Studio 2022 on Windows 11,
and the experiments were conducted by sharing the same programs on the Mac Studio
and Raspberry Pi 4B.

Table 3. Detailed information on the two computing environments.

	Mac Studio	Raspberry Pi 4B
OS	MacOS 14 Sonona	Debian GNU/Linux 11 (bullseye)
CPU	Apple M1 Ultra @3.2 GHz × 20	ARM Cortex-A72 @1.8 GHz × 4
Memory	64 GB	8 GB
C++ Compiler	Apple clang version 15.0.0	gcc version 10.2.1
External Library	OpenCV 4.9.0	OpenCV 4.9.0

4   Results

4.1   Number of Feature Points, Accuracy, and Elapsed Time

The following Fig. 8 and Fig. 9 show the relationship between the number of feature points, generation accuracy, and elapsed time, with variations up to 60,000 points. The sparsity of points in the graph (horizontal axis) is due to the feature point detection method, which results in a nonlinear number of feature points relative to the linear parameter variations. In the case of SURF, although the number of feature points is generally small, the parameters are already set to determine the maximum number of feature points detectable from a single image. To detect more than this, it would be necessary to extend the detection process by dividing the images.

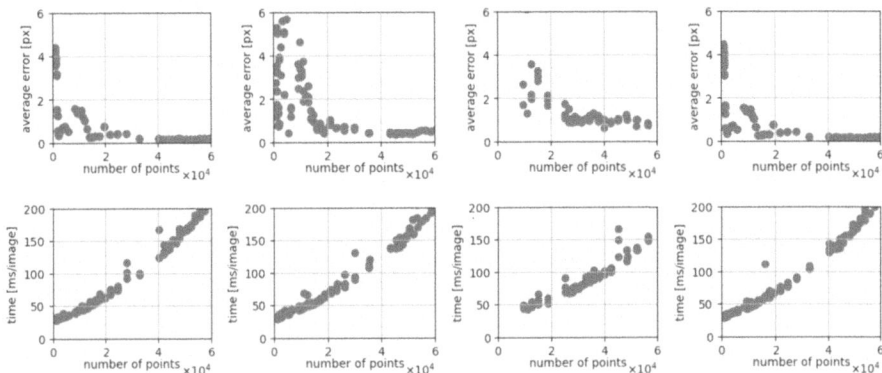

Fig. 8. Results using AKAZE on Mac Studio (repeated three times for each parameter) (from left to right: nominal, brightness fluctuation, contrast fluctuation, and radiation noise).

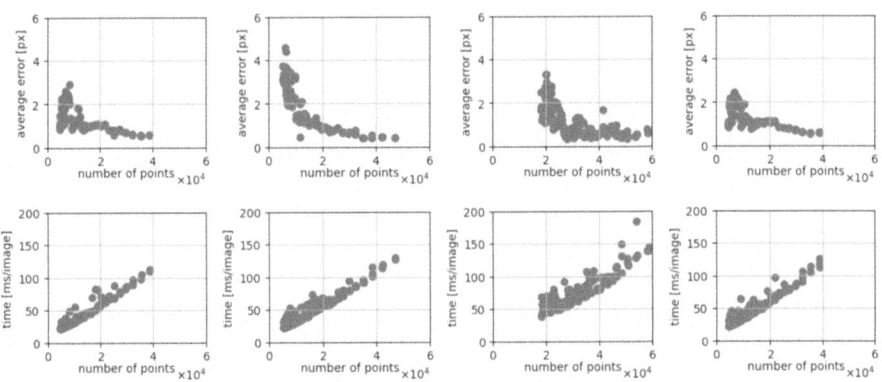

Fig. 9. Results using SURF on Mac Studio (repeated three times for each parameter) (from left to right: nominal, brightness fluctuation, contrast fluctuation, and radiation noise).

Figure 10 shows the number of feature points obtained by adjusting the parameters of the feature point detection methods, AKAZE (akaze_threshold) and SURF

(surf_threshold). By comparing the graphs above, the values of the parameters should be set to achieve the desired number of feature points.

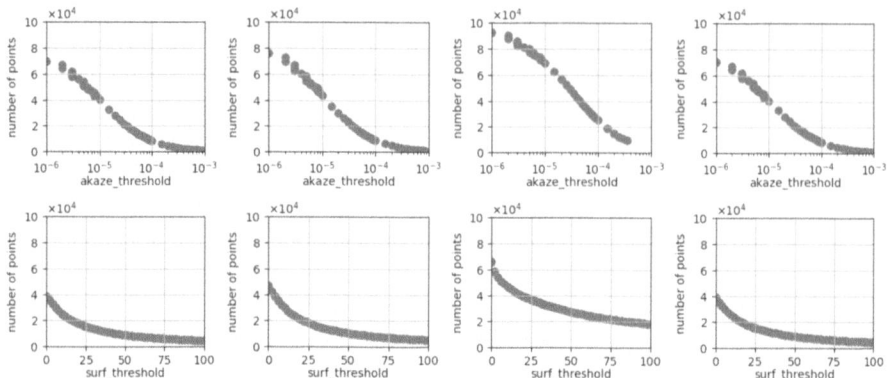

Fig. 10. Relationship between the parameters of the feature point detection method and the number of feature points obtained at that time. Note that the axes of AKAZE are logarithmic. (from left to right: nominal, brightness fluctuation, contrast fluctuation, and radiation noise).

These Fig. 9 and Fig. 10 show an overall trend where error decreases (accuracy improves) as the number of feature points increases, although the degree of improvement varies depending on the disturbance level and feature point detection method. It is essential to achieve optimal performance within resource constraints, understanding that 100% accuracy (0 px error) is not achievable. In the SLIM project, verification utilized images with an acceptable error margin of up to 3 px (approximately 100 m), a requirement met by increasing the number of detected feature points. Regarding the time aspect, validation was carried out using Mac Studio. While this specific time measurement may not be directly relevant now, the observed trends hold significant implications for space probe applications, where time requirements increase with the number of feature points.

Initially, a nominal (no disturbance) case was considered. With AKAZE, extreme accuracy changes were suppressed as the number of feature points approached 20,000, with error stabilizing around 0.2 px. Beyond 40,000 feature points, the required time increased despite no further effect on accuracy improvement, which tended to plateau around this point. AKAZE consistently outperformed SURF in accuracy, although SURF demonstrated a slightly faster computation time.

Now considering luminance anomalies, AKAZE exhibits a wider error distribution compared to SURF when the number of feature points is small. At lower feature point counts, errors can be as high as 6 px, improving to approximately 0.4 px under more optimal conditions due to the inherent randomness in map generation and feature point detection. Retaining high-quality feature points leads to excellent results with RANSAC, but errors can still be significant, depending on the quality of the detected feature points. These trends diminish as the number of feature points approaches 20,000, with the method showing limits to further accuracy improvement, stabilizing around 0.4 px.

Next, we discuss contrast fluctuations, which pose a significant challenge for feature point detection methods. These methods rely on changes in luminance to identify features, but datasets with varying luminance rates make it difficult to consistently identify the same feature points on a map. Even with over 20,000 feature points, an error of approximately 1 px persisted without further improvement. In this aspect, SURF performed better despite the significant variations, maintaining an average error of less than 1 px.

Despite using the same range of parameter changes, SURF detected more feature points due to its sensitivity to luminance changes, resulting in more pixels being identified as features. However, these additional feature points may not be valid compared to those detected in ordinary images. Elimination of such invalid feature points could be effective in improving accuracy in future studies.

Finally, we discuss radiation noise, which refers to damaged pixels in space exhibiting a brightness value of 0 or 255. While the actual occurrence probability is higher in space than on Earth, the frequency remains relatively low. Thus, the results are almost the same as those of nominal conditions, with accuracy improving as the number of feature points increases. The stabilization trend observed beyond 40,000 feature points mirrors that of nominal conditions. To evaluate the algorithm's robustness, further experiments are necessary to explore trends under increased occurrence probability scenarios.

Based on the overall results, it was concluded that AKAZE is a better feature-point detection method for map generation and self-position estimation. This is because the results for at least three types of disturbance datasets—brightness fluctuation and radiation noise—in addition to nominal conditions, which are disturbances that can occur in space, are superior to those of SURF. It is also important to note that the most frequently captured scenes in space are nominal images with no disturbance, and the superior performance using the dataset is also noteworthy.

Regarding parameter settings, it was observed that accuracy did not improve significantly beyond approximately 40,000 feature points for any disturbance type. Therefore, we conclude that parameter

$$akaze_{threshold} = 10^{-5} \tag{4}$$

which is the parameter used for acquiring 40,000 feature points, provides a sufficient balance between accuracy and speed.

4.2 Operation Under Low Computational Resource Environment

With limited resources, we examined the performance when using AKAZE_threshold in Eq. (4). The results are shown in the following Fig. 11.

In the Fig. 11, we confirmed that the algorithm works properly in all cases on a Raspberry Pi 4B. The parameters were set to detect 40,000 feature points in all tests, and the accuracy matched that achieved when running on the Mac Studio.

However, the results show that the minimum time required per image is 1400 ms, which is very long. As presented in Table 3, the Raspberry Pi 4B runs at 1.8 GHz. It should be noted that FPGAs, which are used in computers for space probes, operate at even lower frequencies, typically around 100 MHz. To enable the algorithm to run on

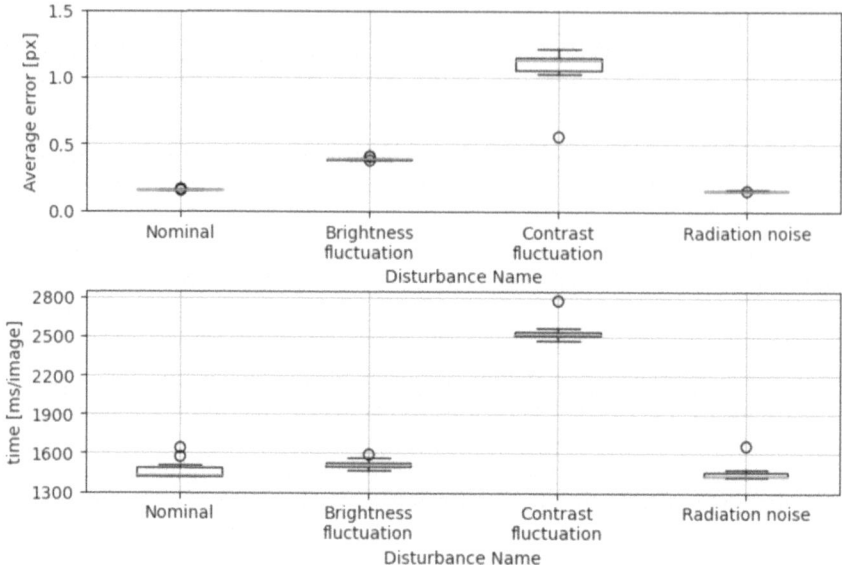

Fig. 11. Results of 10 iterations on Raspberry Pi 4B with a fixed AKAZE_threshold of 10^{-5}.

computers with one order of magnitude lower performance, it needs to be made lighter and faster while maintaining the same level of accuracy.

The following Fig. 12 shows the displacement of feature points using four types of datasets.

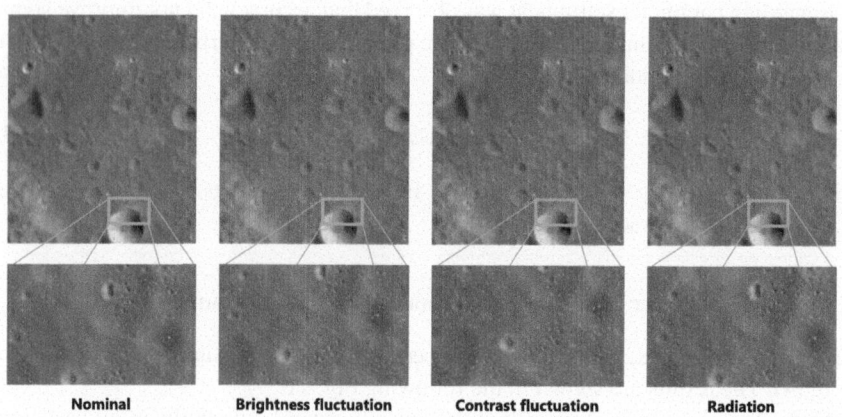

Fig. 12. Misalignment of how much the positions of the feature points have shifted. The green arrows indicate the vector from the original position to the incorrect position. (Color figure online)

Figure 12 shows that the errors were uniformly distributed among the nominal, brightness, and radiation noise. This is because a large number of feature points are used in each map-generation stage, which reduces the influence of feature points with large errors. In the contrast fluctuation, however, the amount of error is larger than in the others, and the error increases toward the bottom of the map. This is due to the accumulation of errors in estimating each position.

5 Conclusion

In this study, we demonstrated a method to achieve accurate self-position estimation in space for landings on unknown celestial bodies using a general-purpose feature point detection method. As a result, an error of 0.2 px ($= 6$ m) was achieved in the nominal case without disturbance, and 1 px ($= 30$ m) was achieved even in the presence of disturbance, particularly with AKAZE. These results meet the technical requirements for landing accuracy within the order of 100 m, as referenced in the SLIM project. In addition, we successfully identified optimal parameters and applied these values to increase speed and reduce the number of features while maintaining accuracy, particularly in environments with limited computing resources. This highlights the feasibility of balancing accuracy with a reduced processing load. Currently, we are exploring the potential of applying this method to space applications.

In the future, we plan to further improve the landing accuracy and tolerance to space-specific disturbances, as well as further reduce the required resources by making the algorithm lighter. We also aim to further increase processing speed and accuracy in low-computational-resource environments, such as the Raspberry Pi and space-grade FPGA.

References

1. Yano, H., et al.: Touchdown of the Hayabusa spacecraft at the Muses sea on Itokawa. Science 312(5778), 1350–1353 (2006)
2. Sei-ichiro, W., et al.: Hayabusa2 mission overview. Space Sci. Rev. 208, 3–16 (2017)
3. Satoshi, U. et al.: A study on guidance technique for precise lunar landing. Proceedings of 31st ISTS (2017)
4. JAXA. https://global.jaxa.jp/press/2024/01/20240125-1_e.html. Accessed 13 Apr 2024
5. Ishida, T. et al.: Optimization and accuracy evaluation of vision-based navigation with craters. Japan Soc. Aeronaut. Space Sci. (2017). (in Japanese)
6. Waragai, Y., et al.: Towards accurate spacecraft self-location estimation by eliminating close craters in camera-shot image. In: the 15th International Symposium on Artificial Intelligence, Robotics and Automation in Space (i-SAIRAS 2020) (2020)
7. Takadama, K. et al.: Artificial intelligence for spacecraft location estimation based on craters, pp. 160–189. Artificial Intelligence for Space (2023)
8. Itoh, M., Kamata, H.: A study of self-position estimation method by lunar explorer by selecting corresponding points utilizing gauss-newton method, Methods and Applications for Modeling and Simulation of Complex Systems. pp. 334–346. Springer Nature Singapore, Singapore (2022). https://doi.org/10.1007/978-981-19-9198-1_25
9. Miyazaki, H. et al.: A study on correspondence point search algorithm using bayesian estimation and its application to a self-position estimation for lunar explorer. In: Hassan, F., Sunar, N., Mohd Basri, M.A., Mahmud, M.S.A., Ishak, M.H.I., Mohamed Ali, M.S. (eds.) AsiaSim 2023. CCIS, vol. 1912, pp. 60–70. Springer, Singapore (2024). https://doi.org/10.1007/978-981-99-7243-2_6

10. Tanji, H., et al.: Study on variational Bayesian self-localization of lunar lander based on image matching. IEEJ Trans. Electron. Inf. Syst. Inst. Electr. Eng. Japan **144**(2), 110–120 (2024)
11. Miura, H., Kamata, H.: Study on visual navigation for high-accuracy landing on unknown celestial bodies. IEICE Proc. Ser. **76**(A2L-41), 60–63 (2023)
12. Miura, H., Kamata, H.: Study on generating a map by feature points for high accuracy landing on unknown celestial bodies. In: 2024 RISP International Workshop on Nonlinear Circuits, Communications and Signal Processing, pp. 488–491 (2024)
13. Alcantarilla, P.F. et al.: Fast explicit diffusion for accelerated features in nonlinear scale spaces. In: British Machine Vision Conference (BMVC), pp.13.1–13.11 (2013)
14. Leonardis, A. et al.: Surf: speeded up robust features. LNCS 3951, vol. Part I, pp. 404–417. ECCV 2006, Springer-Verlag Berlin Heidelberg 2006 (2006)
15. Ohara, S., et al.: Study on high precision matching between captured image with disturbance and map image assuming application to image navigation. In: Proceedings of 63rd Space Sciences and Technology Conference (2019). (In Japanese)
16. Cover, T., Hart, P.: Nearest neighbor pattern classification. IEEE Trans. Inf. Theory **13**(1), 21–27 (1967)
17. David, L.: Distinctive image features from scale-invariant keypoints. Int. J. Comput. Vision **60**, 91–110 (2004)
18. Fischler, M.A., et al.: Random sample consensus: a paradigm for model fitting with applications to image analysis and automated cartography. Commun. ACM **24**(6), 381–395 (1981)
19. Golub, G.H., et al.: Singular value decomposition and least squares solutions. Numerische Mathematik **14**(5), 403–420 (1970)

FedLGN: Federated Lightweight Graph Neural Network for Recommendation

Yuchun Tu, Xiao Song[✉], Songsong Liu, Yong Li, and Kaiqi Gong

School of Cyber Science and Technology, Beihang University, Beijing 100191, China
songxiao@buaa.edu.cn

Abstract. Recommender systems have become a key part of our daily digital experiences, powering personalized content discovery across various online platforms. However, the increasing privacy concerns and the distributed nature of user data pose significant challenges for traditional centralized recommendation models. Existing federated recommendation models often suffer from high computational complexity, which makes them unsuitable for client devices with limited computational and storage resources. To address these challenges, we propose a FedLGN framework in this work. FedLGN leverages the power of graph neural networks to capture complex user-item interactions, while the federated learning paradigm ensures that the model can be trained collaboratively across multiple client devices without the need to share sensitive user data. Meanwhile, our framework also enables efficient federated optimization, reducing the communication overhead and improving the overall system scalability. The experiment results demonstrate the convergence efficiency and communication cost of federated learning while maintaining recommendation accuracy, enhancing the practicality of the system. Additionally, we explore the recommendation performance of the proposed framework with different numbers of clients.

Keywords: Recommendation System · Federated Learning · Privacy Protections · Graph Neural Networks

1 Introduction

The rapid growth of digital data and the increasing demand for personalized recommendations have led to the widespread adoption of recommender systems across a variety of applications, from e-commerce and entertainment to social media and education [1]. Among the recommendation approaches, collaborative filtering (CF) has emerged as a prominent technique, leveraging user-item interaction data to uncover latent preferences and provide tailored suggestions [2]. Traditional centralized CF approaches, however, face significant challenges in modern data-driven environments. As user data becomes increasingly fragmented and distributed across different devices or organizations, directly centralizing the entire dataset for model training raises concerns over data privacy and incurs substantial communication overhead [3]. To address these issues, federated learning has been proposed as a promising paradigm for building collaborative filtering recommender systems [4].

© The Author(s), under exclusive license to Springer Nature Singapore Pte Ltd. 2024
S. Saito et al. (Eds.): AsiaSim 2024, CCIS 2170, pp. 217–227, 2024.
https://doi.org/10.1007/978-981-97-7225-4_17

Federated learning enables the collaborative training of machine learning models while keeping the user data decentralized and private [5]. In the context of recommender systems, federated CF frameworks involve multiple client devices or servers, each of which independently trains a local recommendation model on its own data. These local models are then aggregated into a global model through a federated optimization algorithm, such as FedAvg [6], without the need to share the underlying user data.

Existing works on federated CF have explored various approaches, including federated matrix factorization [7], federated deep learning-based recommenders [8], and graph neural network-based models in the federated setting [9]. These methods have demonstrated the ability to preserve user privacy while achieving competitive recommendation performance compared to their centralized counterparts.

Despite the promising potential of federated recommender systems, there are several challenges that need to be addressed. These include the development of efficient federated optimization algorithms, the handling of non-i.i.d. (independent and identically distributed) data distributions across clients, and the design of effective aggregation strategies to combine local models [3]. Addressing these challenges can lead to significant advancements in the field of federated recommender systems, paving the way for more personalized, privacy-preserving, and effective recommendation services. This work aims to address these challenges by proposing a novel Federated lightweight graph neural network (FedLGN) framework that leverages the strengths of both federated learning and collaborative filtering, while explicitly considering the computational and storage limitations of client devices or servers. Overall, our main contributions to this paper are listed as follows:

- We propose a federated lightweight graph neural network for item recommendation, which can enable heterogeneous data collaboration across different devices or users, improving the generalization performance of the model.
- We integrate the federated framework with LightGCN, which can leverage the graph structure data while preserving user privacy and enhancing the recommendation performance.
- We apply a lightweight graph neural network model with low computational complexity to the federated learning scenario. It significantly improves the convergence efficiency and communication cost of federated learning, while maintaining recommendation accuracy, enhancing the practicality of the system.

The structure of this paper is as follows. First, we introduce related works on the recommendation system and the federated collaborative filtering recommendation in Sect. 2. After that, Sect. 3 presents our proposed FedLGN framework and LightGCN. Section 4 demonstrates the recommendation performance of FedLGN for item recommendation under a federated setting. Finally, Sect. 5 concludes the paper by summarizing the key findings and proposing future research directions.

2 Preliminaries

2.1 Recommendation System

At the core of recommendation systems lies the challenge of accurately modeling user interests and predicting their preferences. Researchers have explored a variety of techniques, from collaborative filtering, which harnesses user-item interactions [10], to content-based approaches that analyze item attributes [11], and hybrid methods that combine multiple data sources [12]. The rapid advancements in machine learning, particularly in the fields of deep learning and graph neural networks, have further enhanced the capabilities of recommendation systems, allowing them to capture complex patterns and relationships within the data [13, 14].

Liang et. al. [15] uses variational autoencoders to model user-item interaction data, effectively capturing the complex latent factor structure and achieving excellent recommendation performance. GCMC [16] transforms the matrix completion problem into a node prediction problem in graph neural networks, effectively capturing the structural information between users and items. KGAT [17] utilizes graph attention mechanisms to model the entities and relations in the knowledge graph, significantly improving the performance and interpretability of the recommender system.

2.2 Federated Collaborative Filtering Recommenation

Federated learning has emerged as a promising paradigm to train machine learning models on decentralized data, while preserving user privacy and reducing communication overhead [4]. In the context of recommender systems, federated learning approaches have been explored to address the challenges of data fragmentation and privacy concerns. Several recent works have investigated federated recommendation models.

FedMF [7] proposed a federated matrix factorization framework, where each client trains a local model and the global model is aggregated using the FedAvg algorithm. FedRec [8] introduced a federated deep learning-based recommender system, utilizing client-side feature extraction and server-side model aggregation. FedNCF [18] proposed utilizing the state-of-the-art Neural Collaborative Filtering (NCF) [19] approach to extract clients' data feature under federated setting. Moreover, LightFR [20], PFedRec [21] and P-GCN [22] have attempted to design new federation architectures that consider the client's limited compute and storage resources to reduce client overhead. The Federated lightweight graph neural network (FedLGN) framework in this work builds upon these advancements, leveraging the simplicity and effectiveness of the Light-GCN model [23] in a federated learning setting to deliver a privacy-preserving and communication-efficient recommender system.

3 Methodology

3.1 Problem Definition

In recommender systems, user-item interaction data is often distributed across different devices or servers. Directly centralizing the entire dataset on a central server for training has many issues, such as data privacy concerns and high communication costs.

Federated learning provides a new paradigm to address this problem, where the global recommendation model is collaboratively trained while preserving the local data on each client.

In this work, we consider a cross-platform scenario where multiple participants, each holding their local dataset. We aim to train a universal recommendation model using a horizontal federated learning framework, without collecting clients' local datasets. The problem definition is as follows:

Federated Recommendation Definition. Given K Clients, each possessing their local dataset D_k containing user-item interaction information $M_k \in \mathbb{R}^{n \times m}$ and rating matrix $\mathcal{R} \in \mathbb{R}^{n \times m}$, where n is the total number of user at *Client$_i$* and m is the total number of item. Our objective is to train a global recommendation model $\mathcal{F}$ that leverages data from all Clients for recommendation without transmitting raw user data. Thus, the federated recommendation problem can be formulated as the following optimization problem:

$$\min_{F} \sum_{k=1}^{K} \mathcal{L}(F, D_k) \tag{1}$$

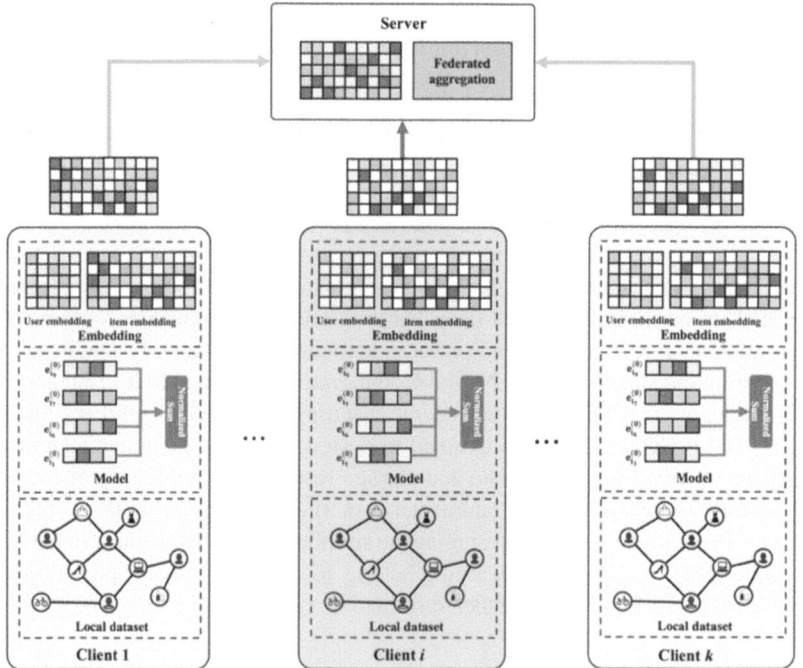

Fig.1. The proposed framework of federated lightweight graph neural network.

3.2 FedLGN Framework

In this section, we introduce the proposed framework of the federated lightweight graph neural network (FedLGN) which is shown in Fig. 1. Unlike centralized recommender systems, which typically require the collection of all user interaction data for centralized training, recommendation models in federated architectures can be jointly trained using user interaction data locally to multiple clients. This ensures that the local dataset is not transmitted elsewhere for the purpose of privacy protection. We consider the federated recommendation algorithm in cross-platform scenarios.

In FedLGN, the coordination server is responsible for initializing, collecting, processing, and sending all item embeddings. In the case of a federated aggregation, for example, the server needs to collect all item embedding from each client. Then operate collected all item embedding according to FedAvg and send new item embedding to each client. After each client obtains the new item embedding, it employs the local model to compute the rating matrix and the value of Bayesian Personalized Ranking loss to the backward optimizing model. The overall proposed FedLGN is shown in Algorithm 1. Due to not constructing a complete user-item interaction matrix, neither the server nor each client can infer more than their own observations.

Algorithm 1: Federated lightweight graph convolution network (FedLGN)

Federated aggregation:

  Initialization: total number of items m, the dimension of latent embedding d, the global training rounds T, generate local dataset and trainer D_i, LGNTrainer$_i$, $i \in Clients$, initialize item embedding $I_S \in \mathbb{R}^{m \times d}$ at Server

  for $epoch\ round\ t = 1, \cdots, T$ **do**

      for each client$_i \in C$ in parallel **do**

      $I_{client_i} \leftarrow$ Server sends item embedding I_S to $client_i$

      $I'_{client_i} \leftarrow$ LGNTrainer$_i(I_{client\ i}, D_i)$

      client$_i$ send its $I'_{client\ i}$ to server for aggregation

      $I_S \leftarrow \frac{1}{|Clients|} \sum_{i=1}^{|Clients|} I'_{client\ i}$ (server aggregation)

      end

  end

LGNTrainer(I, D):

  Initialization: total number of users n, initialize user embedding $U_i \in \mathbb{R}^{n \times d}$, the local epochs E

  downloading item embedding $I'_{client\ i}$ from Server

  for $epoch\ round\ e = 1, \cdots, E$ **do**

      $e_u \leftarrow AGG(e_v, e_u), e_v \in \mathcal{N}(u)$

      $r \leftarrow E_u \cdot E_v$

      Compute $\mathcal{L}oss$ with Eq.(5)

      AdamOptimizer($\mathcal{L}$, model)

  end

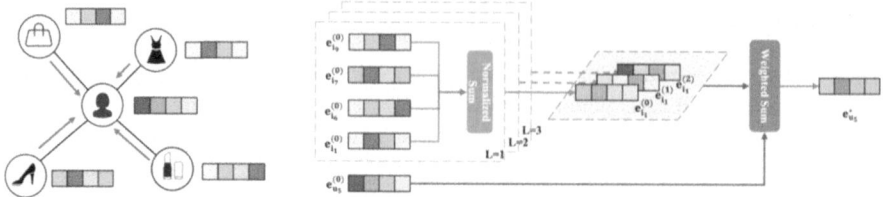

Fig. 2. The multi-layer graph convolution operation of LightGCN.

3.3 LightGCN

Among the state-of-the-art recommendation models, graph convolutional networks (GCNs) have demonstrated impressive performance by effectively capturing complex user-item interactions. Considering the model complexity and computational overhead, we adopt the LightGCN as our recommendation method in our federated framework. The model simplifies the GCN architecture while maintaining good recommendation accuracy. The key idea of LightGCN is to remove the non-linear activation functions and weight matrices commonly used in GCNs, retaining only the essential graph convolution operation. Thus, LightGCN has low complexity and computational overhead.

Specifically, LightGCN constructs a user-item bipartite graph $\mathcal{G} = (\mathcal{U} \cup \mathcal{V}, \mathcal{E})$, where $\mathcal{U}$ and $\mathcal{V}$ represent the sets of users and items, respectively, and $\mathcal{E}$ denotes the user-item interactions. The model learns the embedding representations of users and items through a multi-layer graph convolution process, as illustrated in Fig. 2. The core operation in Light-GCN is the graph convolution, which is formulated as

$$e_i^{l+1} = \sum_{j \in \mathcal{N}(i)} \frac{1}{\sqrt{|\mathcal{N}(i)||\mathcal{N}(j)|}} e_j^l \qquad (2)$$

where e_j^k represents the embedding of node i at the l layer, and $\mathcal{N}(i)$ denotes the set of neighboring nodes of i.

The final user and item embeddings are obtained by averaging the embeddings across all layers:

$$e_i = \frac{1}{K+1} \sum_{l=0}^{K} e_i^l \qquad (3)$$

Then, the rating between user and item are computed by dot product:

$$r_{uv} = e_u \cdot e_v \qquad (4)$$

The model is trained to minimize the Bayesian Personalized Ranking (BPR) loss:

$$\mathcal{L} = -\sum_{(u,i) \in \mathcal{O}^+} \log \sigma(r_{ui}) - \sum_{(u,j) \in \mathcal{O}^-} \log(1 - \sigma(r_{uj})) \qquad (5)$$

where $\mathcal{O}^+$ and $\mathcal{O}^-$ denote the sets of positive and negative user-item pairs, respectively, and r_{ui} represents the predicted score between user u and item i.

4 Experiments

4.1 Experimental Settings

Datasets: We run the proposed FedLGN on three real-world datasets: MovieLens-100K [24], MovieLens-1M [24], and Lastfm-2K [25], to evaluate its effectiveness. They are widely utilized for algorithm benchmarking and validation. The ML-1M dataset, sourced from the MovieLens platform, comprises one million ratings from 6,040 users on 3,952 movies, offering rich user-item interaction data and demographic information. Similarly, ML-100K, an earlier version of MovieLens, contains 100,000 ratings from 943 users on 1,682 movies, serving as a compact yet foundational dataset for collaborative filtering research. Additionally, Lastfm-2k originates from Last. fm and presents implicit feedback data, consisting of listening histories from 1,874 users, where interactions are binary (listened or not listened).

Table 1. Statistics of the utilized datasets.

Datasets	Users	Items	Relations	Sparsity
MovieLens-1M	6040	3592	1000182	95.81%
MovieLens -100k	943	1682	99981	94.70%
Lastfm-2k	1874	17612	92774	99.72%

In reference to [18], we excluded users with less than 10 interactions in each dataset. Table 1 shows the statistics of the utilized datasets. In our federated learning framework setting, we equally split the datasets into n clients (n = 5,10,20).

Implementation Details: In our proposed model, we adopt FedAvg as the federated aggregation method and utilize the Adam [26] optimizer. The learning rate is fixed at 0.001. Additionally, we set the latent dimension $d = 64$ and the neural network layers $L = 3$. The weight parameters in our proposed FedLGN are initialized by Gaussian initialization. Moreover, we sample positive and negative samples in a 1:1 ratio. To ensure that the model converges, we set the global training rounds T to 60 and the local epoch E to 10. All experiments in this work are run on the device (Ubuntu 20.04, Python3.8, Pytorch 1.11.0, RTX 4090 (24 GB),12 vCPU Intel(R) Xeon(R) Platinum 8352V CPU @ 2.10 GHz).

Evaluation Metrics: We evaluate the model performance with top-k Recall (recall@k) and Normalized Discounted Cumulative Gain (NDCG) metrics ($k = 10$) (ndcg@k). Recall measures the proportion of relevant items successfully recommended out of all items the user prefers, serving as a vital metric for evaluating recommendation systems. On the other hand, NDCG (Normalized Discounted Cumulative Gain) combines the relevance and ranking of recommended items, providing a comprehensive assessment of system performance, particularly in higher-ranked recommendations.

4.2 Experimental Results

In this section, we present the experimental results obtained from training and testing at three datasets. These results provide insights into the effectiveness and scalability of our proposed approach across different datasets and scenarios. LightGCN, a state-of-the-art GNN-based method at centralized scenarios, is selected to compare with our proposed method. In addition, we conduct experiments by dividing different numbers of clients to compare the proposed model's recommendation performance under different scale data.

Table 2. Recommendation performance of LightGCN and FedLGN on three datasets in testing.

Methods		MoviesLens-100K		MoviesLens-1M		Lastfm-2K	
		recall	ndcg	recall	ndcg	recall	ndcg
Light-GCN		0.2305	0.4262	0.1621	0.4027	0.1979	0.2367
FedLGN	Client-5	0.2087	0.3554	0.1577	0.3812	0.1665	0.2018
	Client-10	0.1586	0.3091	0.1480	0.3383	0.1489	0.1763
	Client-20	0.1329	0.2493	0.1438	0.3426	0.1389	0.1702

The experimental results are recorded in Table 2, which also shows the recommendation performance of different clients on each metric. The training loss and testing metrics of curves are shown in Fig. 3–4, respectively.

We further plot the training loss curves of LightGCN and FedLGN per 5 epochs on MovieLens and Lastfm-2k, which are shown in Fig. 3. As can be seen, both models demonstrate fast convergence due to their lightweight design. Moreover, the test data (including recall and ndcg) of experiments in different datasets is plotted as shown in the Fig. 4–5. In a federated architecture, they illustrate that the more clients there are on the same dataset, the worse the model recommendation. This statement is in line with our vision that a reduced amount of training data for the model leads to worse recommendation performance.

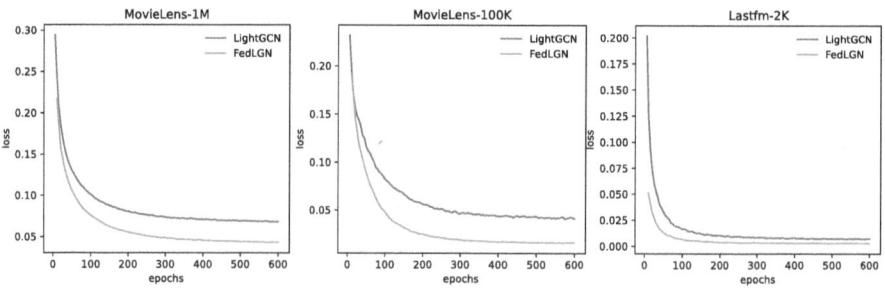

Fig. 3. Loss curves of LightGCN and FedLGN.

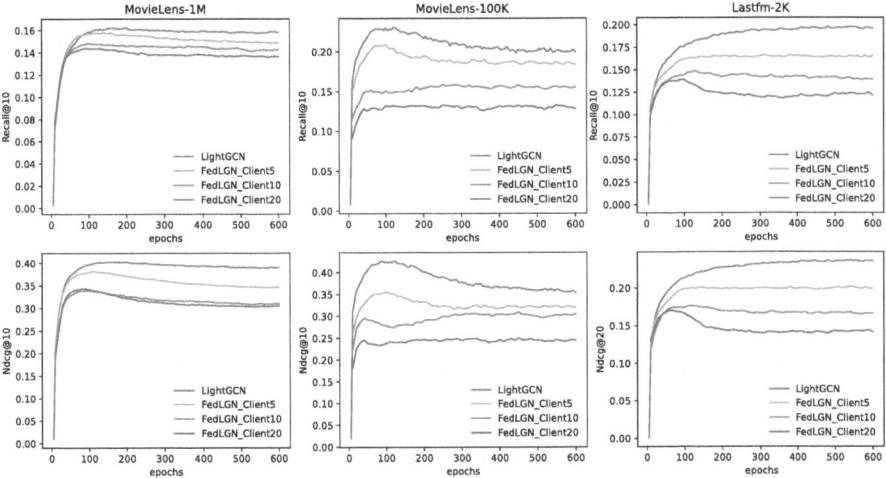

Fig. 4. Recall and ndcg curves of 3-layer LightGCN and FedLGN.

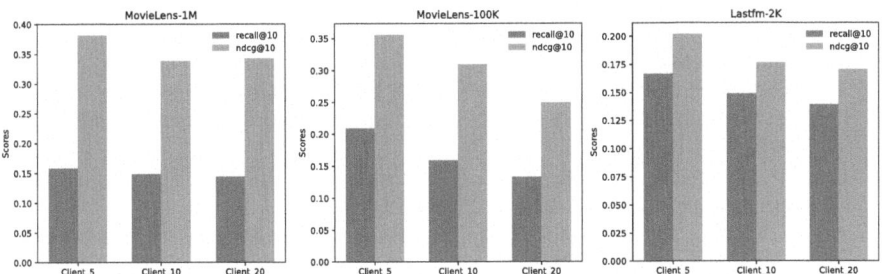

Fig. 5. Performance of LightGCN, FedLGN at different clients.

5 Conclusion

In this work, we present a federated lightweight graph neural network framework (FedLGN) for item recommendations, which coordinates servers and clients to jointly train recommendation models. Due to not transmitting raw user data to the server, our framework can protect user privacy. Meanwhile, the lightweight recommended model reduces computational complexity and enhances the communication efficiency of clients under a federated setting. Lastly, we explore the effect of different numbers of clients on model recommendation performance during training.

Despite the efficiency and effectiveness of our FedLGN, our work has a lot to explore in the future. Firstly, we adopt the FedAvg method as our federated aggregation method, which aggregates all client item embeddings equally. Thus, it is essential to design more advanced federated aggregation methods. After that, negative samples are randomly sampled to train the recommendation model. Considering the impact of samples, exploring new sampling methods is also a vital direction to improve recommendation performance.

Acknowledgments. This work was supported by the National Key Research and Development Program of China under Grant 2023YFB3308200 and Beijing Natural Science Foundation (L233005).

Disclosure of Interests. No potential conflict of interest was reported by the author(s).

References

1. Li, Z., et al.: FedCORE: federated learning for cross-organization recommendation ecosystem. IEEE Trans. Knowl. Data Eng. (2024)
2. Adomavicius, G., Tuzhilin, A.: Toward the next generation of recommender systems: A survey of the state-of-the-art and possible extensions. IEEE Trans. Knowl. Data Eng. **17**(6), 734–749 (2005)
3. Li, T., Sahu, A.K., Talwalkar, A., Smith, V.: Federated learning: challenges, methods, and future directions. IEEE Sig. Process. Mag. **37**(3), 50–60 (2020)
4. Kairouz, P., et al.: Advances and open problems in federated learning. Found. Trends® Mach. Learn. **14**(1–2), 1–210 (2021)
5. McMahan, B., Moore, E., Ramage, D., Hampson, S., y Arcas, B.A.: Communication-efficient learning of deep networks from decentralized data. Artif. Intell. Stat. (2017)
6. McMahan, H.B., Moore, E., Ramage, D., y Arcas, B.A.: Federated learning of deep networks using model averaging. arXiv preprint arXiv:1602.05629 2, 2 (2016)
7. Chai, D., Wang, L., Chen, K., Yang, Q.: Secure federated matrix factorization. IEEE Intell. Syst. **36**(5), 11–20 (2020)
8. Jiang, Y., Konecny, J., Rush, K., Kannan, S.: Improving federated learning personalization via model agnostic meta learning. arXiv preprint arXiv:1909.12488 (2019)
9. Wu, C., Wu, F., Cao, Y., Huang, Y., Xie, X.: Fedgnn: Federated graph neural network for privacy-preserving recommendation. arXiv preprint arXiv:2102.04925 (2021)
10. Sarwar, B., Karypis, G., Konstan, J., & Riedl, J.: Item-based collaborative filtering recommendation algorithms. In: Proceedings of the 10th International Conference on World Wide Web (2001)
11. Pazzani, M.J., Billsus, D.: Content-based recommendation systems. The adaptive web: methods and strategies of web personalization (2007)
12. Burke, R.: Hybrid recommender systems: Survey and experiments. User Model. User-Adapt. Interact. **12**, 331–370 (2002)
13. Zhang, S., Yao, L., Sun, A., Tay, Y.: Deep learning based recommender system: a survey and new perspectives. ACM Comput. Surv. (CSUR) **52**(1), 1–38 (2019)
14. Wang, X., He, X., Wang, M., Feng, F., Chua, T.S.: Neural graph collaborative filtering. In: Proceedings of the 42nd International ACM SIGIR Conference on Research and Development in Information RetrievalIn (2019)
15. Liang, D., Krishnan, R.G., Hoffman, M.D., Jebara, T.: Variational autoencoders for collaborative filtering. In: Proceedings of the 2018 World Wide Web conferencein (2018)
16. Rianne van den, B., Kipf, T.N., Welling, M.: Graph convolutional matrix completion. arXiv preprint arXiv:1706.02263 (2017)
17. Wang, X., He, X., Cao, Y., Liu, M., Chua, T.S..: Kgat: knowledge graph attention network for recommendation. In: Proceedings of the 25th ACM SIGKDD International Conference on Knowledge Discovery & Data Mining (2019)
18. Perifanis, V., Efraimidis, P.S.: Federated neural collaborative filtering. Knowl.-Based Syst. **242**, 108441 (2022)

19. He, X., Liao, L., Zhang, H., Nie, L., Hu, X., Chua, T.S.: Neural collaborative filtering. In: Proceedings of the 26th International Conference on World Wide Web (2017)
20. Zhang, H., Luo, F., Wu, J., He, X., Li, Y.: LightFR: lightweight federated recommendation with privacy-preserving matrix factorization. ACM Trans. Inf. Syst. 41(4), 1–28 (2023)
21. Zhang, C., Long, G., Zhou, T., Yan, P., Zhang, Z., Zhang, C., Yang, B.: Dual personalization on federated recommendation. arXiv preprint arXiv:2301.08143 (2023)
22. Hu, P., Lin, Z., Pan, W., Yang, Q., Peng, X., Ming, Z.: Privacy-preserving graph convolution network for federated item recommendation. Artif. Intell. 324, 103996 (2023)
23. He, X., Deng, K., Wang, X., Li, Y., Zhang, Y., Wang, M.: Lightgcn: Simplifying and powering graph convolution network for recommendation. In: Proceedings of the 43rd International ACM SIGIR Conference on Research and Development in Information Retrieval (2020)
24. Harper, F.M., Konstan, J.A.: The movielens datasets: History and context. ACM Trans. Interact. Intell. Syst. (tiis) 5(4), 1–19 (2015)
25. Cantador, I, Brusilovsky, P., Kuflik, T.: 2nd Workshop on Information Heterogeneity and Fusion in Recommender Systems (HetRec). Proc. RecSys
26. Hu, W., Miyato, T., Tokui, S., Matsumoto, E., Sugiyama, M.: Learning discrete representations via information maximizing self-augmented training. In: International Conference on Machine Learning (2017)

Research on Digital Twin Technology for Converter Steelmaking

Xianghui Meng[1], Tan Li[1(✉)], Jin Guo[1], Jingyu Zhu[1], Weining Song[2], Yalan Xing[3], Wu Lv[4], Mu Gu[5], Meng Chen[1], Nanjiang Chen[1], Gang Wu[1], and Haonan Bu[1]

[1] Nanchang University, Nanchang 330031, China
litan@ncu.edu.cn
[2] East China University of Technology, Nanchang, Jiangxi, China
[3] Beijing University of Aeronautics and Astronautics, Beijing, China
[4] Northeastern University, Shenyang, Liaoning, China
[5] China Aerospace Science and Industry Corporation Limited, Beijing, China

Abstract. Digital twin technology integrates data and mechanistic models of the actual converter steelmaking process for full process simulation, improving efficiency, ensuring safety and optimizing operation. As a effective soft-measurment, digital twin not only monitors furnace conditions in real-time steelmaking, but also synchronously simulates extreme conditions and respond to safety hazards. Mechanism-based and data-based modeling is proposed as well as functional-level applications. Mechanism-based modeling starts from the internal physical process of the converter, and comprehensively restores the steelmaking process by combining a fine mechanism model with a data-driven approach. Innovations in data acquisition technology focus on the efficient integration and processing of real-time, multi-source data, providing a solid foundation for the digital twin system. In terms of functional applications, the digital twin technology has demonstrated excellent performance in actual Converter Steelmaking, including improving efficiency, controlling quality and maximizing resource utilization.

Keywords: Converter Steelmaking · Digital Twins · Data Driven

1 Introduction

Conventional converter steelmaking technology suffers from a number of problems, such as manual operation dependency, rigid control systems, unpredictable safety hazards and inefficient resource utilization. However, the emergence of digital twin technology provides a revolutionary solution to these problems. By building virtual models for real-time monitoring, data analysis and prediction, digital twin technology optimizes production efficiency, improves product quality, and reduces energy consumption and production costs, bringing a new way of production to the converter steelmaking industry and driving the modernization and transformation of the industry.

S. Saito et al. (Eds.): AsiaSim 2024, CCIS 2170, pp. 228–243, 2024.
https://doi.org/10.1007/978-981-97-7225-4_18

2 Digital Twin Technology System for Converter Steelmaking

Converter steelmaking is costly, so only a limited number of measurements, mainly KR, TSC and TSO measurements, are usually performed to control costs. However, these measurements cannot fully reflect the complex states and changes inside the converter. In order to achieve fine control, it is necessary to introduce a converter steelmaking digital twin as a closed-loop control to correct and guide it, as shown in Fig. 1. The technology simulates and predicts the converter steelmaking process based on historical data, providing comprehensive guidance from the pre-blowing stage to the steel production stage. Before the start of blowing, the digital twin performs optimization calculations based on raw material conditions and steel output requirements to provide dosing guidance; in the middle of blowing, it guides the blowing operation of online furnaces based on decarburization, dephosphorization, and slagging control, and provides operational guidance through real-time data analysis; in the late stage of blowing, based on the results of the TSC measurements, the digital twin adjusts the blowing parameters to determine the optimal point of stopping the blowing process to ensure the quality of the products; in the In the process of steel output, according to the composition of steel at the end of blowing and the target composition, the digital twin optimizes and calculates the optimal additive amount of alloying materials and provides dosing guidance. By simulating and analyzing the data, the digital twin provides important guidance for the converter steelmaking process, improving production efficiency and product quality.

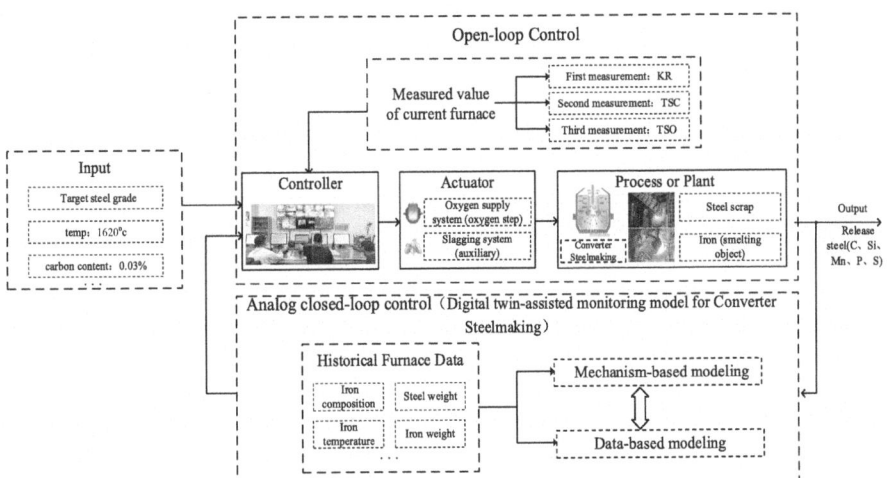

Fig. 1. Digital twins for converter steelmaking as analog closed-loop control.

The converter production process controls the carbon content and temperature of the molten steel in the bath at the end of the blowing process in order to achieve the desired output standard and to ensure the quality of the steel. This process relies on the physical heat of the molten steel in the bath and the heat generated by chemical reactions, and it is necessary to ensure that the carbon content, the content of each metal element and the temperature at the end of the oxygen blowing simultaneously meet the requirements for steel production [1]. The detection of carbon content and temperature in the converter blowing process is divided into traditional sensor-based detection methods, spectral radiation detection, flame image processing and recognition detection, and process data-driven soft measurement methods from the point of view of the measurement principle [2], as shown in Table 1, which compares the advantages and disadvantages of the four methods of carbon temperature/carbon content prediction.

Among them, the process data-driven soft measurement method can make full use of a large amount of real-time data generated in the steelmaking process of the converter, such as oxygen blowing, molten pool temperature, carbon content, etc. These data can comprehensively reflect the state and changes in the steelmaking process, which helps to establish accurate models for endpoint prediction and control. Compared with other methods, this method has higher information and comprehensiveness, which can guide production practice more effectively, and the data cleaning middleware has already cleaned and saved the historical furnace data, so it is more convenient and efficient to choose the process data-driven soft measurement method. Figure 2 shows the endpoint prediction and analysis model using the process data-driven soft measurement method combined with the mechanism model, in which the mechanism model based on thermodynamics and dynamics can provide basic parameter prediction; the process data-driven soft measurement method can obtain the latest data information in time, and make predictions and adjustments according to the real-time situation, to ensure that the parameters of the furnace operate stably within a reasonable range, and to ensure that production meets the expected goals. Literature [3–11] based on a variety of machine learning methods, using different kinds of data processing and analyzing means, respectively, collect the information of converter smelting starting point, smelting end point and so on as model inputs, and establish their respective prediction models. The machine learning methods do not make an in-depth study on the physical and chemical changes of the converter steelmaking process, but construct the prediction model based on massive on-site production data, and the accuracy is very considerable.

The digital twin system of traditional converter steelmaking technology includes data-based modeling, mechanism-based modeling and digital twin functional layer. First, data-based modeling ensures data reliability through high-performance sensor data acquisition and data cleaning middleware, providing a reliable foundation for subsequent modeling and analysis. Second, mechanism-based modeling uses modular design to build models for material analysis, auxiliary monitoring and alloy cost optimization, and combines the mechanism model and data-driven approach to simulate and predict the steelmaking process. Finally, the digital twin functional layer application includes the functions of auxiliary monitoring, material analysis, cost optimization and endpoint analysis of the converter steelmaking process, realizing real-time monitoring and accurate analysis through historical data and intelligent learning to improve production efficiency and the accuracy of management decisions.

Table 1. Comparison of advantages and disadvantages of four carbon temperature/carbon content prediction methods.

Forecasting methodology	advantages and disadvantages
Traditional testing methods	Traditional contact monitoring methods such as sub-lance control and furnace gas analysis[12] are advantageous in converter steelmaking, providing direct measurement of the terminal carbon temperature, increasing the hit rate of predictions, and reducing the number of furnace inversions and labor experience costs. However, the point measurements of the sub-lance technique do not provide continuous information, have limited hit rates, high equipment and maintenance costs, and are not suitable for scenarios such as small to medium sized converters
Spectral radiation detection methods	The spectroradiometric monitoring method realizes real-time on-line monitoring of the flame spectrum at the furnace opening [13], which improves the prediction accuracy and stability, and at the same time allows monitoring of key parameters in the molten steel. However, it requires high-cost specialized instruments and equipments, which are less suitable for small and medium-sized converters, and the operators also need high technical level and professional knowledge
Flame image processing and recognition detection method	The use of non-contact measurement will not interfere with the converter operation, and has the advantages of low cost, high accuracy, and not affected by human subjective factors. However, due to the instability of the flame combustion process and the complexity of the flame image, there are certain difficulties in the characterization and image analysis, which require further research and technical improvement
Process data driven soft measurement approach	Process data-driven soft measurement methods utilize a large amount of data to build models that reflect nonlinear relationships, are real-time and well-maintained, and are useful for predicting the carbon content or temperature at the end point of converter steelmaking [14]. However, it requires reasonable selection and pre-processing of auxiliary variables and online calibration to ensure accuracy, which is time and resource consuming

3   Digital Twin Technology System for Converter Steelmaking

3.1   Mechanism Based Modeling

In complex industrial production such as converter steelmaking, it is crucial to establish accurate and reliable digital twin models. However, the traditional single modeling approach has limitations, so combining three modeling approaches based on mechanism, data and knowledge becomes an effective way. In this paper, the steelmaking process is decomposed into key models such as material analysis, auxiliary monitoring, endpoint prediction and alloy cost optimization model, and by constructing and realizing these models, data information of each link can be systematically processed to realize data-driven intelligent control. This approach can effectively improve the productivity, quality stability and economy of the converter steelmaking process.

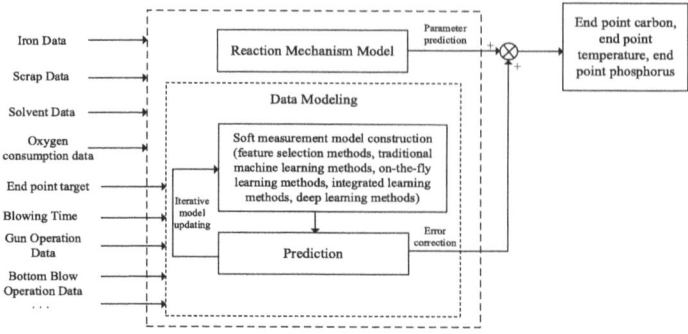

Fig. 2. Endpoint prediction analysis.

Material Analysis Model. The oxidation of impurity elements during converter steel-making is a complex reaction involving several variables coupled with each other. The rate of oxidation of these elements is influenced by a number of factors, including the decomposition pressure of the oxides of each element, the percentage of the element in the molten iron, and the melt pool temperature. This multivariate coupling makes it difficult to accurately measure and model this process [16]. For this complex oxidation reaction process, it is often necessary to combine multiple modeling approaches for analysis and prediction. On the one hand, a mechanism-based modeling approach can be used, i.e., the basic model of the oxidation reaction, including the oxidation rate equations of each element, can be established based on the principles of chemical reactions and the knowledge of thermodynamics. On the other hand, since data and operational variables in the actual production affect the oxidation process, it is also necessary to combine the data-driven modeling approach to build a data model through historical data analysis to reveal the relationship between the variables. The combination of mechanistic and data models can be used to improve the accuracy and applicability of the model by introducing correction terms or fusion methods. For example, the data model can be used to optimize and calibrate the parameters of the mechanistic model, or the advantages of both can be combined using model integration to achieve more accurate prediction and control of the oxidation process. This multi-model integration approach can effectively address the complex and variable problem of oxidation of impurity elements in converter steelmaking and improve the stability and efficiency of the production process.

In the static model of converter steelmaking, elemental oxidation reactions usually involve four elements, carbon, silicon, manganese, and phosphorus, with the following reaction equations [17]:

$$\begin{cases} [Si] + O_2 = (SiO_2) \\ [Mn] + \frac{1}{2}O_2 = (MnO) \\ [P] + \frac{5}{4}O_2 = (PO_{2.5}) \\ [C] + \frac{1}{2}O_2 = (CO) \end{cases} \tag{1}$$

These elemental oxidation reactions are very important chemical reactions in the converter steelmaking process, which directly affect the metal element content, oxygen utilization rate, furnace temperature and other key indicators in the steelmaking process. By considering these reactions in the static model, the dynamic process of elemental oxidation in the furnace can be simulated and predicted, which helps to optimize the steelmaking process and control the reaction conditions in the furnace, thus improving the efficiency of steelmaking and product quality. In the static model, the three most important parameters are: total oxygen blowing, slagging agent addition, and heat balance material addition, and the calculation formulas are as follows:

$$O = \sum_{i=1}^{4} (S_i - T_i) \times A \times m_i \times 0.7 \tag{2}$$

$$A_{\text{slag}} = (R \times (S_2 - T_2) \times A \times 2.84 + A_{\text{Mg}} \times (q_{\text{Sio}_2} - R \times q_{cao})) / (p_{cao} - R \times p_{\text{Sio}_2})$$

$$\tag{3}$$

$$A_{Heat} = (H_{iron} + H_e + H_f - H_{steel} - H_g - H_s - H_r - H_m) / \triangle H_{heat} \tag{4}$$

where, O is the total oxygen blowing; S_i is the in-furnace content of carbon, silicon, manganese and phosphorus respectively; T_i is the out-of-furnace content of the four elements; A is the in-furnace weight of molten iron; m_i is the relative molecular mass required for the four elements to consume one oxygen molecule; A_{slag} is the amount of slagging agent added; q_{sio_2} is the silica content of magnesium material; P_{sio_2} is the silica content of slagging material; R is the target alkalinity; H_e is the elemental oxidizing heat; H_{steel} is the physical heat of molten steel; H_s is the physical heat of soot; H_r is the heat loss; H_m is the physical heat of slagging; and ΔH_{heat} is the physical heat of slagging as the amount of heat absorbed by each kilogram of thermal equilibrium material.

In actual steelmaking, the current furnace data information is first calculated by the converter blowing mechanism model of material balance and heat balance to get the content prediction value of each material, but in the actual converter steelmaking production process, the model parameters, such as oxygen utilization rate, lining erosion, heat loss, etc., are difficult to model, and it is very difficult to make accurate prediction only by the constructed mechanism model. For this reason, the historical furnace data can be taken out through the data cleaning middleware, and methods such as statistics or machine learning can be used to establish a mathematical relationship model between the oxidation rate of each element and key parameters, as shown in Fig. 3. Afterwards, by using the prediction results in the data model as the input or correction term of the mechanism model, the output of the mechanism model can be corrected to improve the accuracy and prediction ability of the model; or the mechanism model and the data model can be used as the components of the integrated model, and the advantages of the two are comprehensively utilized, such as model fusion, model stacking, etc., to obtain more reliable prediction results; or the prediction results of the data model can be used as the basis of real-time monitoring, and compare it with the actual data in order to achieve dynamic optimization and stability of the model.

Auxiliary Monitoring Model. The auxiliary monitoring model for converter steelmaking is a system designed to achieve real-time monitoring and prediction of the furnace

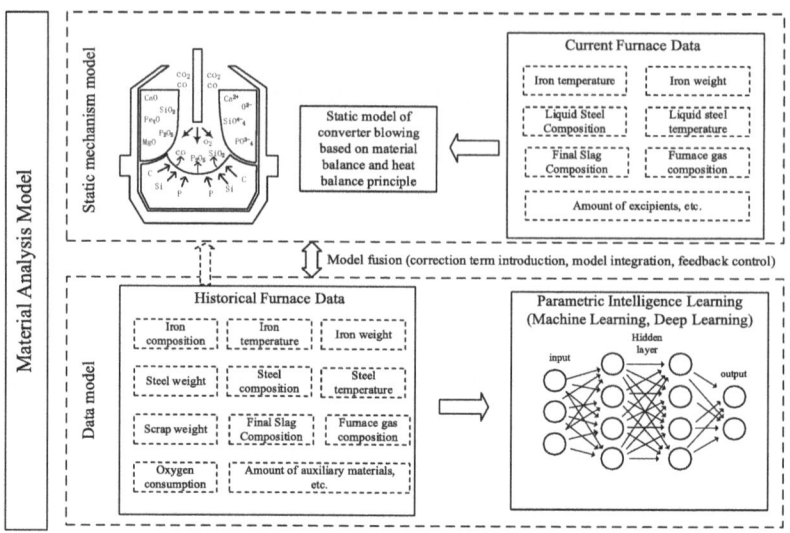

Fig. 3. Material Analysis Model.

process, so as to guide the operation, optimize the process, and improve the productivity and product quality. This model covers a number of aspects, including mechanism model, knowledge planning, data model, etc. These components work together to form a complete monitoring and optimization system, as shown in Fig. 4. The mechanism model is one of the core components of the auxiliary monitoring model. It is based on the principles of reaction thermodynamics and macroscopic kinetics, and models and predicts key parameters such as redox reactions, carbon content changes, temperature distribution, etc. in converter steelmaking. By describing various chemical reactions and material transfer processes in the furnace through mathematical modeling, the trend of various indicators in the furnace can be predicted more accurately. However, due to the mutual coupling of variables, it is not possible to model and analyze all variable parameters, such as the impact pit area, slagging agent melting, etc. It is difficult to model, and the process parameter indicators usually come from the experience of experts, so only by integrating the mechanism model, the data model and the knowledge model can we monitor the current reaction state of the furnace in a comprehensive way and give reasonable guidance.

Data models use historical data and real-time data to learn and analyze, and construct prediction models for the current furnace. These models can be based on statistical methods, machine learning algorithms and other technologies to predict and optimize changes in various parameters. For example, learning the pattern of temperature changes in the furnace based on historical data predicts the temperature trend for the current furnace and provides adjustment suggestions. On the other hand, the knowledge planning part involves guiding and monitoring the current furnace situation. It includes functions such as detecting and recognizing abnormal status, process parameter adjustment suggestions, and combustion status monitoring. By setting reasonable monitoring indexes and judgment conditions, the system can monitor the furnace status in real time, provide

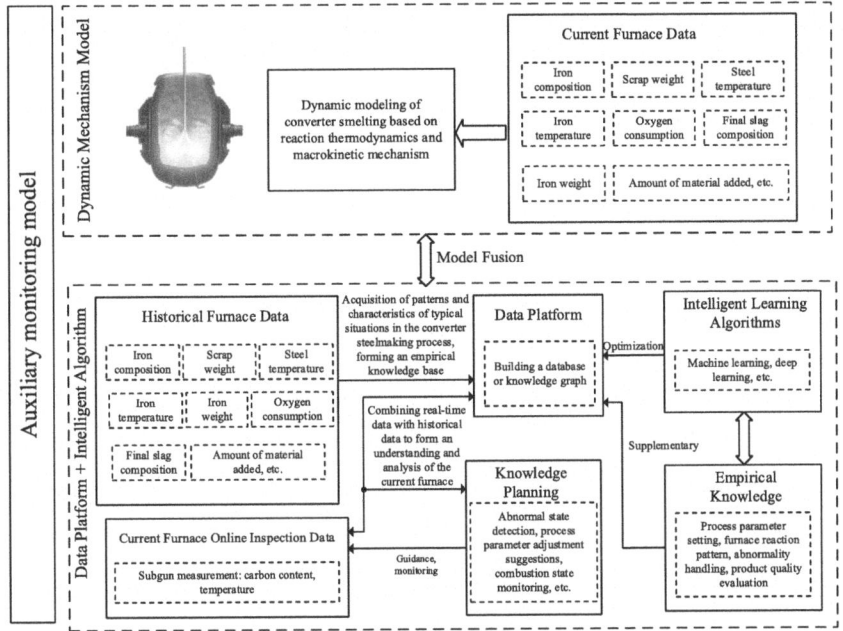

Fig. 4. Auxiliary monitoring model.

operation suggestions and abnormal handling solutions to ensure the stability and safety of the production process.

Finally, the auxiliary monitoring model will integrate the mechanism model, knowledge model and data model by means of intelligent learning technology and knowledge fusion, and comprehensively consider the influence of various factors on the process in the furnace. This fusion can improve the prediction accuracy and practicality of the model, making the system more intelligent and adaptable, and better able to cope with complex production situations and changes. Taken together, the auxiliary monitoring model for converter steelmaking realizes comprehensive monitoring, prediction and optimization of the process in the furnace by combining a variety of technical means and modeling methods, which provides important support and guidance for the production process.

Alloy Cost Analysis Model. In converter steelmaking, a wide range of steel grades are commonly encountered, each with unique process requirements and performance criteria. These differences lead to significant differences in the consumption of steel feed, energy and alloys in the smelting process. In particular, steel feed consumption, as the dominant part of steelmaking costs, varies significantly from process to process. From the perspective of steel grades, there are large differences in steel consumption, oxygen consumption, limestone consumption, etc. between different steel grades, and the impact of these differences on the overall cost of steelmaking can not be ignored. Converter steelmaking cost analysis model can be achieved through a combination of mechanism modeling and knowledge modeling, for the mechanism model, the Lagrange multiplier method to seek the conditional extreme value of the modeling, the objective function is the minimum value of the cost of deoxidation alloying, the constraints can include the range of composition of the steel range limit, the material limit, alloying to add the total amount of formulae for reference are as follows:

$$\min C_{total} = c_1 x_1 + c_2 x_2 + \ldots + c_n x_n \tag{5}$$

$$y_{min,i} \leq \sum_{j=1}^{n} x_j \cdot c_{i,j} \leq y_{max,i}, i = 1, 2, \ldots m \tag{6}$$

$$0 \leq x_j \leq X_{supply,j}, j = 1, 2, \ldots n \tag{7}$$

$$0 < x_1 + x_2 + \ldots + x_n \leq X_{total} \tag{8}$$

where Eq. (5) for the deoxidation alloying cost of the lowest construction of the objective function; Eq. (6) for the range of steel composition limiting constraints, $c_{i,j}$ for the j in the alloy of the i component content; Eq. (7) for the material limiting constraints, $X_{supply,j}$ is the maximum supply of each alloy; Eq. (8) for the total amount of constraints. For the alloying cost model, expert knowledge is also particularly important. Literature [18, 19] achieves the cost-optimal alloy proportioning scheme through cost calculation and classification of steel grades, which can be added as a correction term to the constructed mechanism model to jointly realize the analysis of alloy cost for converter steelmaking and optimize the proportioning strategy.

3.2 Data Based Modeling

Data is the foundation of digital twin, and its accuracy and practicability will directly affect the implementation effect of subsequent model development and application services. The data in the steel production site covers real-time acquisition data, quality inspection data, production operation data, audio and video data, etc. The data sources and structures are diverse and complex. There are many challenges in data acquisition, transmission, and storage, such as Fisher OJ et al. [20] defined the data aspect of digital twin as one of the main challenges in developing simulation models.

Data Acquisition. Data acquisition is a very critical and complex part of the converter steelmaking process, which is related to the monitoring, control and optimization of the whole production process. The quality, quantity, type and other characteristics of data acquisition will directly affect the fidelity of the digital twin. For the real-time data of the converter steelmaking site as well as database, logs and other multi-source heterogeneous data, such as iron data, scrap data, smelting process volume data, melting event data, etc. are generally obtained through the common data acquisition methods in the traditional process industry, including programmable logic controllers (PLC), distributed control systems (DCS), and supervisory control and data acquisition (SCADA) [21], etc., as shown in Fig. 5. In order to more accurately obtain the number of process parameters in converter steelmaking, the use of various types of intelligent sensors will also improve the quality of data acquisition, such as new electrochemical sensors for the determination of phosphorus content in iron water [22].

In digital twin systems, although data collection techniques and unstructured data can provide a wealth of information, there are a number of problems associated with the direct use of such data. First, these data often suffer from noise, missing values, errors, or inconsistencies, which can lead to inaccuracies and unreliability in models. Second, the characteristics of unstructured data make it difficult to apply directly to models and require processing and transformation before they can be effectively utilized by models. In addition, the way data are collected may introduce repetitive, redundant or invalid information, which increases the difficulty of data processing and analysis. Therefore, these drawbacks need to be addressed by data cleaning middleware.

Data Cleaning Middleware.   The data cleaning middleware is a system component responsible for cleaning, organizing, distributing and storing the collected data to obtain the training data required by the model. Among them, the development of interface and specification link is an important step to ensure that data from different data sources are transmitted and exchanged according to a unified format and communication protocol. The data storage module unifies the management of data from different data sources,

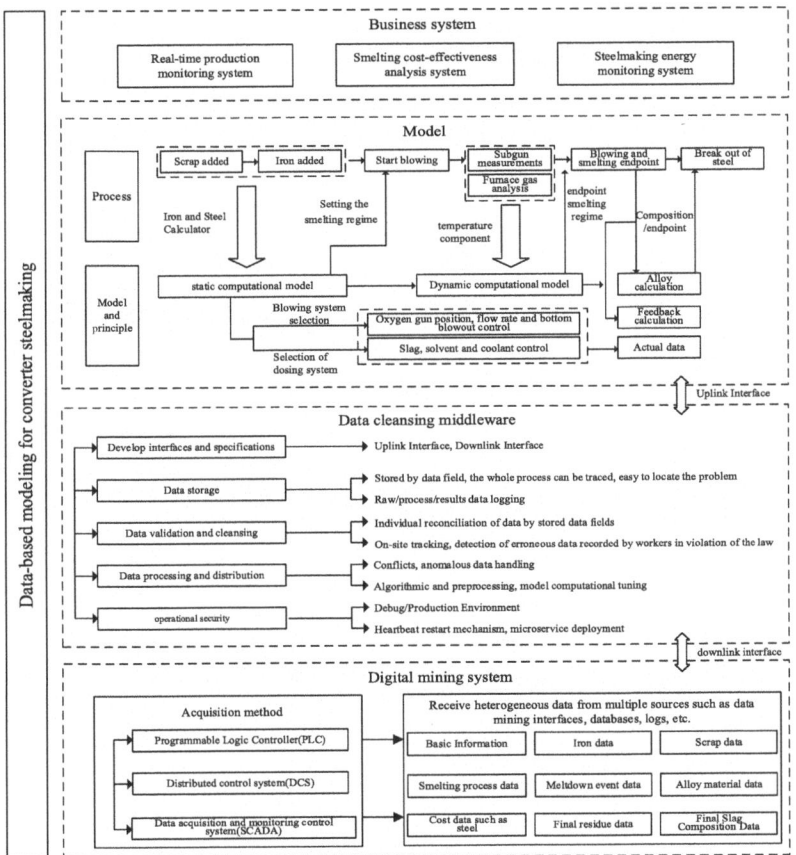

Fig. 5. Data-based modeling for converter steelmaking.

including the selection of appropriate databases or data warehouses for storage and the establishment of data table structures to guarantee the security and reliability of the data. The data verification and cleansing link is dedicated to ensuring data accuracy and consistency, including data validation, de-duplication, error correction, filling of missing values and other operations to improve data quality. The data processing and distribution module processes, calculates, and analyzes the cleaned and verified data, and distributes the results to business systems or users to support data application and decision-making. The operation guarantee module is responsible for monitoring the data processing process, recording operation logs, handling abnormal situations, and taking appropriate measures to ensure system stability. Key technologies such as heartbeat restart mechanism and microservice deployment are applied to automatically restart abnormal services and improve system stability, availability and flexibility.

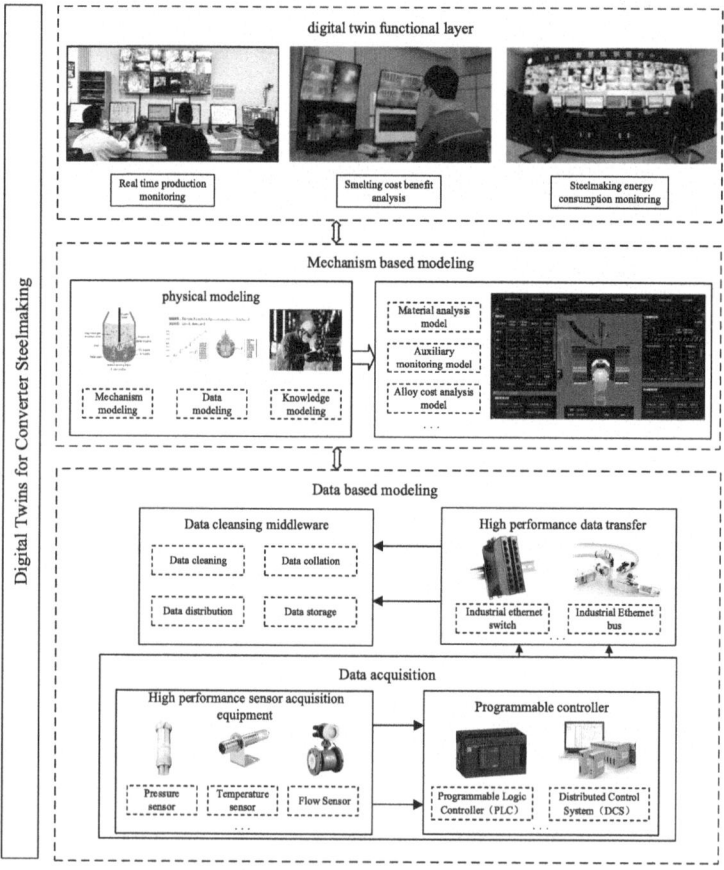

Fig. 6. Digital Twins for Converter Steelmaking.

4 Digital Twin Functional Layer for Converter Steelmaking

The upper layer applications of the converter steelmaking digital twin include systems for real-time production monitoring, smelting cost-benefit analysis, and steelmaking energy consumption monitoring, see Fig. 6. These systems are built on data-based modeling and model-based modeling logic. A reliable data foundation is constructed through the underlying data collection and cleaning middleware, while model-driven and data-driven approaches are fused to form models for material analysis, auxiliary monitoring, and alloy cost optimization. These underlying logics provide a reliable data support and modeling foundation for the upper layer applications, enabling systems such as real-time production monitoring, smelting cost-benefit analysis, and steelmaking energy consumption monitoring to be implemented and operated on top of them, providing comprehensive monitoring, analysis, and optimization support for the steelmaking process.

The real-time production monitoring system is built on top of underlying predictive models such as material analysis models and auxiliary monitoring models. These models provide real-time access to key data in the steelmaking process, such as slag, molten iron, scrap and oxygen usage, as shown in Fig. 7. The material analysis model predicts and analyzes the composition of raw materials to provide guidance for the production process; the auxiliary monitoring model monitors the internal state of the converter and the condition of the furnace lining refractory material. These models form the core of the real-time monitoring system, providing real-time data support and predictive capabilities. The data is processed and analyzed and mapped onto a geometric model for real-time simulation, which visualizes the internal condition of the converter and helps operators understand and master changes in the production process. By comparing with the simulation model, potential problems or anomalies such as wear and tear of the refractory lining or abnormal temperature fluctuations can be detected, and adjustment

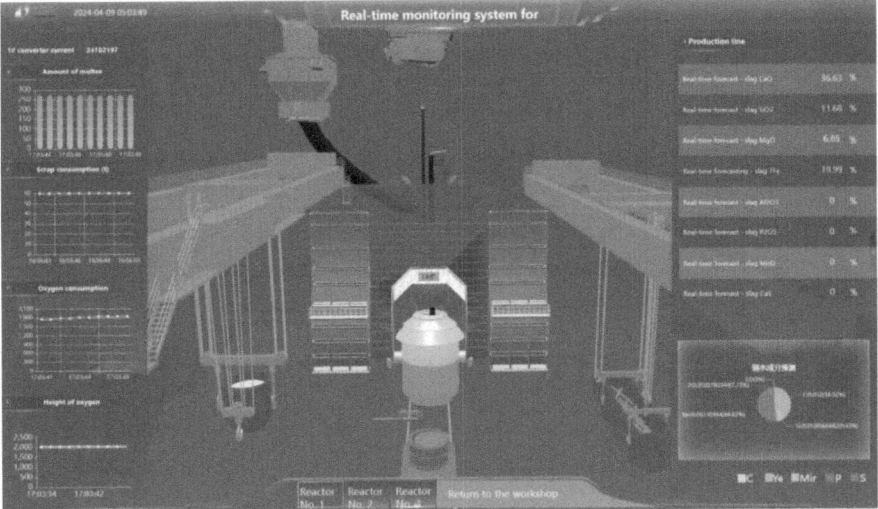

Fig. 7. Converter steelmaking real-time monitoring system.

measures can be taken to ensure production stability and product quality. The system is equipped with early warning and alarm functions to detect abnormalities and send out alarms instantly, reducing production risks and ensuring production safety and stability.

The Smelting Cost-Effectiveness Analysis System is built on an alloy cost analysis model and an auxiliary monitoring model, and is mainly used for real-time monitoring and analysis of various costs in the steelmaking process. The system evaluates the cost-effectiveness of different process solutions and production plans by analyzing alloy

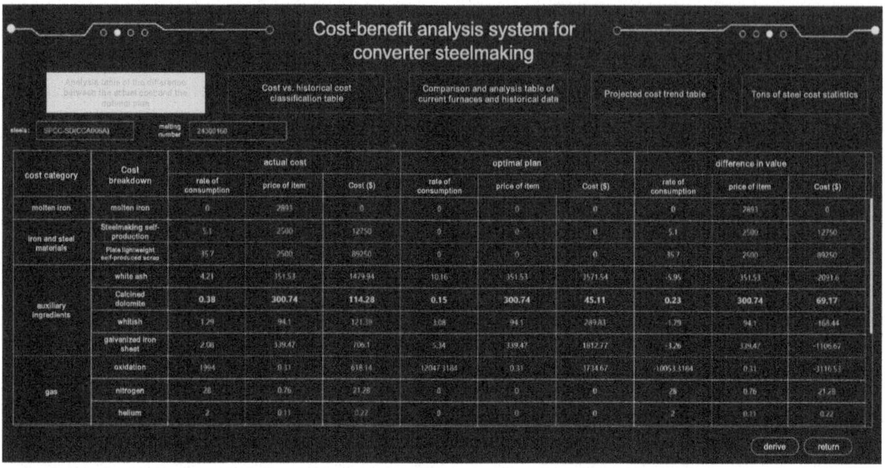

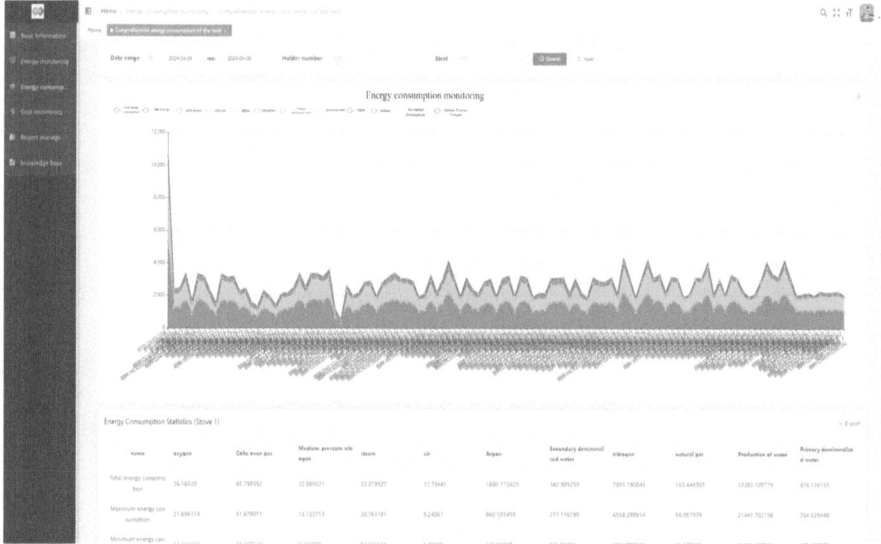

Fig. 8. Converter steelmaking cost-benefit analysis system.

Fig. 9. Energy consumption monitoring system for converter steelmaking.

costs and other factors, in order to optimize resource allocation and improve production efficiency and profitability. The system also includes a variety of functional modules, such as a table for analyzing the difference between the actual cost and the optimal plan, a table for classifying and comparing the cost with the historical cost, a table for comparing and analyzing the current furnace and the historical data, a table for predicting the trend of the cost change, and a table for the statistical table of the cost of ton of steel, etc., as shown in Fig. 8. These modules enhance the system's analyzing and forecasting ability, providing more comprehensive data support and decision-making reference for managers.

The steelmaking energy monitoring system utilizes predictive analysis models, such as material analysis models, to monitor energy consumption in the steelmaking process in real time, including fuels and gases, as shown in Fig. 9. The system evaluates the efficiency of energy utilization and provides suggestions and measures to save energy and reduce consumption. The data is visualized through geometric and simulation models, helping business managers to intuitively understand energy consumption and take appropriate energy-saving measures. By monitoring, analyzing and visualizing energy consumption in real time, the system provides important energy-saving and consumption-reduction support for enterprise managers, helping to reduce energy consumption and improve productivity and economic efficiency.

5   Conclusion

In this paper, the application and benefits of digital twin technology for converter steelmaking are explored in depth by studying its application in the steelmaking process. Firstly, data-based modeling and model-based modeling logic are established, and model bases such as material analysis model, auxiliary monitoring model, and alloy cost optimization model are constructed. These models are validated in practice and provide reliable support for the realization of systems such as real-time production monitoring, smelting cost-benefit analysis, and steelmaking energy consumption monitoring. Through the research in this paper, real-time monitoring and analysis of the converter steelmaking process can be realized to improve the stability and quality of the production process and to reduce the cost of steelmaking by optimizing the material ratios. However, there are still some challenges and problems in converter steelmaking operations, such as operators need more comprehensive auxiliary guidance and support in complex operating environments. Subsequent work can integrate unstructured text data by introducing technologies such as knowledge graph technology and big models to provide operators with more accurate and reliable operation suggestions and decision support, reduce operation errors and costs, realize the intelligence and automation of the steelmaking process, and make greater contributions to the development of the steel production industry.

References

1. Chang, S., Zhao, C., Li, Y., et al.: Multi-channel graph convolutional network based endpoint element composition prediction of converter steelmaking. IFAC-PapersOnLine **54**(3), 152–157 (2021)

2. Liu, H., Shao, B., Yang, L., et al.: Research review and prospect of determination method of blowing endpoint in converter steelmaking. Metall. Autom. **47**(02), 1–15+26 (2023). (in Chinese)

3. Zhang, Y., Zhang, C., Zeng, K., et al.: Research on terminal control model of intelligent mining of flame spectral information of converter mouth in late smelting stage. Ironmaking Steelmaking **48**(6), 677–684 (2021)

4. Wei-min, Y., Jie, Z., Sheng-yong, W., et al.: Converter temperature forecasting based on hybrid recurrent genetic neural network. Control. Eng. **14**(B05), 33–34 (2007). (in Chinese)

5. Liang-tao, Y., Ming, L., Da-yong, Y.: Prediction study of carbon content at the end of converter based on GA-KPLSR. Control. Eng. **24**(5), 923–926 (2017). (in Chinese)

6. Yan, L.T., Li, M., Yang, D.Y.: Prediction of carbon content at end point based on GA-KPLSR in converters. Control Eng. China **24**, 923–926 (2017)

7. He, F., Zhang, L.: Prediction model of end-point phosphorus content in BOF steelmaking process based on PCA and BP neural network. J. Process. Control. **66**, 51–58 (2018)

8. Han, Y., Zhang, C.J., Wang, L., et al.: Industrial IoT for intelligent steelmaking with converter mouth flame spectrum information processed by deep learning. IEEE Trans. Industr. Inf. **16**(4), 2640–2650 (2019)

9. Feng, L., Zhao, C., Li, Y., et al.: Multichannel diffusion graph convolutional network for the prediction of endpoint composition in the converter steelmaking process. IEEE Trans. Instrum. Meas. **70**, 1–13 (2020)

10. Zhang, B., He T., Zang, R., et al.: Study and application of theoretical modeling of material balance and heat balance in steelmaking based on field working conditions. Wide Plate (2021). (in Chinese)

11. Holappa, L.: Historical overview on the development of converter steelmaking from Bessemer to modern practices and future outlook. Mineral Process. Extr. Metall. **128**(1–2), 3–16 (2019)

12. Wang, G.: Research and application of automatic control system for converter steelmaking. Northeastern University (2014). (in Chinese)

13. Xu, L.: Research on the online carbon content measurement method of converter steelmaking endpoint based on the spectral information of furnace mouth flame. Nanjing University of Science and Technology (2011). (in Chinese)

14. Han, M., Zhang, R., Xu, M.: A variable selection algorithm based on improved gray correlation analysis. Control Decis. Making **32**(9), 1647–1652 (2017). (in Chinese)

15. Uhlemann, T.H.J., Lehmann, C.,Steinhilpe, R.R.: The digital twin: Realizing the cyber-physical production system for industry 4.0 Procedia Cirp, 2017, able digital twins: Simulation-based development and operation of complex technical systems. In: IEEE International Symposium on Systems Engineering (ISSE)pp. 1–6 (2016)

16. Qu Li-ping, Q., Yong-yin, B.J., et al.: Intelligent automation system for steelmaking. Control. Eng. **S3**, 17–19 (2007). (in Chinese)

17. Lytvynyuk, Y., Schenk, J., Hiebler, M., et al.: Thermodynamic and kinetic model of the converter steelmaking process. Part 1: The description of the BOF model. Steel Res. Int. **85**(4), 537–543 (2014)

18. Hu, S.,Yang, F.: Optimization of alloy cost in converter steelmaking. Metallurgical Management 2021(03), 101+176 (in Chinese)
19. Sheng, H.: Optimization of steelmaking cost by classification of converter steel grades. Metall. Mater. 41(02), 123–124 (2021). (in Chinese)
20. Watson, N.J., Fisher, O.J., Escrig, J.E., et al.: Considerations, challenges and opportunities when developing data-driven models for process manufacturing systems. Comput. Chem. Eng. 140 (2020)
21. Cheng-dong, L., Tian-chi, Z., Yuan, Z., et al.: Research and application of 3D digital twin virtual factory platform for copper smelter. Metall. Autom. 45(4), 12 (2021). (in Chinese)
22. Wen, T., Yu, J., Jin, E., et al.: A novel electrochemical sensor for phosphorus determination in the high phosphorus liquid iron. J. Market. Res. 9(3), 3530–3536 (2020)

Pool-Based Active Classification Based on Expected Error Reduction with Uncertainty Subsampling

Chako Takahashi[1]([✉]) [iD] and Toki Sato[2,3P] [iD]

[1] Graduate School of Science and Engineering, Yamagata University, 4-3-16 Jonan, Yonezawa, Yamagata 992-8510, Japan
chako@yz.yamagata-u.ac.jp
[2] Graduate School of Information Sciences, Tohoku University, 6-3-09 Aramaki-aza-Aoba, Aoba-ku, Sendai, Miyagi 980-8579, Japan
[3] Faculty of Engineering, Yamagata University, 4-3-16 Jonan, Yonezawa, Yamagata 992-8510, Japan

Abstract. Active learning is a practical machine learning approach useful when labelling data is costly. This study focuses on pool-based active learning, where instances considered most useful for model training are selected from an unlabelled pool set. A well-known label query strategy in pool-based active learning is expected error reduction (EER), which labels instances that are expected to minimise future errors in the model. However, the computational cost of EER grows significantly with the size of the pool set, limiting its practical applicability. To address this issue, we propose to combine EER with uncertainty subsampling, a method that selects queries based on expected error after prefiltering candidate instances using uncertainty. By reducing the size of the unlabelled set for which the expected error is computed, this approach reduces the computational burden of EER while still selecting useful instances. Our experiments show that EER with uncertainty subsampling requires less labelling to achieve a certain performance level than random subsampling. Furthermore, the proposed method allows for a more effective selection of samples that reduce the future error of the model, which is in line with the goal of EER.

Keywords: pool-based active learning · selective sampling · subsampling · uncertainty sampling · expected error reduction

1 Introduction

Active learning is a practical approach to machine learning when labelling data (annotation) incurs high costs. Active learning generates or selects unlabelled data instances that are useful for model training, queries the annotator for their labels, and incrementally increases the size of the labelled training set. In statistical query learning, the annotator is called an *oracle* [5]. When many data

© The Author(s), under exclusive license to Springer Nature Singapore Pte Ltd. 2024
S. Saito et al. (Eds.): AsiaSim 2024, CCIS 2170, pp. 244–255, 2024.
https://doi.org/10.1007/978-981-97-7225-4_19

instances are readily obtainable, pool-based active learning is utilized [11]. Pool-based active learning requires labelling the instances from the unlabelled pool set that are expected to improve the model's performance the most. These instances are used to retrain the model. For example, while many unlabelled images can be easily obtained, e.g. by crawling, labelling them is expensive. Pool-based active learning can be applicable in such a situation. One of the known query selection methods (called *query strategies*) in pool-based active learning is expected error reduction (EER) [3,4,8,10,12]. EER selects instances for labelling that reduce future errors in the model. The procedure of EER proposed by Roy and McCallum [10] is as follows: First, a tentative label is assigned to a particular instance in the pool set. The model is then retrained on the training set that contains this labelled data. Consider the retrained model as a future model and calculate the expected errors on this model. These processes are performed for each instance in the pool set. The computational cost of EER depends on the number of instances in the pool set. Therefore, as the size of the pool set increases, the cost of selecting queries explodes. Several ways to reduce the number of candidate instances in EER include random subsampling from a pool set, outlier prefiltering, and subsampling using other active learning metrics [10]. These methods are essential to suppress the computational cost of EER to a practically usable level. However, their effects on model training have yet to be investigated closely.

In this study, we investigate EER with uncertainty subsampling, which selects queries based on expected error after prefiltering candidate instances of labelling via uncertainty calculations. Uncertainty sampling is one of the typical query strategies in pool-based active learning. Due to its low computational cost, it is often the first choice in many problem settings. EER on uncertainty-subsampled instances is expected to select queries with high uncertainty in the current model and reduce errors in the future model. Numerical experiments show that EER with uncertainty subsampling requires less labelling to achieve a certain performance than EER with random subsampling.

2   Pool-Based Active Learning

The main problem settings in active learning can be categorised into three types [11]: membership query synthesis, which generates data to be labelled; stream-based sampling, which decides whether or not to label incoming unlabelled data in a stream; and pool-based sampling, which selects data to be labelled from a set of unlabelled data set (pool). In this study, we assume a pool-based sampling setting, which is the most commonly considered in machine learning.

We assume that there exists a small labelled initial dataset $\mathcal{D}_{\text{labelled}} = \{(x, y)^{(l)}\}_{l=1}^{n_{\text{init}}}$ as well as an unlabelled dataset $\mathcal{D}_{\text{pool}} = \{x^{(u)}\}_{u=1}^{n_{\text{pool}}}$. The outline of training via pool-based sampling is as follows.

1. Train the model with the small initial training set $\mathcal{D}_{\text{labelled}}$.

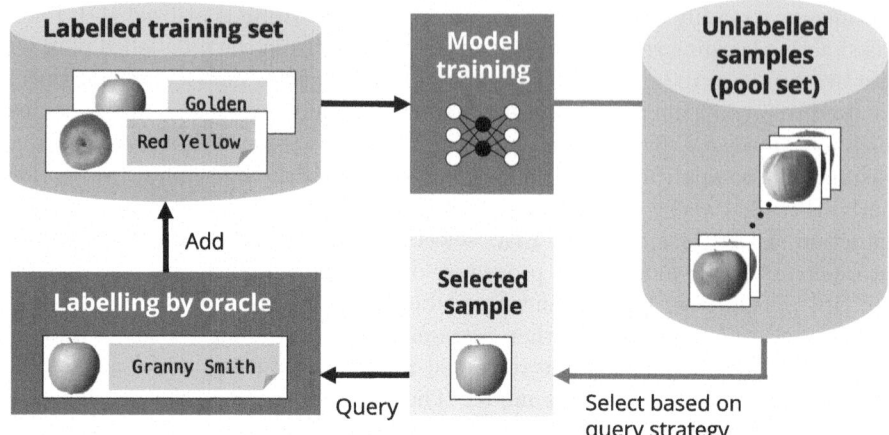

Fig. 1. The outline of pool-based active learning. The apple images and labels are from the Fruits 360 dataset [9].

2. Calculate the utility for each instance in $\mathcal{D}_{\text{pool}}$. The utility is a measure of which data in $\mathcal{D}_{\text{pool}}$ are likely to be labelled to improve the performance of the model. It corresponds to the acquisition function described below.
3. Select the instance with the highest utility in $\mathcal{D}_{\text{pool}}$ as the instance to be labelled next, $\boldsymbol{x}_{\text{queried}}$.
4. Query an oracle for a label for $\boldsymbol{x}_{\text{queried}}$, add $(\boldsymbol{x}_{\text{queried}}, y_{\text{queried}})$ to $\mathcal{D}_{\text{labelled}}$, and remove $\boldsymbol{x}_{\text{queried}}$ from $\mathcal{D}_{\text{pool}}$.
5. Retrain the model with the current training set $\mathcal{D}_{\text{labelled}}$ and return to Step 2.

Figure 1 shows a graphical outline of pool-based active learning.

In active learning, the strategy for selecting the next instance to be labelled is called the *query strategy*. The subsequent sections describe the query strategies relevant to this study, including uncertainty sampling and EER.

2.1   Uncertainty Sampling

Uncertainty sampling is one of the typical query strategies. Uncertainty sampling quantifies the uncertainty of predictions for unlabelled instances in the current model and labels instances with high uncertainty.

In classification, the following three acquisition functions are often used as primary measures of uncertainty [6,11]. In the following, P_θ is the model with parameter θ and $P_\theta(y \mid x)$ is the probability that data x is classified into class y by model P_θ.

Least Confident Sampling. Select the sample that the current model is least confident in predicting. Namely, the sample with the smallest value of the maximum predictive probability,

$$x^*_{\text{lcf}} = \underset{x}{\text{argmin}}\, P_\theta(\hat{y} \mid x) \tag{1}$$
$$= \underset{x}{\text{argmax}}\, (1 - P_\theta(\hat{y} \mid x)),$$

is selected, where $\hat{y} = \text{argmax}_y P_\theta(y \mid x)$.

Margin Sampling. Select the sample with the smallest margin as

$$x^*_{\text{mar}} = \underset{x}{\text{argmin}}\, (P_\theta(\hat{y}_1 \mid x) - P_\theta(\hat{y}_2 \mid x)). \tag{2}$$

Margin is the difference between the class probability with the highest predictive probability, $P_\theta(\hat{y}_1 \mid x)$, and the class probability with the second highest predictive probability $P_\theta(\hat{y}_2 \mid x)$).

Entropy Sampling. Select the sample with the highest entropy. Using Shannon entropy

$$H_\theta(Y \mid x) = - \sum_{y \in \mathcal{L}} P_\theta(y \mid x) \log P_\theta(y \mid x) \tag{3}$$

as the utility, where $\mathcal{L}$ is the set of possible labels, the sample selected by entropy sampling is denoted as

$$x^*_{\text{ent}} = \underset{x}{\text{argmax}}\, H_\theta(Y \mid x). \tag{4}$$

2.2 Expected Error Reduction

A common objective in machine learning is to obtain models that can predict well on unknown data, i.e. that have high generalisability. Expected error reduction (EER) is a query strategy that tries to achieve this objective more directly. In contrast to the above three uncertainty sampling methods, which focus on the predicted probability of unlabelled instances in the current model, EER queries the sample that minimises the expectation of errors in the future model.

The labels of unlabelled instances can only be known once they are queried. The inputs that the model will receive in the future are also unknown. Therefore, it is not possible to make any predictions about them. To address these two problems faced in calculating the expected error, Roy and McCallum proposed to use the posterior distribution of the model [10]. Specifically, we assign a tentative label to an unlabelled instance, add a pair of the instance and the label to the training data, and then retrain the model. Moreover, we assume unknown future inputs are approximated by a pool set and calculate the expected error for the pool set on the new model. The sample with the smallest expected error is then selected.

This study considers log-loss (also called cross-entropy loss), widely used in multi-class classification, as the loss function. EER selects the sample with the smallest expected value of log-loss as

$$
\begin{aligned}
x^*_{\mathrm{EER}} &= \operatorname*{argmin}_{x} \mathbb{E}_{Y|\theta,x} \left[\sum_{x' \in \mathcal{D}_{\mathrm{pool}}} \mathbb{E}_{Y|\theta^+,x} \left[-\log P_{\theta^+}(y \mid x') \right] \right] \\
&= \operatorname*{argmin}_{x} \sum_{y \in \mathcal{L}} P_\theta(y \mid x) \left(\sum_{x' \in \mathcal{D}_{\mathrm{pool}}} \left(-\sum_{y' \in \mathcal{L}} P_{\theta^+}(y' \mid x') \log P_{\theta^+}(y' \mid x') \right) \right) \\
&= \operatorname*{argmin}_{x} \sum_{y \in \mathcal{L}} P_\theta(y \mid x) \sum_{x' \in \mathcal{D}_{\mathrm{pool}}} H_{\theta^+}(Y \mid x'),
\end{aligned}
\tag{5}
$$

where θ^+ denotes the model parameter after the retraining using a labelled set $\mathcal{D}_{\mathrm{labelled}} \cup (x_t^{(u)}, y_t)$. Here $x_t^{(u)}$ is a candidate sample in $\mathcal{D}_{\mathrm{pool}}$ and y_t is a tentative label for $x_t^{(u)}$ (i.e., we assume the hypothetical oracle response for x_t as y_t).

Algorithm 1. Query selection by expected error reduction [10]

Input: $\mathcal{D}_{\mathrm{labelled}} = \{(x, y)^{(l)}\}_{l=1}^{n_{\mathrm{init}}}$: initial training set, θ: model parameters trained with the initial training set, $\mathcal{D}_{\mathrm{pool}} = \{x^{(u)}\}_{u=1}^{n_{\mathrm{pool}}}$: pool set (unlabeled instances).
Output: $x^*_{\mathrm{EER}} \in \mathcal{D}_{\mathrm{pool}}$.
1: **for all** x in $\mathcal{D}_{\mathrm{pool}}$ **do**
2:　　**for all** y in $\mathcal{L}$ **do**
3:　　　Compute $P_\theta(y \mid x)$; predictive probabilities of x on the current model.
4:　　　Retrain P_θ using $\mathcal{D}_{\mathrm{labelled}} \cup (x, y)$ and obtain a new model P_{θ^+}.
5:　　　**for all** x' in $\mathcal{D}_{\mathrm{pool}}$ **do**
6:　　　　Compute $H_{\theta^+}(Y \mid x')$.
7:　　　**end for**
8:　　　Compute $P_\theta(y \mid x^{(u)}) \sum_{x' \in \mathcal{D}_{\mathrm{pool}}} H_{\theta^+}(Y \mid x')$.
9:　　**end for**
10:　Compute $\sum_{y \in \mathcal{L}} P_\theta(y \mid x^{(u)}) \sum_{x' \in \mathcal{D}_{\mathrm{pool}}} H_{\theta^+}(Y \mid x')$; an expected error for x.
11: **end for**
12: **return** $x^*_{\mathrm{EER}} = \operatorname*{argmin}_{x} \sum_{y \in \mathcal{L}} P_\theta(y \mid x) \sum_{x' \in \mathcal{D}_{\mathrm{pool}}} H_{\theta^+}(Y \mid x')$.

We present the algorithm of the EER in Algorithm 1. To select one query using Algorithm 1, $O(n_{\mathrm{pool}}^2)$ is required. In many cases n_{pool} is not small. Thus it is not practical to use this algorithm simply to select a query.

3   Expected Error Reduction with Uncertainty Subsampling

To cope with the significant computational cost of EER, we propose a method that combines uncertainty subsampling and EER. We show the algorithm of the proposed method in Algorithm 2. The proposed method uses uncertainty sampling to create a subsampled pool set $\mathcal{D}'_{\text{pool}}$ and compute expected errors using $\mathcal{D}'_{\text{pool}}$.

Algorithm 2. Query selection by expected error reduction with uncertainty subsampling

---

Input: $\mathcal{D}_{\text{labelled}} = \{(x, y)^{(l)}\}_{l=1}^{n_{\text{init}}}$: initial training set, θ: model parameters trained with the initial training set, $\mathcal{D}_{\text{pool}} = \{x^{(u)}\}_{u=1}^{n_{\text{pool}}}$: pool set (unlabeled instances), UC: uncertainty measure, $n_{\text{subsampled}}(< n_{\text{pool}})$: the number of subsampled instances.
Output: $x^*_{\text{UCEER}} \in \mathcal{D}_{\text{pool}}$.
1: Initialise a set of subsampled instances $\mathcal{D}'_{\text{pool}} = \emptyset$.
2: **for all** x in $\mathcal{D}_{\text{pool}}$ **do**
3:    Compute the utility of x based on an uncertain measure $\text{UC}(x)$.
4: **end for**
5: Add the top $n_{\text{subsampled}}$ most uncertain instances in $\mathcal{D}_{\text{pool}}$ to $\mathcal{D}'_{\text{pool}}$.
6: Run Algorithm 1 using $\mathcal{D}'_{\text{pool}}$ as a pool set and get a selected query $x^*_{\text{UCEER}} = x^*_{\text{EER}}$.
7: **return** x^*_{UCEER}.

In contrast to many other active learning studies, the proposed method does not modify the formulation of utilities. As a measure of uncertainty, another measure of uncertainty can also be used, as well as Eqs. (1), (2), and (4). The advantage of this method is simplicity of implementation. It requires only the addition of an uncertainty subsampling part.

4   Numerical Experiments

This section demonstrates active learning methods using EER with uncertainty subsampling through several settings. We compare these methods with uncertainty sampling and EER with random subsampling.

4.1   Settings

Datasets. For this experiment, we use three multi-class classification datasets, ranging from small to medium size, shown in Table 1. The parameters in Table 1 are respectively: n_{initial} is the number of initial training data, n_{pool} is the size of the pool set and n_{query} is the total number of queries. Each query labels one instance in this experiment, and thus n_{query} denotes the number of data added. $n_{\text{subsampled}}$ is the number of unlabelled instances for which the EER is calculated, filtered by subsampling, and n_{test} is the size of the test set.

Table 1. A summary of the datasets and experimental setup used in the experiments.

Datasets	# of classes	# of features	# of instances	n_{init}	n_{pool}	n_{query}	$n_{subsampled}$	n_{test}
Wine [1]	3	13	178	18	71	20	20	89
Vehicle [7]	4	18	846	80	343	100	20	423
Digits [2]	10	64	1797	50	848	100	20	899

Models. We use two basic classifiers: logistic regression classifiers and support vector machines (SVMs) with linear kernels. We used the implementations included in scikit-learn for both models. The regularisation parameter C, which adjusts the strength of the L_2 regularisation applied to the SVM classifiers, was set to $C = 0.025$ in all experiments.

Methods. The query selecting methods for comparison in this section are as follows:

1. Random sampling (rnd): Selects a query randomly from a pool set.
2. Least confident (lcf): Selects a query according to Eq. (1).
3. Margin sampling (mar): Selects a query according to Eq. (2).
4. Entropy sampling (ent): Selects a query according to Eq. (4).
5. EER with random sampling ($rnd+EER$): Subsamples $n_{subsampled}$ samples randomly from the pool set and create a subsampled set $\mathcal{D}'_{pool}$. Select the sample with the lowest expected error in $\mathcal{D}'_{pool}$ as a query.
6. EER with least confident subsampling ($lcf+EER$): Selects a query by Algorithm 2 with least confident subsampling.
7. EER with margin subsampling ($mar+EER$): Selects a query by Algorithm 2 with margin subsampling.
8. EER with entropy subsampling ($ent+EER$): Selects a query by Algorithm 2 with entropy subsampling.

The experiments do not consider EER with a whole pool set because of its high computational cost. Note that $rnd+EER$, which performs EER with a randomly reduced pool set, can be considered a simplified alternative to full EER.

4.2   Results

This section shows the cross-entropy losses and classification accuracy curves for each dataset and model on the test set. Each result is an average of over 20 trials. The split between the initial training set, the pool set and the test set is different for each trial.

Comparison of EER with Uncertainty Subsampling with Uncertainty Sampling. Figures 2, 3 and 4 show the losses and classification accuracies against the number of queries for the wine, vehicle and digits datasets, respectively. In each figure, (a) is the result of the logistic regression classifier, and (b)

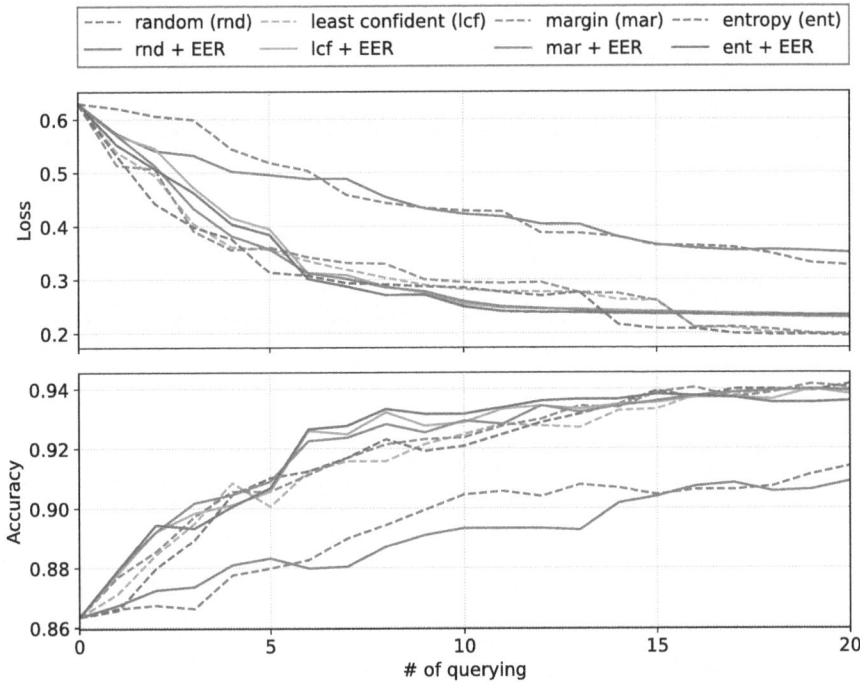

(a) Loss and accuracy curves in logistic regression classifiers.

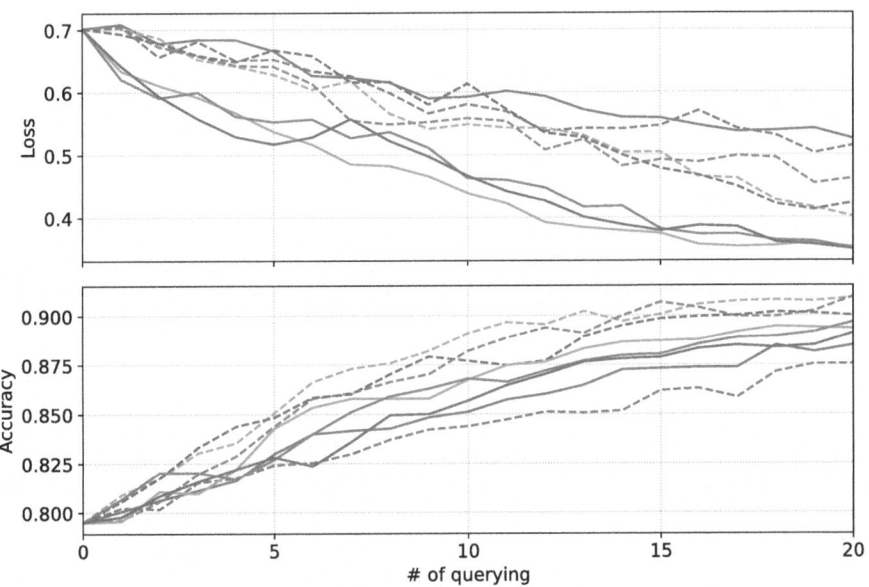

(b) Loss and accuracy curves in support vector machine classifiers.

Fig. 2. Loss and accuracy curves for the number of queries on the UCI wine dataset.

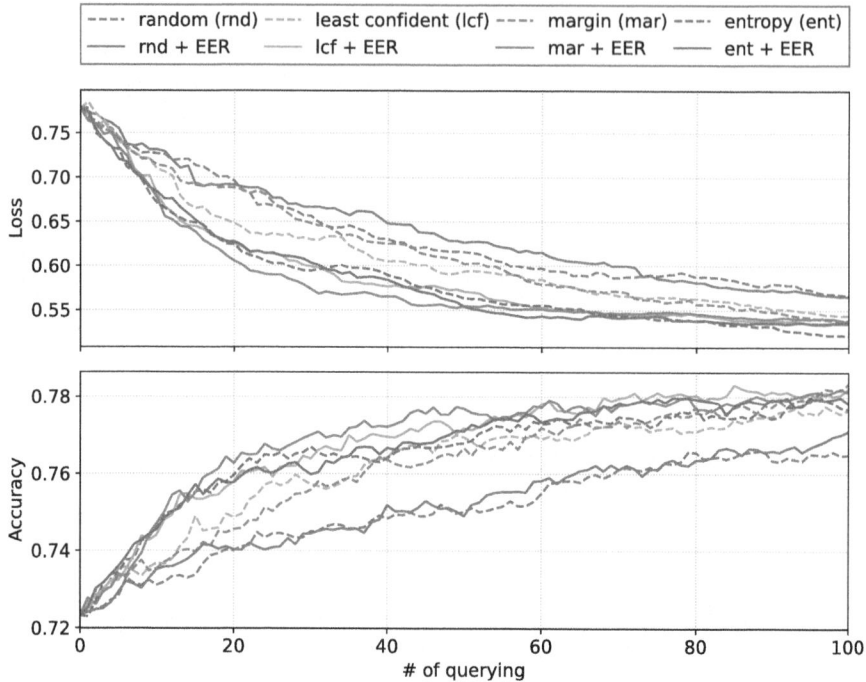

(a) Loss and accuracy curves in logistic regression classifiers.

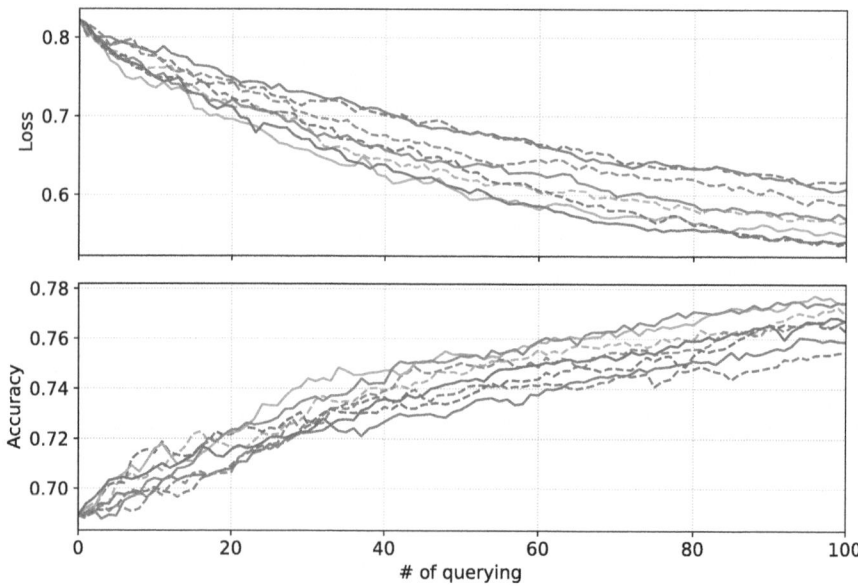

(b) Loss and accuracy curves in support vector machine classifiers.

Fig. 3. Loss and accuracy curves for the number of queries on the UCI vehicle dataset.

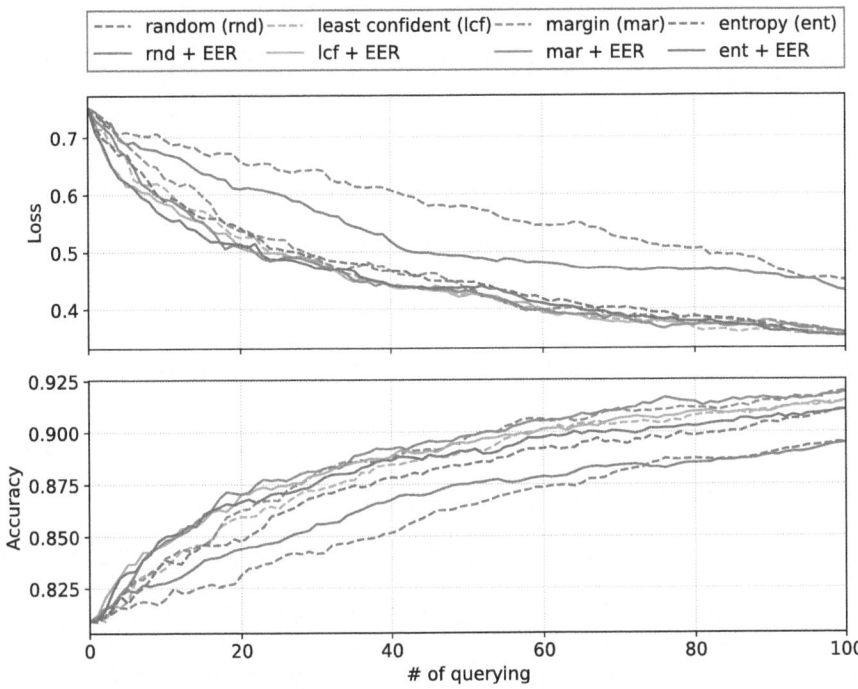

(a) Loss and accuracy curves in logistic regression classifiers.

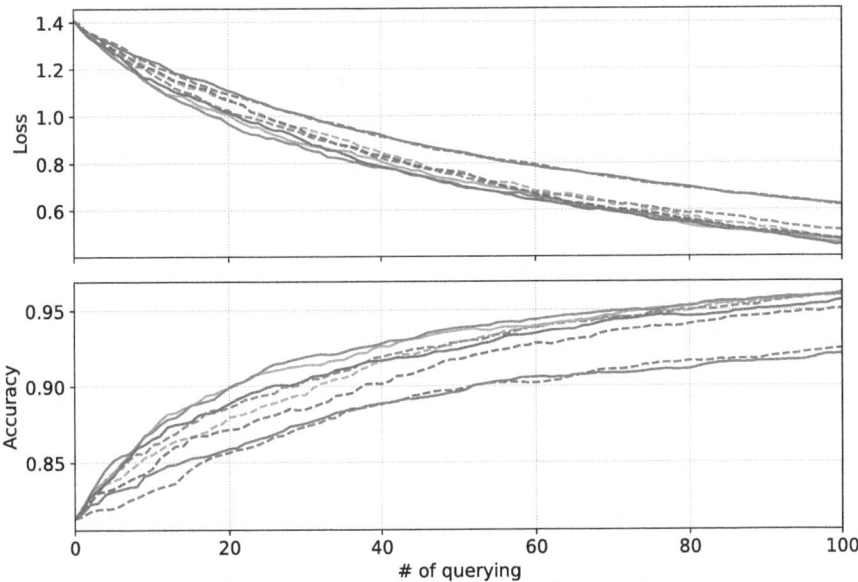

(b) Loss and accuracy curves in support vector machine classifiers.

Fig. 4. Loss and accuracy curves for the number of queries on the UCI digits dataset.

is the result of the SVM classifier. The value on the horizontal axis in each figure is the number of queries. When the number of queries is 0, the results are shown for training on n_{init} initial datasets in Table 1. Focusing on the loss in each of Figs. 1, 2 and 3, the method that combines uncertainty subsampling and EER reduces the losses faster than the method that selects samples by uncertainty alone. In particular, *lcf+EER* can reduce losses faster than *lcf*. We can see similar results for *mar* and *ent* (although sometimes this is not necessarily observed, such as the losses at the several initial queries in Fig. 2(a) and the losses' relationship between *ent* and *ent+EER* in Fig. 3(a)). We also observed that when the loss decreases faster in the method that combines uncertainty subsampling and EER than in the method that selects samples by uncertainty alone, accuracy also increases faster (exceptions were also observed, as in Fig. 2(b)).

Comparison of Uncertainty Subsampling with Random Subsampling. Next, we compare *rnd+EER* with the methods that combine uncertainty subsampling and EER (*lcf+EER*, *mar+EER* and *ent+EER*) in Figs. 2, 3 and 4. Query selections by *lcf+EER*, *mar+EER* and *ent+EER* select samples that reduce loss faster and improve accuracy faster than *rnd+EER*. Assuming that *rnd+EER* is regarded as a simplified alternative to full EER, as described in the previous section, these results imply that queries to obtain a model with a smaller future error (here cross-entropy loss), which is the goal of full EER, are more likely to be achieved by uncertainty subsampling.

5   Conclusion

This study investigates a query selection method that combines uncertainty subsampling and EER. The proposed subsampling method reduces the computational cost of EER by reducing the size of the unlabelled set based on an uncertainty measure. The proposed method can select, without modifying the acquisition function, samples that have high uncertainty for the current model and for which the future error of the model is expected to be smaller if the sample is used for training. Experimental results on several small datasets show that the proposed method can reduce the labelling required to train models that achieve a certain level of accuracy compared to random subsampling. Furthermore, the results suggest that uncertainty subsampling facilitates a more effective selection of samples that reduce the model's future error, which is the objective of EER.

This study focuses on EER when cross-entropy loss (log-loss) is used as the loss function in multi-class classification. Uncertainty subsampling can be easily applied to EERs with 0/1 loss, i.e., minimising expectations of $\sum_{x \in \mathcal{D}_{\text{pool}}} \mathbb{E}_{Y|\theta,x} [y \neq \hat{y}]$, used in binary classification without modifying the acquisition function itself. We use the earliest EER algorithm [10] that has been used in our experiments. Other practical EER algorithms have also been proposed, such as MM+M [4]. Future studies will detail the effectiveness of uncertainty subsampling on binary classification tasks and different EER algorithms.

In our experiments, we used relatively small datasets in size, number of classes and features, and basic machine learning models. In more advanced situations, such as deep learning, the critical point is how the computational cost of EER can be reduced in batch active learning, where multiple samples are labelled in a single query. Effective uncertainty subsampling strategies for training deep models are also one of the topics for future research.

Acknowledgments. This work was supported by JSPS KAKENHI (Grant Number JP21K17804).

Disclosure of Interests. The authors have no competing interests to declare that are relevant to the content of this article.

References

1. Aeberhard, S., Forina, M.: Wine. UCI Machine Learning Repository (1991). https://doi.org/10.24432/C5PC7J
2. Alpaydin, E., Kaynak, C.: Optical Recognition of Handwritten Digits. UCI Mach. Learn. Repository (1998). https://doi.org/10.24432/C50P49
3. Fu, W., Wang, M., Hao, S., Wu, X.: Scalable active learning by approximated error reduction. In: Proceedings of the 24th ACM SIGKDD International Conference on Knowledge Discovery & Data Mining. pp. 1396–1405. KDD 2018, Association for Computing Machinery, New York, NY, USA (2018). https://doi.org/10.1145/3219819.3219954
4. Guo, Y., Greiner, R.: Optimistic active learning using mutual information. In: Proceedings of the 20th International Joint Conference on Artifical Intelligence, pp. 823–829. IJCAI 2007, Morgan Kaufmann Publishers Inc., San Francisco, CA, USA (2007)
5. Kearns, M.: Efficient noise-tolerant learning from statistical queries. J. ACM **45**(6), 983–1006 (1998). https://doi.org/10.1145/293347.293351
6. Monarch, R., Munro, R., Manning, C.: Human-in-the-Loop Machine Learning: Active Learning and Annotation for Human-centered AI. Manning (2021)
7. Mowforth, P., Shepherd, B.: Statlog (Vehicle Silhouettes). UCI Mach. Learn. Repository. https://doi.org/10.24432/C5HG6N
8. Mussmann, S., Reisler, J., Tsai, D., Mousavi, E., O'Brien, S., Goldszmidt, M.: Active learning with expected error reduction. arXiv preprint arXiv:2211.09283 (2022)
9. Oltean, M.: Fruits 360 dataset. Mendeley Data, V1 (2018). https://doi.org/10.17632/rp73yg93n8.1
10. Roy, N., McCallum, A.: Toward optimal active learning through sampling estimation of error reduction. In: Proceedings of the Eighteenth International Conference on Machine Learning, pp. 441–448. ICML 2001, Morgan Kaufmann Publishers Inc., San Francisco, CA, USA (2001)
11. Settles, B.: Active Learning. Synthesis Lectures on Artificial Intelligence and Machine Learning. Morgan & Claypool, California (2012)
12. Zhang, Y., Wang, Y., Cai, W., Zhou, S., Zhang, Y.: From theory to practice: efficient active cost-sensitive classification with expected error reduction. In: Proceedings of the 2017 SIAM International Conference on Data Mining (SDM), pp. 153–161 (2017). https://doi.org/10.1137/1.9781611974973.18

Networks and Complex Systems

Network Robustness Assessment via Edge Criticality Evaluation: Improvement of Bridgeness and Topological Overlap Methods by the Iterative Metrics Re-estimation

Hamza Ejjbiri$^{(\boxtimes)}$ and Vasily Lubashevskiy ⓘ

Tokyo International University, 4-42-31 Higashi-Ikebukuro, Toshima City, Tokyo, Japan
ejjbiri.hamza@gmail.com, vlubashe@tiu.ac.jp

Abstract. Detection of significant edges maintaining the connectivity in complex networks is essential in many applications such as attack vulnerability analysis, the spread of epidemic diseases, and information spreading patterns discovery. There are many existing methods enabling us to evaluate the criticality ranking of links in networks, which are based on straightforward algorithms and topological features of analyzed graphs. In this paper, we propose an improvement of well-known algorithms for the edge criticality ranking in complex networks, bridgeness, and topological overlap. The two resulting methods, iterative bridgeness, and iterative topological overlap are both funded on the principle of iterative metrics re-estimation. To compare methods, we conducted a numerical simulation and decomposition of three real-world benchmark networks. The resulting algorithms are compared with the original methods and demonstrated an increase of efficiency of up to 40% for the iterative bridgeness and up to 11% for the iterative topological overlap.

Keywords: Network Robustness · Edge Criticality · Topological Overlap · Bridgeness

1 Introduction

The advent and evolution of complex networks have marked a crucial step forward in our understanding of various natural and social systems. Whether it's the biological networks [1] that drive life processes within organisms, the social networks [2] that shape human interactions and the dissemination of ideas, or the technological networks [3] that support our critical infrastructures and information exchanges, network science provides an essential lens through which we can understand the dynamics that govern these complex systems. As our dependence on these networks increases, whether for the maintenance of social, economic, or ecological functions, the need to understand their functioning [4], robustness, and vulnerability [5] to internal or external disruption becomes a compelling priority.

© The Author(s), under exclusive license to Springer Nature Singapore Pte Ltd. 2024
S. Saito et al. (Eds.): AsiaSim 2024, CCIS 2170, pp. 259–270, 2024.
https://doi.org/10.1007/978-981-97-7225-4_20

Network science, as a multidisciplinary field, covers a wide range of theories, models, and analytical approaches designed to unlock the structure and dynamics of networks. By examining networks from a variety of perspectives, be it the topology of their structure, the strength and meaning of their links, or their behavior in the face of change [6], researchers can identify patterns and principles that transcend the specificities of each system. This holistic approach is crucial for developing strategies to improve the resilience and efficiency of networks [7], with the aim of building more robust and adaptative systems capable of withstanding and recovering from challenges and crises of the modern world.

To understand the significance of links within networks, several metrics have been explored in the scientific literature [8], including edge betweenness centrality [9], degree product [10], diffusion importance [11], bridgeness [12, 13], topological overlap [14] and k-path centrality [15]. These methods offer diverse perspectives on the importance of edges within networks, ranging from their role in information communication to their contribution to community structure. Our study, while recognizing the value of these metrics, focuses specifically on two of them: "Bridgeness" and "Topological Overlap," which represent typical methods focusing on the topology of networks. By focusing on these aspects, we aim to unveil the underlying mechanisms that contribute to the stability and flexibility of networks, thus offering new insights into how to manage these complex systems to optimize their resilience.

At the heart of the networks that structure our interactions and exchanges are linked "edges" whose value stretches beyond their mere connection. Among them, the notion of "Bridgeness" emerges as a measure of their ability to unite different sections of a network. These links serve as crucial channels that not only facilitate the flow of information but also ensure continuity in communication, even in the event of disruptions. Bridgeness is based on the idea that an edge that effectively connects cliques of fully interconnected nodes is vital to the integrity of the network. When an edge has a high level of "Bridgeness," it often indicates a strategic point whose reinforcement or preservation can considerably increase the overall robustness of the network [12].

The concept of "Topological Overlap" represents a metric that helps us understand how links between nodes in a network are resilient in the face of disturbances. Based on the number of neighbors shared between two nodes, this metric gives an idea of the multiplicity of possible communication paths. The higher the value, the more likely it is that the network will continue to function efficiently, even if certain links disappear. This notion is fundamental in assessing the reliability of a network and its aptitude to maintain itself despite any issues that may arise [14].

This research builds on these two concepts to reveal the key structural elements that contribute to network stability. By examining how "Bridgeness" and "Topological Overlap" influence the cohesion and flexibility of networks, we can better understand how these systems adapt and evolve in response to change. By adopting a rigorous analytical approach inspired by advanced methodologies such as "Deep Link Entropy" [16] and "Improved Link Entropy" [17], which uses the principle of iteration to re-evaluate metrics after each edge removal, our study aims to offer a dynamic perspective on how networks respond to disturbances.

In addition to this iterative perspective, applied to the concept of "Bridgeness" and "Topological Overlap", we adopt the S_{rgc} measurement, the area undet the curve of the largest connected component [20, 21], to assess the impact of selective edge removal on network stability. This allow us to target links vital for connectivity and to evaluate the effectiveness of our methods. S_{rgc} reduction guides us in understanding network robustness and improving our strategies for enhancing system resilience in the face of change and disruption.

This paper is structured to progressively illustrate our approach; after explaining our pioneering method, we present its application on a variety of reference networks, such as the Dolphins [18], Football [9], and Jazz [19] networks. Details of the method used to test the effectiveness of our approach will be given, followed by the presentation of experimental results, which reveal a substantial improvement in the approach. The final discussion will place our findings in a wider context and conclude with the implications and prospects of our work.

This research enriches the field of network science by providing valuable insights into the mechanisms underlying the robustness and adaptability of networks. By identifying the key elements that contribute to network stability and exploring ways in which they can be optimized, our work paves the way for significant advances in network design and management, offering innovative solutions to enhance their sustainability and face the challenges of the modern world.

2 Methods

2.1 Bridgeness

Bridgeness is an index that quantifies the central role an edge plays in maintaining the connectivity of a network. This index is particulary relevant in contexts where the assessment of links strength is limited by the lack of access to personal or historical data, as is often the case in social networks.

The structure of a network is often characterized by cliques, groups of interconnected nodes forming a complete sub-graph. A clique is determined by its number of nodes (k), and an edge is considered all the more central as it connects larger cliques. This, an edge that links two distinct cliques performs an essential function as a bridge between thes two substructures, facilitating interactions within the network.

To explain this mathematically, the "Bridgeness" B_e of an edge e is calculated using the following formula:

$$B_e = \frac{\sqrt{S_x S_y}}{S_e}$$

where S_x represents the size of the largest clique, including node x at one end of the edge, excluding cliques that include node y at the other end, and S_y is similarly defined for node y, excluding those containing x, and S_e is the size of the largest clique to which the edge itself belongs.

To illustrate this concept, let's take this example: assume an edge connects a five-member clique to a four-member clique, and the same edge belongs to a clique of length

2. Applying the "Bridgeness" Formula, we would obtain the following:

$$B_e = \frac{\sqrt{5 * 4}}{2}$$

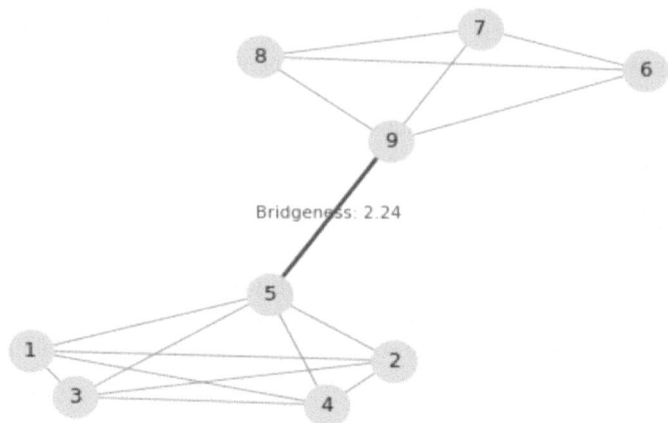

Network Diagram with Bridgeness Calculation

Fig. 1. An example for Calculating Bridgeness

This calculation shown above in Fig. 1 would indicate that the edge in point has a "Bridgeness" of 2.24, underlining its role as a major axis of passage between the two sets of nodes, which has a direct impact on the structure of the network and its ability to maintain connectivity between its different segments. Edges with significant "Bridgeness" are strategic: their removal would result in significant network fragmentation, demonstrating their importance in the stability and robustness of the overall system.

2.2 Topological Overlap

Topological Overlap is a relevant index used to measure the similarity between the social environments of two nodes within a network. It is used to quantify, in topological terms, the proximity of relationships based on the number of shared friends. This index, first introduced to analyze the strength of links in mobile telecommunications networks, is proving to be a reflection of the strength of social links.

Formally, the topological overlap between two nodes i and j, denoted O_{ij}, is given by the following formula:

$$O_{ij} = \frac{n_{ij}}{((k_i - 1) + (k_j - 1) - n_{ij})}$$

where n_{ij} is the number of common neighbors between node i and j, and k_i and k_j represent the respective degrees of nodes i and j, meaning the number of connections they possess. This measure is based on the probability that neighbors at both ends of an edge are identical, thus reflecting repetitive local structures within the network, whether at the core or the margin.

Where there are no common neighbors between two nodes, the topological overlap is zero, indicating that the edge between them could represent a bridge between two distinct communities. Conversely, a topological overlap of one means that all friends of i are also friends of j, implying a very strong connection, usually within the same clique or community.

Therefore, the topological overlap is a measure based on the probability that the neighbors of two connected nodes are interrelated, which can provide information about the local structures of the network regardless of its overall shape.

2.3 Iterative Approach

In order to analyze the robustness of networks by applying " Bridgeness" and "Topological Overlap," we adopt a systematic process that progressively alters the structure of the network. This procedure begins with the calculation of these metrics for all edges, followed by the identification and removal of the most important edge according to the specified metric. With each removal, the size of the largest connected component R_{gc} is recalculated to reflect the new topological conditions.

The effectiveness of this iterative process is evaluated by comparing the size of R_{gc} relative to the proportional number of edges removed. A more pronounced decrease in R_{gc} suggests more impactful edge removal and, therefore, a more efficient quantification method. To ensure consistent evaluation across different networks, two normalizations are commonly applied: the R_{gc} is scaled relative to the number of nodes, and the number of removed edges is normalized relative to the total number of edges in the original network. This approach provides a clear visual and quantitative means of comparing the effectiveness of methods in terms of overall connectivity across the network.

The use of R_{gc} as a criterion owes its success to its clarity: the faster the value of R_{gc} decreases with increasing of ρ -the percentage of the edges removed relative to the total number of edges in the network- the more significant the edges removed are considered to be. This criterion makes it possible to rank methods clearly according to their efficiency with a curve $R_{gc:1}$ (ρ) less than $R_{gc:2}$ (ρ) for any $\rho > 0$, meaning that the former method is more efficient than the latter. This methodical approach focuses on the general characteristics of the network topology rather than the details of the network structure at the microscopic level.

Furthermore, for more accurate numerical comparisons, the area under the curve S_{rgc} representing this decrease in the largest connected component is calculated, providing a performance metric where a smaller area indicates a more effective method.

3 Data and Numerical Experiments

To examine how the introduction of an iterative approach improves the effectiveness of "Bridgeness" and "Topological Overlap," we set up an in-depth numerical analysis. We selected three representative networks [9, 18, 19] that are varied in terms of size and link density. These networks, regularly used in benchmark studies, serve as a test case to evaluate the progress made by our iterative method compared with traditional measures of "Bridgeness" and "Topological Overlap."

3.1 Dolphins Network

The Dolphins network, a social system comprising 159 interactions between 62 dolphins living near Doubtful Sound in New Zealand, provides an ideal study ground for examining the effects of measures on network connectivity. The graph shows a significant fluctuation in the size of the largest connected component R_{gc} when sequentially removing edges from the network. This allows us to examine the effect of various methods used. These include Fig. 2, representing the circular display of the Dolphins network, and Fig. 3, detailing the decomposition process within the same network, respectively.

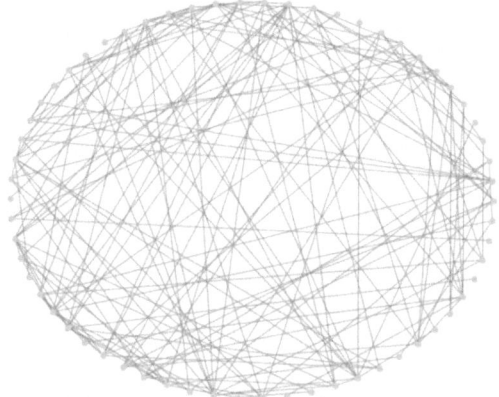

Fig. 2. Circular display of the Dolphins network graph

The illustration shows significantly that the iterative approach to Bridgeness and Topological Overlap leads to an accelerated decrease in R_{gc}, much faster than the classical approaches. This indicates that the iterative methods, which adapt their calculation after each removal, offer a more reactive and accurate perspective on the relative importance of connections within the network. The rate at which R_{gc} decreases in iterative versions underlines their potential for assessing network robustness and predicting the effects of possible disturbance.

3.2 American College Football Network

The American football network, capturing the contests between Division IA teams during the autumn 2000 regular season, stands out as a large-scale case study for our study. With its 115 nodes connected by 613 edges, the network lends itself to a detailed breakdown of the structural robustness via measures of Bridgeness and Topological overlap. These

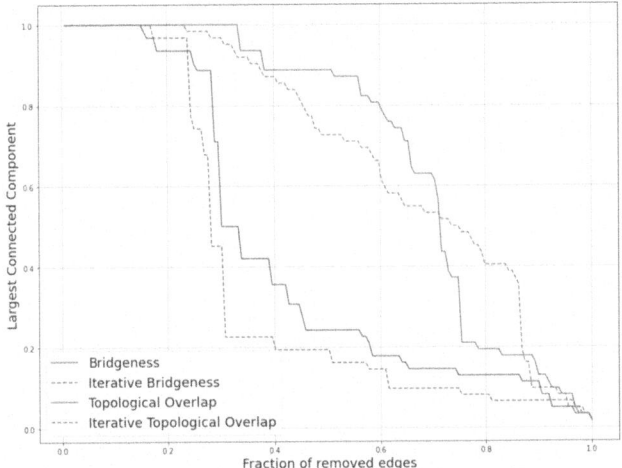

Fig. 3. Display of classic and iterative edge removal in the Dolphins network: Drop of the R_{gc} as a function of deleted edges

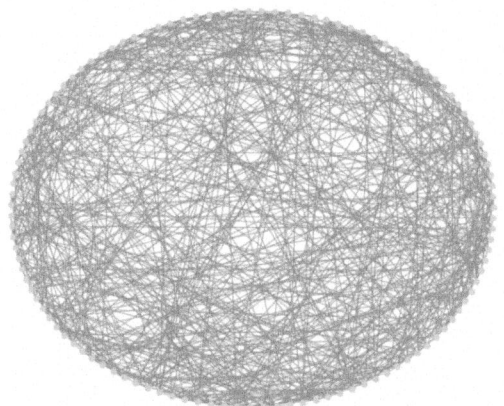

Fig. 4. Circular display of the Football network graph

are supported by Fig. 4, which shows the circular display of the network, and Fig. 5, which details the decomposition process within the network.

The illustration shows that, although the results of the classical and iterative Bridgeness are broadly similar for the majority of edge deletions, a notable distinction emerges in the last 20% of deletions. At this point, the curve corresponding to the iterative method shows a steeper fall than that of the classical method. Regarding the iterative Topological Overlap, it shows a more gradual decrease in R_{gc}, suggesting that this method subtly captures structural variations as the network evolves. These insights underline the relevance of iterative approaches to understanding and maintaining the structural robustness of a complex network.

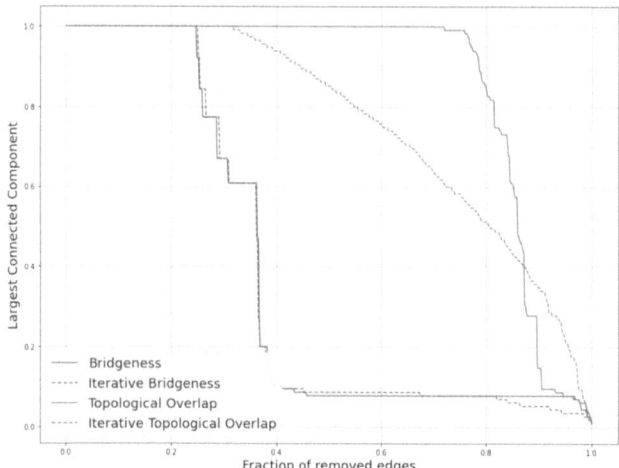

Fig. 5. Display of classic and iterative edge removal in the Football network: Drop of the R_{gc} as a function of deleted edges

3.3 Jazz Network

The Jazz Network, reflecting collaborations between members of the Jazz world, with its 198 nodes woven together by 2742 edges, offers an enriching perspective for examining connectivity mechanisms. This complex structure allows us to apprehend the network's resilience through measures of bridgeness and topological overlap, which can be viewed from both static and dynamic angles. Figure 6 shows the circular display of the network, and Fig. 7 details the decomposition process within the network, respectively.

Fig. 6. Circular display of the Jazz network graph

The illustration shows that the iterative "Bridgeness" method causes a steeper reduction in R_{gc} compared to the classic approach, highlighting a sharper skill in identifying

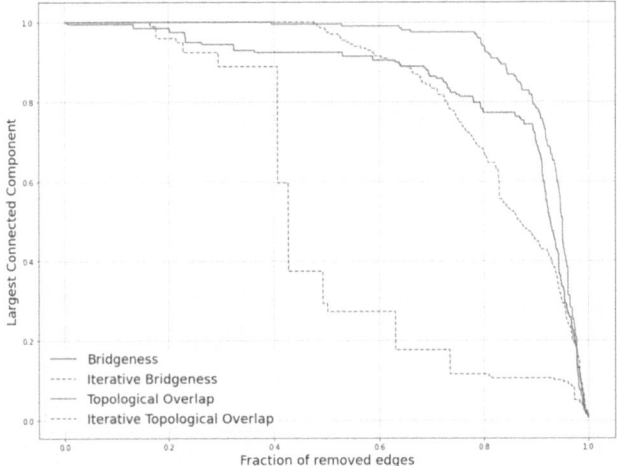

Fig. 7. Display of classic and iterative edge removal in the Jazz network: Drop of the R_{gc} as a function of deleted edges

key edges that support the network. At the same time, iterative Topological overlap induces a more moderate but sustained drop in R_{gc}, suggesting an ability to finely distinguish evolutions in network structure. This distinction highlights the significant advantage of iterative methods in understanding and preserving the structural integrity of complex social networks.

4 Discussion

In our comparative analysis, we begin by establishing a fundamental understanding of the networks under study by detailing their statistical parameters, as summarized in Table 1. This preliminary data provides a quantitative basis from which to assess the network's structural complexities. The networks analyzed vary in size and complexity, as indicated by the number of nodes (N) and connections (edges, E). The degree distribution of each network is represented by the average degree ($\langle k \rangle$) and the maximum degree K_{max}, which highlight connectivity patterns within the networks. We also evaluate the degree of heterogeneity calculated as $H_k = \langle k2 \rangle / \langle k \rangle 2$, which measures the variance of connectivity between nodes. In addition, degree assortativity (r) provides insight into the propensity of nodes to link with other nodes with similar or dissimilar degrees. Finally, the clustering coefficient (C) measures the degree to which nodes tend to group together to form interconnected triplets.

Following the foundational analysis presented in Table 1, we proceed to examine the comparative strengths of iterative and classic methods through the analysis of distinct networks, as shown in the following Table 2. Focusing on the bridgeness and topological overlap indicators, we have integrated S_{rgc} dimension to assess the impact of targeted edge elimination on network stability. The table shows the performance of the methods according to the area under the curve, revealing the effectiveness of the iterative and

Table 1. Basic statistics of the three networks.

Networks	N	E	$\langle k \rangle$	K_{max}	H_k	r	C
Dolphins	62	159	5.1290	12	1.3268	−0.0436	0.2590
Football	115	613	10.6609	12	1.0069	0.1624	0.4032
Jazz	198	2742	27.6970	100	1.3951	0.0202	0.6175

classic approaches in relation to the proportional number of edges removed. A smaller era under the curve suggests a more impactful edge removal method, reinforcing the efficiency of the iterative process.

Table 2. S_{rgc} integral area: Comparison of 'Bridgeness' and 'Topological Overlap' measurements in classic and iterative contexts for the three networks

Network	Bridgeness	Iterative Bridgeness	Topological Overlap	Iterative Topological Overlap
Dolphins	0.4260	0.3636	0.6941	0.6796
Football	0.3794	0.3775	0.8594	0.7577
Jazz	0.8538	0.5062	0.9250	0.8413

A closer look at the numbers in Table 2 shows that the iterative method demonstrates its superiority in evaluating edge criticality. The improvement of brightness by the iterative approach decreases the space under the curve by up to 40%, with an average improvement of 18%, considering the noted benchmark networks. The iteration of topological overlap enhanced the ranking of link criticality up to 11%, with an average improvement of 7%. This finding is of crucial importance, underlining the ability of the iterative approach to identify and neutralize the edges whose absence is most determined to network integrity.

Adopting this iterative approach opens up innovative perspectives for optimizing network resilience. By incorporating progressive edge removal measures, this methodology reveals points of robustness that classic approaches might otherwise overlook. The approach adopted proves to be a valuable tool for practitioners seeking to design networks that are resilient to disturbance while retaining optimal functionality.

The enhanced efficiency in the evaluation of edge criticality and detection of a more rapid decomposition sequence has been obtained with the price of the increased computational complexity of the approach. Considering the fact that the essence of the iterative approach is in the recalculation of the topological features of the network after each individual edge removal, we can consider the resulting complexity as N times the computation complexity of the foundational method, where N is the number of edges in the original network. The present increment of complexity can be shortened by the reduction of the re-evaluation frequency, but it might result in a decrement in the efficiency of edge criticality assessment. The balance between the frequency of the informational update, the computation complexity of the approach, and the resulting efficiency must be the subject of further investigation.

This detailed analysis centered around the above results represents only an initial step in a comprehensive assessment of network robustness. It is conceivable that the trends observed could guide the development of targeted intervention strategies, where critical edges could be strengthened or reorganized to improve overall resilience. However, the scope and the applicability of the findings require further investigations based on a wider range of networks and outage scenarios.

5 Conclusion

The quest to identify critical edges within complex networks is a fundamental line of research in network science. Our study has taken an innovative route, adopting an iterative method to examine these key components. This iterative approach, in application to the concepts of Bridgeness and Topological overlap, enabled us to assess the reaction of networks to specific edge deletions, reflecting their capacity for adaptation and resilience.

The present work and numerical experiments demonstrated that the proposed iterative approach excels and outperforms traditional methods, revealing its potential to accurately assess link criticality. Analysis revealed that our method makes the conventional algorithms more efficient, notably by demonstrating a significantly faster decrease in the largest connected component. The numerical experiment of this approach on the Dolphins, Football, and Jazz networks highlighted its ability to identify links whose removal would have a significant impact on the overall network structure.

Furthermore, it should be emphasized that the effectiveness of existing methods for network evaluation is well-established and recognized. Our contribution, through the application of the iterative approach, aims to enrich the range of available methods by proposing an enhancement of it in certain specific network contexts. This approach is part of a spirit of collaboration and ongoing dialogue with existing work, paving the way for an enriched exploration of complex dynamics within different types of networks. Our aim is to contribute to the diversification of evaluation techniques for a more nuanced and adapted understanding of the networks we study.

This research lays a foundation stone for network science, going beyond conventional measurements and offering a deep understanding of network topology. In conclusion, the current study not only provides a framework for the assessment of network robustness but also opens a dialogue on the practical application of this knowledge for the design of resilient systems capable of sustaining and operating effectively in the face of future challenges.

Acknowledgments. This study was partially funded by the Personal Research Fund of Tokyo International University and by the Japan International Cooperation Agency "JICA" ABE Initiative Scholarship.

Disclosure of Interests. The authors have no competing interests to declare that are relevant to the content of this article.

References

1. Newman, M.E.J.: Spread of Epidemic Disease on Networks. Phys. Rev. E, Stat. Phys. Plasmas Fluids Relat. Interdiscip. Top. **66**(1), 016128 (2002)
2. Pei, S., Muchnik, L., Tang, S., Zheng, Z., Makse, H.A.: Exploring the complex pattern of information spreading in online blog communities. PLoS ONE **10**(5), e0126894 (2015)
3. Wan, R., Xiong, N., Loc, N.T.: An energy-efficient sleep scheduling mechanism with similarity measure for wireless sensor networks. Hum-. Centric Comput. Inf. Sci. **8**(18), 1–22 (2018)
4. Newman, M.E.J.: The structure and function of complex networks. SIAM Rev. **45**(2), 167–256 (2003)
5. Xia, Y., Hill, D.J.: Attack vulnerability of complex communication networks. IEEE Trans. Circuits Syst. II Express Briefs **55**(1), 65–69 (2008)
6. Albert, R., Barabási, A.-L.: Statistical mechanics of complex networks. Rev. Mod. Phys. **74**(1), 47–97 (2002)
7. Louzada, V.H.P., Daolio, F., Herrmann, H.J., Tomassini, M.: Generating robust and efficient networks under targeted attacks. In: Król, D., Fay, D., Gabryś, B. (eds.) Propagation Phenomena in Real World Networks Intelligent Systems Reference Library, vol. 85, pp. 215–224. Springer, Cham (2015). https://doi.org/10.1007/978-3-319-15916-4_9
8. Qian, Y., Li, Y., Zhang, M., Ma, G., Lu, F.: Quantifying edge significance on maintaining global connectivity. Sci. Rep. **7**(1), 45380 (2017)
9. Girvan, M., Newman, M.E.: Community structure in social and biological networks. Proc. Natl. Acad. Sci. **99**(12), 7821–7826 (2002)
10. Wang, W.X., Chen, G.R.: Universal robustness characteristic of weighted networks against cascading failure. Phys. Rev. E, **77**(2), 026101 (2008)
11. Liu, Y., Tang, M., Zhou, T., Do, Y.: Improving the accuracy of the k-shell method by removing redundant links: from a perspective of spreading dynamics. Sci. Rep. **5**, 45380 (2015)
12. Cheng, X.Q., Ren, F.X., Shen, H.W., Zhang, Z.K., Zhou, T.: Bridgeness: a local index on edge significance in maintaining global connectivity. J. Stat. Mech: Theory Exp. **2010**(10), P10011 (2010)
13. Wu, A.K., Tian, L., Liu, Y.Y.: Bridges in complex networks. Phys. Rev. E **97**, 012307 (2018)
14. Onnela, J.-P., et al.: Structure and tie strengths in mobile communication networks. Proc. Natl. Acad. Sci. U.S.A. **104**(18), 7332–7336 (2007)
15. De Meo, P., Ferrara, E., Fiumara, G., Ricciardello, A.: A novel measure of edge centrality in social networks. Knowl.-Based Syst. **30**, 136–150 (2012)
16. Ozaydin, S.Y., Ozaydin, F.: Deep link entropy for quantifying edge significance in social networks. Appl. Sci. **11**(23), 11182 (2021)
17. Lubashevskiy, V., Ozaydin, S.Y., Ozaydin, F.: Improved link entropy with dynamic community number detection for quantifying significance of edges in complex social networks. Entropy **25**(2), 365 (2023)
18. Lusseau, D., Schneider, K., Boisseau, O.J., Haase, P., Slooten, E., Dawson, S.M.: The bottlenose dolphin community of doubtful sound features a large proportion of long-lasting associations. Behav. Ecol. Sociobiol. **54**, 396–405 (2003)
19. Gleiser, P.M., Danon, L.: Community structure in jazz. Adv. Complex Syst. **6**, 565–573 (2003)
20. Lubashevskiy, V., Lubashevsky, I.: Evolutionary approach for detecting significant edges in social and communication networks. IEEE Access **11**, 58046–58054 (2023)
21. Lubashevskiy, V.: Entropy-based approaches of edge significance quantification in complex networks: detection of link vulnerabilities using static, dynamic and group-focused methods. In: Skiadas, C.H., Dimotikalis, Y., (eds.) 16th Chaotic Modeling and Simulation International Conference, Springer Proceedings in Complexity. Springer Cham (2024)

Edge Criticality Evaluation in Complex Structures and Networks Using an Iterative Edge Betweenness

Zihao Gao$^{(\boxtimes)}$, Hamza Ejjbiri, and Vasily Lubashevskiy ⓘ

Tokyo International University, Tokyo, Japan
gzihao2000x@gmail.com, vlubashe@tiu.ac.jp

Abstract. Identifying the edges that played critical roles in understanding the structure and robustness of the network. These key edges facilitate the uninterrupted connection between different network components, and their detection can prevent potential cascading collapses. Various traditional methods that are based on straightforward algorithms can enable us to evaluate the criticality of edges. A widely used method for assessing the importance of these connections is the edge betweenness method. This technique measures the criticality of edges by identifying the number of shortest paths passing through a specific edge between any pair of nodes. In our study, we are focused on enhancing the Edge Betweenness method by applying the iterative principle to refine the measurement of edge criticality, which significantly improved the efficiency of network analysis. We applied this improved method to three different benchmark networks: Zachary's Karate Club Network, American Football College Network, and Power Grid Network. The results showed improved efficiency over other state-of-the-art methods, with performance increases of 11.17%, 39.23%, and 90.63% for each network, respectively. These results underline the effectiveness of the Iterative Edge Betweenness method in detecting edges critical to the robustness and stability of complex networks.

Keywords: Complex Networks · Edge Significance · Edge Betweenness

1 Introduction

In the contemporary era, complex networks play an important role in enhancing the functionality of modern society, permeating every facet of daily life. These networks, ranging from communication networks that can provide more efficient means of communication [1] to power supply networks ensuring stable power transmission [2] and transportation [3] systems to facilitating easier travel, have become indispensable for people. In addition to that, complex networks also assist scientific analysis, such as protein DNA biological systems [4] for the DNA structures and political networks [1] for research and the spread of ideas. As the complex network gets deeper into various sectors, it is necessary to conduct thorough research on the robustness to guarantee the operation.

© The Author(s), under exclusive license to Springer Nature Singapore Pte Ltd. 2024
S. Saito et al. (Eds.): AsiaSim 2024, CCIS 2170, pp. 271–284, 2024.
https://doi.org/10.1007/978-981-97-7225-4_21

Recently, network science has attracted much attention as a multidisciplinary area. With the works of the last decades, it includes a broad range of theories and methods. The patterns of complex networks can be uncovered through the examination from different perspectives, such as the topological structures or the criticality value of the links. We can regard them as a basis to enhance the robustness, preventing the network from cascading failures or even collapse. There are multiple directions in the network science. Some researchers are focused on finding the optimization approaches by prediction of new connections in the networks [5–8]. Some researchers focus on the detection of the resilience of the networks by evaluating the influential nodes [9–13]. Some researchers are devoted to detecting the critical edges of the network and getting the result of network connectivity after link removal to evaluate the robustness of the network. While past research on network robustness has primarily concentrated on node removal, further study into link removal is still necessary. This study will address this gap by focusing on the removal of edges rather than nodes. This direction is of interest to the present research.

Various methods have been developed to evaluate the criticality of the edges. Most of these methods can be classified into two types of algorithms: straight-forward and integral optimization. Straightforward algorithms are the algorithms that focus on analyzing the features of network topology. They assign specific properties to independent edges and get the criticality values [20]. Straightforward algorithms include the Edge Betweenness [17], the Degree Product [16], the Bridgeness [18], the Diffusion Importance [19], the Topological Overlap [1], the K-path centrality [20], and the Link Entropy [21]. These methods exhibit different performance advantages across network types. For example, in assortative networks, the Degree Product method excels, while in denser networks, the Bridgeness method is more advantageous [21].

In contrast, integral optimization algorithms adopt a holistic perspective rather than the assignment of properties to individual links; these algorithms are in favor of observing the macro-level characteristics. The Evolutionary Approach [22] is a kind of integral optimization algorithm that leverages genetic algorithm principles to evaluate the edge criticality, demonstrating superior efficiency to straightforward algorithms such as improved Link Entropy from the perspective of the final criticality ranking. However, the Evolutionary Approach is much more demanding in terms of computational time and capacity.

Analyzing recent advantages in network science, we have noticed that two new straightforward algorithms have been developed: Deep Link Entropy [14] and Improved Link Entropy [15]. Both of these methods were rooted in Link Entropy and enhanced their efficiency via iterative re-estimation of the topological features after each edge removal. In this paper, we decided to apply the same principle of step-by-step re-evaluate of the network structure using another criticality approach, the Edge Betweenness [17].

We named the resulting method the iterative edge betweenness, and the result has been compared with the standard edge betweenness, degree product, and Bridgeness methods. We conducted a numerical simulation of edge criticality ranking and decomposition of three well-known real-life benchmark networks: Zachary's Karate Club [22], American College Football [21], and Power Grid [21]. We will compare the efficiency of the four methods to examine the effectiveness of the Iterative Edge Betweenness method in the following sections.

2   Methods

The Iterative Edge Betweenness method is tested on the decomposition of three known benchmark networks and compared with the foundational algorithms, such as Edge Betweenness, Degree Product, and Bridgeness. The selection of the noted algorithms is based on the following logic: the recent research demonstrated that Link Entropy, Deep Link Entropy, and Improved Link Entropy have a sufficient number of parameters that might vary the outcome of the analysis, such as a multiplication coefficient for neighboring links entropy (Chi), number of assumed communities, and method of communities detection [23]. So, for the present work, we selected three known and widely used benchmark algorithms. However, we acknowledge that for further research, the number of benchmark networks and the number of benchmark algorithms should be extended.

In this section, we will first introduce three methods for network decomposition: the Edge Betweenness, the Degree Product, and the Bridgeness. These methods are used to evaluate the criticality widely in different networks. The Edge Betweenness method is the method that we want to improve in this research. For the other two methods, we will use them to compare the decompose efficiency with the Edge Betweenness. The details of the iterative edge betweenness method will also be explained after introducing these three methods.

2.1   Edge Betweenness

Edge Betweenness is a method that counts the number of shortest paths between one pair of nodes through the specific edge. This algorithm was developed by Linton Freeman; it uses Edge Betweenness centrality to evaluate the criticality of edges [17]. The edge, which is frequently evaluated as the shortest path between various pairs of nodes, has relatively high criticality because it also means that the edge can connect different parts of the network.

Step 1. For a given undirected graph, $G = (N, E)$, where N is the set of nodes, and E is the set of edges. For two random nodes s, t included in N; we denote σ_{st} as the total number of shortest paths between N and E. It reflects the different paths that information or flows in the network.

To find the shortest paths between N and E, we will use the Floyd-Warshall algorithm. The algorithm will initialize a matrix X based on the distance of the path between one pair of nodes (s, t); for the nodes without a direct connection, the distance of the path is infinite. The algorithm will select a random node k in the network and try to find a shorter path between s and t. If node k can help to find a shorter path, the algorithm will update the shortest path between i and j.

Step 2. For the shortest paths between s and t, which go through an edge E, we denote it as $\sigma_{st}(E)$ as the factor to discover the criticality of the edge. The algorithm will evaluate each edge in the network. The $\sigma_{st}(E)$ reflects the reliability of E for the connection between N and E. If $\sigma_{st}(E)$ has a higher value than the other edges connected between s and t, it also means that edge E has higher criticality.

Step 3. After calculating σ_{st} and $\sigma_{st}(E)$, we need to know the proportion of the shortest paths between s and t, which go through an edge E to all of the shortest paths between s and t. If $\sigma_{st}(E) / \sigma_{st}$ has a greater value than the other edges, it reflects that the edge plays an important role in maintaining the connection between s and t. In other words, E has high criticality.

The formula of this method is the following:

$$Betweenness_E = \sum\nolimits_{s \neq t} \frac{\sigma_{st}(E)}{\sigma_{st}} \tag{1}$$

When the network is fragmented, $\sigma_{st} = 0$ when the fragmentation occurs. It is assumed that, in cases where nodes s and t are disconnected, $\frac{\sigma_{st}(v)}{\sigma_{st}}$ is set to zero for the calculation.

2.2 Degree Product

This is a method utilizing the Degree Product method to describe cascading-failure-induced disasters in weighted networks [16].

$$Degree_E = k_x k_y \tag{2}$$

The degrees of node x and y are denoted by k_x and k_y, respectively. The Degree Product uses the degree of each node, which refers to the number of edges connected to the node. It functions effectively in assortative networks in which nodes are high degree connected [21].

2.3 Bridgeness

Bridgeness method uses Bridgeness to quantify the significance of edges in sustaining network connectivity [18].

$$Bridgeness_E = \frac{\sqrt{S_x S_y}}{S_E} \tag{3}$$

In the Bridgeness method, there are two endpoints, x and y, and edge E is included, and the size of a node's or edge's clique E is the largest clique size that includes either the edge or the node. A clique is a subset of a graph in which each pair of nodes is directly connected. This method performs well on dense networks in which similar or relevant nodes are likely to interconnect, such as documents and social networks [21].

2.4 Iterative Edge Betweenness

In normal criticality evaluation, the Edge Betweenness method only evaluates the criticality of edges in the network for once at the beginning, then the edges will be ranked based on their criticality and removed by the sequence. However, with changes in the network topology, the criticality of edges in the current network may also change. Therefore, it is necessary to re-evaluate the criticality after each removal of the edge to suit the current network structure.

As we checked the results from the other researchers, we found examples of "Deep Link Entropy" [14] and "Improved Link Entropy" [15] by applying the iteration principle to improve the efficiency of the Link Entropy method. These methods improved the efficiency of detecting network structures in complex networks by applying the iterative principle. The iterative principle helps these methods to analyze the network structure dynamically, which is crucial to evaluate the criticality of edges accurately. The results of Deep Link Entropy and Improved Link Entropy showed a significant improvement compared to Link Entropy [14] for detecting edge criticality [15]. Based on this discovery, we are going to apply the same iteration principle to the Edge Betweenness method to verify if it can also improve the decomposition efficiency of the Edge Betweenness method.

In our research, we will do the process as Fig. 1 shows.

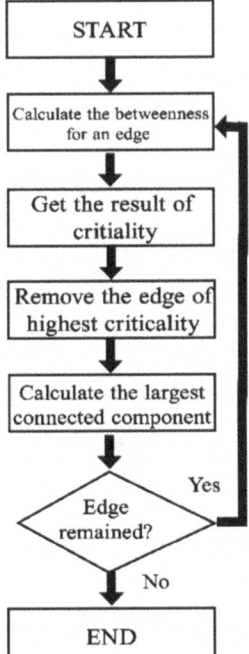

Fig. 1. Flow chart of the whole process for iterative Edge Betweenness.

We will use the Edge Betweenness method to evaluate criticality for all edges at the beginning. Afterward, we will rank them in descending order and select the edge which has the highest criticality. It will be regarded as the most important edge for maintaining connectivity and removing it from the original network. We will calculate R_{gc} and ρ values after this edge's removal and record them in a list. The data of the new network, which is the network that has already removed the edge from the original network, will be saved in a separate new list. Then, we will execute the above process repetitively until the whole network is completely decomposed.

3   Algorithm Performance Measurement

We decompose a complex network by removing the edges. The sequence of edge removal should follow the rank of the criticality of each edge in descending order. The algorithm's decompose efficiency is determined by the size of the largest connected component and the number of removed edges, which is usually denoted as R_{gc} and ρ. It is the value that is obtained by calculating the size of connected components divided by the fraction of the removed edges to all edges. The initial value of R_{gc} is 1 since each node can reach other nodes successfully, and ρ is 0 since there is no edge removed yet.

At first, the evaluation algorithm will traverse all edges in the network and evaluate the criticality of each of them. This process will result in criticality for all edges in a list; thus, we can remove the edges based on rank. Secondly, we will remove the edge that has the highest criticality on the list. When the network decomposes completely, the value ρ will become 1, and R_{gc} will become 0, eventually. The reason for using $R_{gc}(\rho)$ criteria is attributed to its intelligibility. The highest criticality edge has the greatest impact on network connectivity; removal of the edges may affect the communication efficiency of the network. The greater the criticality of the removed edges, the more abruptly the value of R_{gc} decreases as ρ grows, which also reflects the decomposition efficiency.

Based on the $R_{gc}(\rho)$ criteria, we applied the S_{rgc} value as the accurate measurement of the decomposition efficiency. Intuitively speaking, S_{rgc} is the value that calculates the area under the curve $R_{gc}(\rho)$. This criterion emphasizes the general features of network topology instead of the specific aspects of the network's structure at the micro-level [21]. The network decomposition is denoted as a sequence of discrete values $\{\rho_i\}(i \in [0, m])$ with the spacing $1/m$. The formula of the S_{rgc} is shown in the following:

$$S_{rgc} = \frac{1}{m} \sum\nolimits_{i=0}^{m} R_{gc}(\rho_i) \tag{4}$$

The S_{rgc} represents a distinct approximation of the theoretical continuous integral of $R_{gc}(\rho)$ across the interval from 0 to 1 [21]. We acknowledge that the network decomposition is more efficient while the S_{rgc} value is smaller. The S_{rgc} value of different methods will be shown in the discussion part of the paper to compare the decomposition efficiency more accurately.

4   Data and Numerical Modeling

To show the performance among the original methods such as Edge Betweenness, Degree Product, and Bridgeness and iterative Edge Betweenness, we will use three popular benchmark networks, Zachary's Karate Club Network, American Football Network, and Power Grid Network with R_{gc} (ρ) criteria to do the numerical modeling for these four methods. To show the structural properties of these benchmark networks, we listed the basic statistical data in Table 1.

Table 1. Basic statistics of three benchmark networks. Structural properties include the number of nodes (N), the number of edges (E), the average degree ($<k>$), the maximum degree(k_max), the degree heterogeneity (H_k), and the clustering coefficient (C).

Networks	N	E	$<k>$	k_{max}	H_k	C
Karate	34	78	4.5882	17	1.6933	0.5706
Football	115	613	10.6609	12	1.0069	0.4032
Power	4941	6594	2.6691	19	1.4504	0.0801

Zachary's Karate Club Network
This network is often used in testing various methods of network analysis. It is the network of friendships that binds the 34 karate club members at a US university together, comprising 78 edges and 34 nodes, which represents a small size network. We will show the circular layout of Zachary's Karate Club Network in Fig. 2.

In Fig. 3, we will compare Zachary's Karate Club Network's decomposition. R_{gc} (ρ) curves of Edge betweenness, Iterative Edge Betweenness, Degree Product, and Bridgeness are shown in the graph.

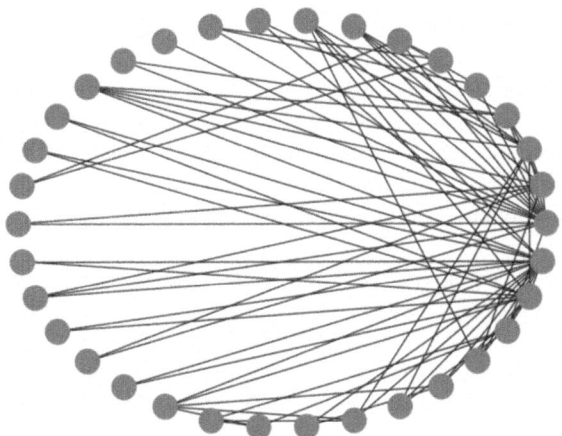

Fig. 2. This figure presents Zachary's Karate Club network in a circular layout. This network has relatively low density, indicating that not every node interacts with the other nodes.

According to the graph, compared to the curve of Edge Betweenness, the curve of iterative Edge Betweenness has a more significant drop, which also indicates that it has better efficiency. However, the curves of Iterative Edge Betweenness and Bridgeness are similar. It is hard to figure out which one is better. To compare the efficiency accurately, we will calculate the S_{rgc} value for Zachary's Karate Club Network's decomposition in the discussion part. In a small-size network, the iterative Edge Betweenness may not outperform well than the other three methods. In the next part, we will use the American College Football Network, a medium-sized network for decomposition with these four methods.

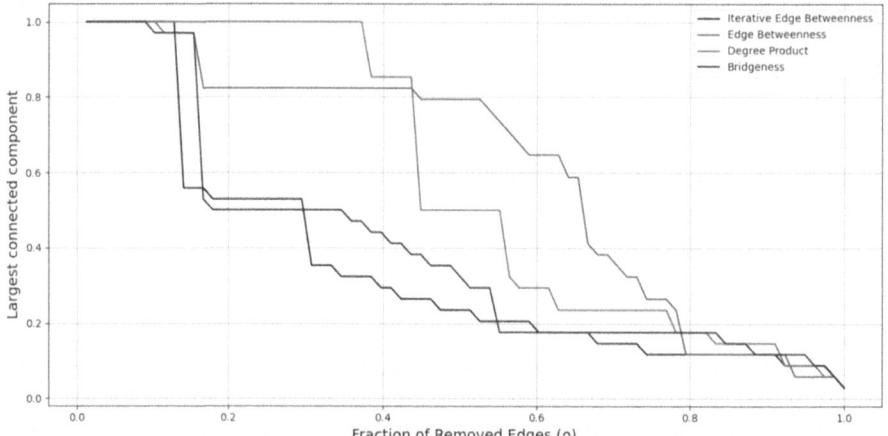

Fig. 3. This figure draws a graph along with Zachary's Karate Club Network's decomposition as determined by four methods: the Edge Betweenness (pink line), the Degree Product (green line), the Bridgeness (blue line), and Iterative Edge Betweenness (black line). The graph shows how the proportion of nodes ρ in the largest connected component of R_{gc} changes as we remove edges. (Color figure online)

American College Football Network

This is a network of American College Football games among Division IA colleges. The data of this network was collected from the standard season in 2000's autumn. It can be considered an exemplar of medium-scale networks, comprising 115 nodes and 613 edges. The circular layout of the American College Football Network will be shown in Fig. 4.

In Fig. 5, we will decompose the American College Football network and compare the R_{gc} (ρ) curves of Edge Betweenness, Iterative Edge Betweenness, Degree Product, and Bridgeness.

We can see that the Iterative Edge Betweenness method has dropped at the earliest time among these four methods. It shows that the Iterative Edge Betweenness method not only outperforms the decomposition efficiency of the Edge Betweenness method but is also better than the Degree Product and Bridgeness methods. To compare the efficiency accurately, we will calculate the S_{rgc} value for American College Football's

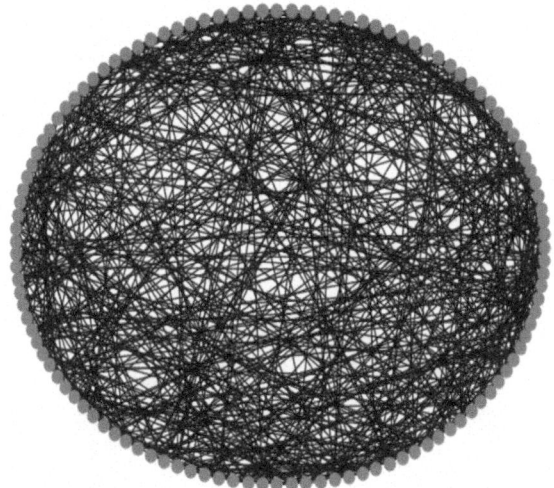

Fig. 4. This figure presents the American College Football network in a circular layout. This network has lots of connections.

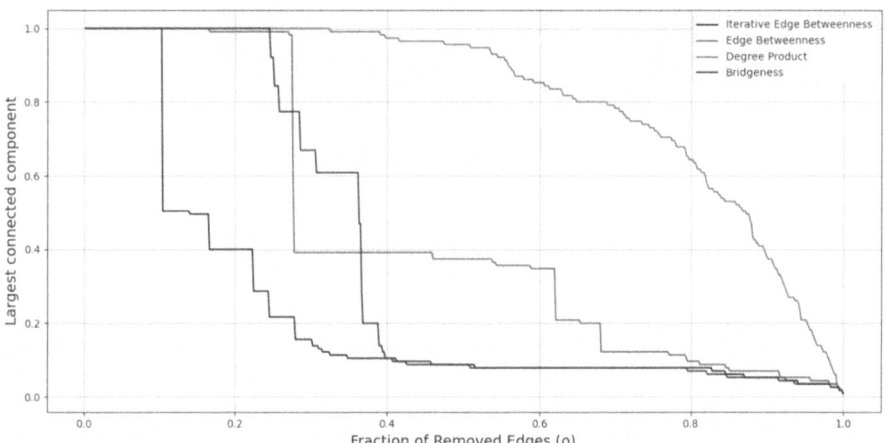

Fig. 5. This figure draws a graph along with American College Football's decomposition as determined by four methods: the Edge Betweenness (pink line), the Degree Product (green line), the Bridgeness (blue line), and the Iterative Edge Betweenness (black line). The graph shows how the proportion of nodes ρ in the largest connected component of R_{gc} changes as we remove edges. (Color figure online)

decomposition in the discussion part. From the result of this decomposition, we can see that the Iterative Edge Betweenness has more obvious efficiency advantages in medium-size networks. In addition to the above two benchmark networks, we also want to do network decomposition with these methods on networks with large sizes.

Power Grid Network

Power Grid Network as an undirected, unweighted representation of the topology of the Western States Power Grid of the United States. This network consists of 4941 nodes and 6594 edges, whose size is larger than Zachary's Karate Club Network and American College Football Network. This network was used as one of the benchmark networks with unique characteristics to compare the performance of different methods. This is a network that considers the flow and distribution of resources. We will show the circular layout of the Power Grid Network in Fig. 6.

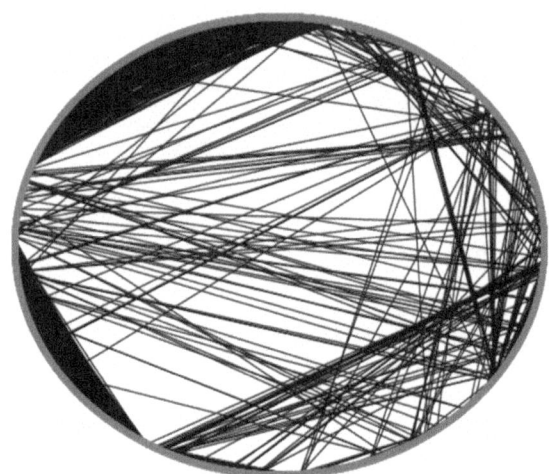

Fig. 6. This figure presents the Power Grid Network in a circular layout. Some edges in this network are close to each other, which reflects that this network has more obvious clustering than the other networks.

In Fig. 7, we will decompose the Power Grid Network and compare the R_{gc} (ρ) curves of Edge betweenness, Iterative Edge Betweenness, Degree Product, and Bridgeness.

According to the graph of the Power Grid Network's decomposition, the curve of the Iterative Edge Betweenness method has dropped dramatically at the beginning, and the largest connected component value goes down to 0 rapidly, which means that the efficiency is much better than the other three methods. To make this a little bit more intuitive, we will calculate the S_{rgc} value for the Power Grid Network's decomposition in the discussion part. From the result of this decomposition, we can see that the advantage of the efficiency of Iterative Edge Betweenness has become more significant in large-size networks.

5   Discussion

We have decomposed three benchmark networks of different scales of nodes and edges with Zachary's Karate Club network, the American College Football network, and the Power Grid network. The results are shown in Fig. 3, Fig. 5, and Fig. 7. These figures

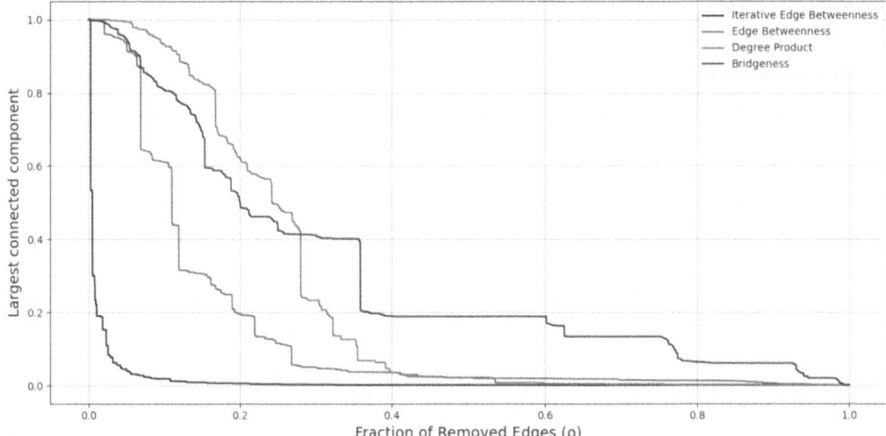

Fig. 7. This figure draws a graph along with the Power Grid network's decomposition as determined by four methods: the Edge Betweenness (pink line), the Degree Product (green line), the Bridgeness (blue line), and Iterative Edge Betweenness (black line). The graph shows how the proportion of nodes ρ in the largest connected component of R_{gc} changes as we remove edges. (Color figure online)

show the impact of removing edges on the size of the largest connected component in R_{gc} for the Edge Betweenness, Degree Product, Bridgeness, and Iterative Edge Betweenness which are used in network decomposition.

Through the research, we have found that the Iterative Edge Betweenness method outperforms the other methods. The R_{gc} curve of decomposition efficiency of iterative Edge Betweenness has performed a more pronounced drop. To compare the efficiency accurately among different methods, we will calculate S_{rgc} for the decomposition of each benchmark network with these methods. The data of S_{rgc} values will be shown in Table 2.

Table 2. This table presents the S_{rgc} of three benchmark networks. These values enable us to compare the decomposition efficiency among different methods.

Methods	S_{rgc} in the Karate network	S_{rgc} in Football network	S_{rgc} in Power network
Iterative Edge Betweenness	0.34766	0.22829	0.01326
Edge Betweenness	0.60294	0.44442	0.14153
Degree Product	0.56825	0.81237	0.24180
Bridgeness	0.39140	0.37567	0.30766

According to the table, the lower S_{rgc} of iterative Edge Betweenness reflects that it has better efficiency of network decomposition in three benchmark networks with different sizes and structures. It is not only better than the Edge Betweenness method, but also better than the Degree Product method, and Bridgeness method.

We selected the method that has the lowest S_{rgc} value other than Iterative Edge Betweenness in each benchmark network and used it to show the percentage of efficiency increment of the Iterative Edge Betweenness method. In Zachary's Karate Club, Bridgeness has the S_{rgc} value 0.39140, which has the best efficiency other than the Iterative Edge Betweenness method. With the S_{rgc} value 0.34766, the Iterative Edge Betweenness method exceeds Bridgeness's S_{rgc} value in 11.17%. In the American College Football Network, Bridgeness has the S_{rgc} value 0.37567, which has the best efficiency other than the Iterative Edge Betweenness method. The Iterative Edge Betweenness method exceeds Bridgeness's S_{rgc} value by 39.23%. In the Power Grid network, Edge Betweenness has the S_{rgc} value 0.14153, which has the best efficiency other than the Iterative Edge Betweenness method. The Iterative Edge Betweenness method exceeds the Edge Betweenness's S_{rgc} value by 90.63%. With these values, we can assume that as the network's size increases, the superiority of the Iterative Edge Betweenness method also becomes more significant. It should come from the fact that the more complex and big the network – the harder it is to define the best decomposition sequence from the initial topological metrics. Correspondingly, the iterative reassessment of those features should provide a much better result. However, this assumption verification might be the purpose of the next research.

The Iterative Edge Betweenness method exhibits significant limitations, particularly in terms of running times. Compared to the other traditional method, Iterative Edge Betweenness needs much time to identify the criticality of all edges in the network since it recalculates the shortest paths between every pair of nodes for each edge within the network. In addition to this, since the Iterative Edge Betweenness method needs to be run after each edge removal, it needs much more time than the Edge Betweenness method. Consequently, as network size increases, so does computational time. Therefore, for a time-sensitive decomposition task, this method may not be the optimal choice.

6   Conclusion

The study of edge criticality in complex networks is an essential pillar of network science. A variety of methods have been developed to address this issue more efficiently across networks of different sizes and types. Our research has been innovated by adopting an iterative approach applied to the Edge Betweenness method to evaluate the edge criticality, which can help evaluate the robustness of the network.

The iterative principle has already been shown to be feasible by applying it to the edge betweenness method. It demonstrated superior efficiency by significantly improving network decomposition compared to traditional methods such as Edge Betweenness, Degree Product, and Bridgeness. Notably, it succeeded in obtaining lower S_{rgc} values, proving its potential for optimizing network analysis performance. Through numerical tests carried out on various networks such as the Zachary's Karate Club Network, the American College Football Network, and the Power Grid Network, we confirmed that the iterative Edge Betweenness approach offers superior results in terms of critical edge identification, outperforming traditional methods.

Our research aims not to undermine the effectiveness of traditional methods but to highlight that the iterative principle could provide additional benefits to traditional methods in specific networks. Indeed, this research enriches the existing corpus by demonstrating that methodological innovation can bring substantial benefits to the assessment of network robustness. We hope our findings will inspire other researchers to explore new methods for improving network analysis and to consider new research directions in this dynamic field.

In conclusion, our study not only offers an alternative method for evaluating edge criticality in complex networks but also opens up an enriching dialogue on the practical application of these techniques for the design and optimization of networks in the face of future challenges. It lays a solid foundation for future explorations in network science, underlining the importance of continuing to diversify and refine evaluation tools to understand better and manage the complexity of our world's interconnected systems.

Acknowledgments. This study was partially funded by the Personal Research Fund of Tokyo International University.

Disclosure of Interests. The authors have no competing interests to declare that are relevant to the content of this article.

References

1. Onnela, J.-P., et al.: Structure and tie strengths in mobile communication networks. Proc. Natl. Acad. Sci. **104**(18), 7332–7336 (2007)
2. Wang, Z., He, J., Nechifor, A., Zhang, D., Crossley, G.: Identification of critical transmission lines in complex power networks. Energies **10**(9), 1294 (2017)
3. Ghosh, S., Banerjee, A., Sharma, N., Agarwal, S., Ganguly, N.: Statistical analysis of the Indian railway network: a complex network approach. Acta Phys. Pol. B Proc. Suppl. **12**(2), 123 (2011)
4. Sathyapriya, R., Vijayabaskar, M.S., Vishveshwara, S.: Insights into protein–DNA interactions through structure network analysis. PLoS Comput. Biol. **4**(9), e1000170 (2008)
5. Ma, X., Sun, P., Wang, Y.: Graph regularized nonnegative matrix factorization for temporal link prediction in dynamic networks. Physica A **496**, 121–136 (2018)
6. Jamali, A.A.,Kusalik, A., Wu. F.X.: NMTF–DTI: a nonnegative matrix tri factorization approach with multiple kernel fusion for drug target interaction prediction. IEEE/ACM Trans. Comput. Biol. Bioinform. **20**(1), 586–594 (2023)
7. Tofighy, S., Charkari, N.M., Ghaderi, F.: Link prediction in multiplex networks using intralayer probabilistic distance and interlayer co-evolving factors. Physica A: Stat. Mech. Appl. **606**, 128043 (2022)
8. Yin, Y., Wu, Y., Yang, X., Zhang, W., Yuan, X.: SE–GRU: structure embedded gated recurrent unit neural networks for temporal link prediction. IEEE Trans. Netw. Sci. Eng. **9**(4), 2495–2509 (2022)
9. Lv, J., Yang, B., Yang, Z., Zhang, W.: A community-based algorithm for influence blocking maximization in social networks. Clust. Comput. **22**, 5587–5602 (2017)
10. Salavati, C., Abdollahpouri, A., Manbari, Z.: Ranking nodes in complex networks based on local structure and improving closeness centrality. Neurocomputing **336**, 36–45 (2019)

11. Dai, Z., Li, P., Chen, Y., Zhang, K., Zhang, J.: Influential node ranking via randomized spanning trees. Physica A: Stat. Mech. Appl. **526**, 120625 (2019)
12. Zhao, G., Jia, P., Zhou, A., Zhang, B.: InfGCN: identifying influential nodes in complex networks with graph convolutional networks. Neurocomputing **414**, 18–26 (2020)
13. Yang, H., An, S.: Critical nodes identification in complex networks. Symmetry **12**(1), 123 (2020)
14. Ozaydin, S.Y., Ozaydin, F.: Deep link entropy for quantifying edge significance in social networks. Appl. Sci. **11**(23), 11182 (2021)
15. Lubashevskiy, V., Ozaydin, S.Y., Ozaydin, F.: Improved link entropy with dynamic community number detection for quantifying significance of edges in complex social networks. Entropy **25**(2), 365 (2023)
16. Wang, W., Chen, G.: Universal robustness characteristic of weighted networks against cascading failure. Phys. Rev. E **77**(2), 026101 (2008)
17. Girvan, M., Newman, M.E.J.: Community structure in social and biological networks. Proc. Natl. Acad. Sci. **99**(12), 7821–7826 (2002)
18. Cheng, X., Ren, F., Shen, H., Zhang, Z., Zhou, T.: Bridgeness: a local index on edge significance in maintaining global connectivity. J. Stat. Mech. Theory Exp. **10**, P10011 (2010)
19. Liu, Y., Tang, M., Zhou, T., Do, Y.: Improving the accuracy of the k-shell method by removing redundant links from a perspective of spreading dynamics. Sci. Rep. **5**, 13172 (2015)
20. De Meo, P., Ferrara, E., Fiumara, G., Ricciardello, A.: A novel measure of edge centrality in social networks. Knowl.-Based Syst. **30**, 136–150 (2012)
21. Qian, Y., Li, Y., Zhang, M., Ma, G., Lu, F.: Quantifying edge significance on maintaining global connectivity. Sci. Rep. **7**(1), 45380 (2017)
22. Lubashevskiy, V., Lubashevsky, I.: Evolutionary approach for detecting significant edges in social and communication networks. IEEE Access **11**, 58046–58054 (2023)
23. Lubashevskiy, V.: Entropy-based approaches of edge significance quantification in complex networks: detection of link vulnerabilities using static, dynamic and group-focused methods. In: Skiadas, C.H., Dimotikalis, Y., (eds.) 16th Chaotic Modeling and Simulation International Conference, Springer Proceedings in Complexity. Springer Cham (2024)

Risks of Supply Chain Disruption and Market Concentration: Constructing Conceptual Models of Transaction Structures in Supply Chain Networks

Toko Sasaki[1,2]([✉]) and Akira Nagamatsu[2]

[1] Faculty of Business and Informatics, Niigata University of International and Information Studies, Niigata, Japan
tohko@nuis.ac.jp
[2] Graduate School of Engineering, Tohoku University, Sendai, Japan

Abstract. Gaining competitive advantage and market share is essential for business continuity and growth. However, under conditions of increased uncertainty (e.g. economic, geopolitical, technological, and environmental risks), the risk of supply chain disruptions also increases. Thus, market concentration and the risk of supply chain disruption are closely related, and how to balance supply chain resilience (information and technology sharing for risk dispersion) and competitiveness (price, quality, and delivery advantages) is an important issue. This study is to construct a conceptual model that captures phenomena such as supply fluctuations (disruptions and delays) and changes in trading relationships (overlaps and concentrations) that occur in supply chain networks (SCNs), and to visualize how supply chain disruption and market concentration affect SCNs by constructing two types of conceptual models: the SCNs disruption model and the SCNs overlap model. Although, it is very important to analyze SCNs with large and complex structures in more detail, this study considers it is also important to construct simple models, and combining them, quickly analysis.

Keywords: Supply Chain Network · Risk · Disruption · Market Concentration · Conceptual Model

1 Introduction

Supply chain networks (SCNs) are large and complex structures formed by many manufacturers and suppliers, which change significantly depending on the environment surrounding them and the passage of time. Transactions between companies and the flow of management resources constitute the links in the network, which interact with each other to form the entire supply chain. It is important to understand the SCN as a kind of phenomenon, rather than as a mere aggregate of organizations and companies. This is because changes in one element can affect other elements, which in turn can have a significant impact on the network as a whole. It can also be an important perspective in terms of risk management, productivity improvement, and strategy formulation.

© The Author(s), under exclusive license to Springer Nature Singapore Pte Ltd. 2024
S. Saito et al. (Eds.): AsiaSim 2024, CCIS 2170, pp. 285–298, 2024.
https://doi.org/10.1007/978-981-97-7225-4_22

For example, the supply chain serving Japan's automobile industry is a complex network of interlocking assembler–supplier relationships. This keiretsu structure historically has enabled Japan's automobile manufacturers to remain lean and flexible while exercising a degree of control over supply akin to vertical integration [1]. Japan's automobile industry had resembled a pyramid composed of several tiers of suppliers beneath main assembly groups or single assemblers [2].

However, during the 1990s, Japan's automobile industry faced serial crises, including the collapse of Japan's "bubble" economy, the Yen's appreciation against the US dollar, the 1995 Kobe earthquake. In response to these crises, Japan's automobile industry accelerated its shift to overseas production, and promoted establishment of a global supply network. Some Japanese companies supply parts to assemblers outside their keiretsu, numerous lower-tier suppliers provide their products across their keiretsu, and some major parts manufacturers supply the entire Japan's automobile industry. The supply of first-tier and second-tier parts (functional components) is decentralized among several suppliers, whereas the supply of lower-tier parts (simple components) is centralized in one company that uses specialized process technology [3]. In other words, while the lower tiers have become more concentrated in specific suppliers which have a competitive advantage in quality and price, the suppliers in first or second tier become more dispersed. Thus, Japan's automobile industry became a diamond-like structure.

In the pyramidal structure, when a disaster occurs and a supplier in the lower tier are affected, the damage is limited within the keiretsu group. However, in the diamond structure, the damage will spread throughout the automobile industry. The supply of components from lower-tier suppliers was interrupted during the earthquakes of 2007 and 2011, and the impact affected entire Japan's automobile industry. Thus, in the diamond structure of centralized lower-tier suppliers production stoppages at suppliers' plants affect nearly every Japan's automobile manufacturer.

To begin with, in the 1980s, before the focus on supply chain risk and vulnerability, Kraljic (1983) expressed supply uncertainty as the term 'supply risks' in his article where he used a matrix based on complexity of supply market and importance of purchasing to classify items into four categories: strategic items, bottleneck items, procurement leverage items, and non-critical items [4]. Subsequently, natural disasters such as earthquakes, floods, economic crisis, and pandemics have disrupted supply chain, it has become more important building supply chain risk management and creating resilient supply chain. For instance, Christopher and Peck (2004) suggested three categories (internal to the firm, external to the firm but in-ternal to the supply chain network, and external to the network) of risk which can be further sub-divided to produce a total of five categories (process risk, control risk, demand risk, supply risk, and environmental risk) [5]. Yossi and Rice described how resilient companies build flexibility into each of five essential supply chain elements: the supplier, conversion process, distribution channels, control systems, and underlying corporate culture. And they illustrated how building flexibility in these supply chain elements not only bolsters the resilience of an organization but also creates a competitive advantage in the marketplace [6]. Tang (2006) reviewed various quantitative models for managing supply chain risks, and related various supply chain risk management strategies examined in the research literature with actual practices [7]. Moreover, Wagner and Bode (2006) investigated the relationship between supply chain vulnerability and supply chain risk, supply chain characteristics such as a firm's dependence on certain customers and suppliers, the degree of single sourcing, or reliance on global supply sources are relevant for a firm's exposure to supply chain risk [8]. In terms of more empirical research using models, Klibi and Martel (2012) proposed a supply chain risk modeling approach to support the generation of plausible future scenarios including extreme events [9].

Some studies have attempted to understand the real structure of SCNs by using complex network analysis and social network analysis approaches using large-scale real data, especially from the automobile industry. For example, Kito (2013) used an automotive information platform to identify 2,196 suppliers of Toyota, and then used complex network science to characterize the Toyota supply network and thoroughly analyze its structure. This analysis made it possible to understand the robustness of this network, which could not be ascertained before [10]. Furthermore, Kito (2015) utilized the 'Survey of Production and Distribution of 200 Automotive Parts (2012)' by I.R.C. Co., Ltd., extracting twelve automotive manufacturers (assemblers) and 561 domestic suppliers for 202 automotive parts, and then investigated the structural heterogeneity of real-life supply networks in the automobile industry, through complex network analysis of large-scale empirical data. This study suggested that the structure of SCNs reflects not only the diversity of the characteristics of parts and the diversity of supplier manufacturing capabilities but also the diversity of automotive manufacturers' strategies [11]. Additionally, Okamoto (2016) utilized three points in time (1993, 2004, and 2013) from 'The Reality of the Japanese Automotive Parts Industry' by I.R.C. Co., Ltd., to extract inter-manufacturer inducement networks and conducted an analysis of supplier overlap. As a result, the overlap of suppliers simultaneously cooperating with multiple cooperative associations has significantly increased, suggesting a trend towards closer inter-manufacturer relationships for each vehicle type [12].

These previous studies on SCNs, supply chain disruptions, or supply chain risks can be broadly divided into two approaches: one that interprets phenomena qualitatively [4–6, 8] and the other that evaluates phenomena quantitatively [7, 9–12]. Furthermore, these quantitative approaches can be further divided into complex network analysis [10–12], articles review analysis [7], and simulation analysis [9].

It is very important to analyze SCNs with large and complex structures in more detail. Additionally, since these structures change over time and in different environments, analysis requires more data and time. However, disasters do not wait for these results. Therefore, this research considers it also important to construct simple models and use them for quickly and frequent analysis. The purpose of this research is to construct a conceptual model that quickly captures phenomena such as supply fluctuations (disruptions and delays) and changes in trading relationships (overlaps and concentrations) that occur in SCNs. This will help to understand and identify potential risks in advance. Furthermore, the conceptual model developed through this process is useful for balancing supply chain risks and competitive advantages and developing strategies to adapt to environment changes.

2 The Outline of the Conceptual Model for Transaction Structure of SCNs

2.1 The Area of the Conceptual Model

In general, SCNs can be divided into two main areas: the area from manufacturing plants through wholesale and retail to consumers, and the area from the supply of raw materials through component plants and semi-finished product assembly plants to manufacturing plants (see Fig. 1). This study focuses on the latter area.

2.2 Structures of the Conceptual Model

The Typical and Conceptual Models

It will explain the difference between the typical hierarchical structure commonly used to represent SCNs (see Fig. 2(a)) and the structure of the conceptual model developed in this study (see Fig. 2(b)). In the case of the typical model (an example), the number of suppliers in each tier is equal to the square of the tier number. In addition, each supplier receives orders from only one supplier one tier higher, so it is not possible to represent orders receipt from multiple suppliers. Assume that the manufacturer and all suppliers place an order the same number of units to two suppliers one tier lower, all suppliers in the same tier would receive orders for the same number of units.

On the other hand, in this conceptual model, the number of suppliers in each tier is equal to the tier number plus one. In addition, as each supplier can receive orders from two suppliers one tier higher, it is possible to represent the variation in transactions within the same tier, even if the above assumptions are made.

The Pyramid and Diamond Structures. The conceptual model comprises two types of structures: the pyramid structure and the diamond structure, each with five tiers of suppliers below the manufacturer (see Fig. 3). The pyramid structure has one manufacturer with 20 suppliers, while the diamond structure has one manufacturer with 12 suppliers. It shows that at the lower tiers of the diamond structure, demand is concentrated on a few suppliers with price and quality advantages and is becoming increasingly oligopolistic.

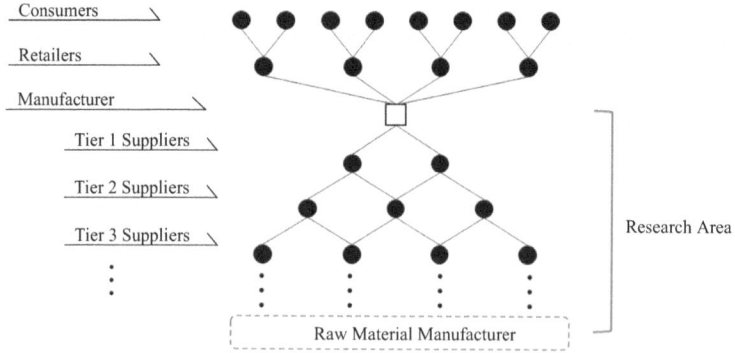

Fig. 1. The Area of the Conceptual Model

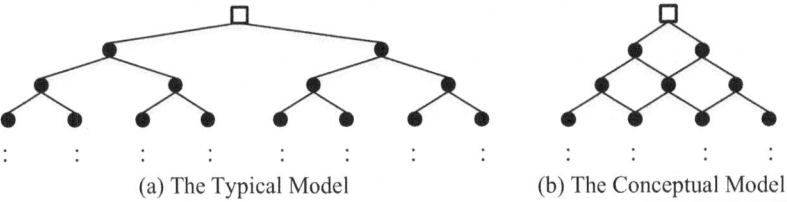

(a) The Typical Model   (b) The Conceptual Model

Fig. 2. The Structures of Typical and Conceptual Models

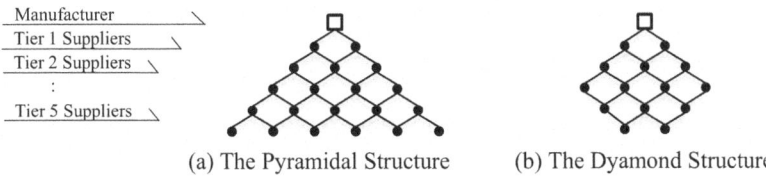

(a) The Pyramidal Structure   (b) The Dyamond Structure

Fig. 3. The Pyramid and Dyamond Structure

The Supplier Number. This model uses the supplier number. The tens digit in the number are the tier number, and the ones digit are the serial number. For example, no. 54 is the Tier 5 supplier within the pyramid structure, positioned fourth from the left within that tier, and no. 44 is the Tier 4 supplier, and located fourth from the left. The diamond structure uses the same tier numbers as the suppliers in the pyramid structure so that the corresponding suppliers can be identified (see Fig. 4).

2.3   Simulation Methods of the Conceptual Model

Simulation Tool. This conceptual model was built using Arena, a discrete event simulator, which enables the modeling of random events and fluctuations within systems. Arena is a versatile simulation tool that can be used across various industries and processes. In addition, this capability facilitates the creation of scenarios for assessing risks and efficiencies. By modeling SCNs with Arena, it becomes possible to evaluate and assess the impact of demand and supply fluctuations, transport delays, and to optimize inventory levels.

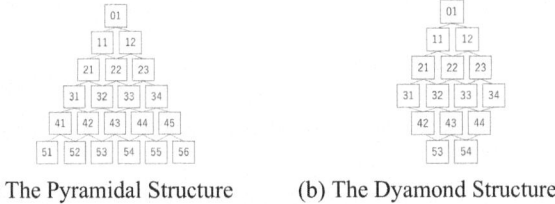

(a) The Pyramidal Structure   (b) The Dyamond Structure

Fig. 4. Supplier Numbers

Demand and Supply. The model consists of a simple demand structure (see Fig. 5), creating two types of entities: demand and supply. Demand entities create from the manufacturer at the top of the structure. This manufacturer places orders with the two Tier 1 suppliers, which in turn place orders with the two suppliers one tier lower, finally arriving the Tier 5 suppliers. The ratio of orders to two suppliers is given as a variable. Supply entities create from the Tier 5 suppliers at the bottom of the structure. Each supplier produces the number of units ordered and supplies to suppliers one tier higher. The time taken for supply (production time and transport time) is set as the lead time.

In this model, demand and supply are represented as 'units'. 'Units' can refer to quantities, time, frequency, etc., depending on the product, goods, or services.

Order Processing Method. This model adopts the backorder method. This method allows orders to be accepted even when inventory is insufficient. Orders placed during shortages are dispatched at a later date when the inventory becomes available.

In this model, the inventory levels of each supplier are update using the number of units assigned to the arrived entities.

3   The SCNs Disruption Model

3.1   The Outline of the SCNs Disruption Model

Scenarios. In this model, four scenarios were established for each structure (see Tables 1 and 2). The P1 and D1 scenarios assume that no disruptions occur. The P2 and D2 scenarios assume that supplier no. 54 stops production during emergencies such as natural disasters. The P3 and D3 scenarios assume that transport delays on no.54 to no.44 section is caused by traffic disruptions such as traffic jams and road closures. The P4 and D4 scenarios assume that the transport delays on all sections are caused by problems related to reduced transport capacity, such as "the 2024 problem". In Japan, there are concerns about various problems that will arise as a result of the cap on drivers' working hours starting in April 2024. One major concern is that transport capacity, particularly for long-distance transport, will be reduced.

Parameters. The parameters for each scenario are also shown in Tables 1 and 2. In all scenarios of both structures, the demand units per day from the manufacturer are 500 each, and the order ratio to two suppliers is 50% each. In the case of the pyramid structure, the supply units per day at suppliers no. 51, no. 52, no. 53, no. 54, no. 55, no. 56 are 16, 78,156, 156, 78, 16. In the case of the diamond structure, the supply units at suppliers no. 53 and no. 54 are both 250. However, in the P2 and D2 scenarios where supplier no. 54 stops production, the supply units of supplier no. 54 are set to 0. The lead times indicated for production time and transport time are one day each, except in the case of disruption of the section. The lead times for P3, D3, P4, and D4 in the section disruption scenario are set to two days, as it is assumed that alternative transport means and routes can be secured.

In the simulation experiments, the replication length is set to one hundred days. To accommodate lead time and initial inventory setup, observations for the first four days from the start of execution were excluded, results from the subsequent 96 days were used for analysis.

Fig. 5. The Basic Structure of Demand

Table 1. Scenarios of the SCNs Disruption (Pyramidal Structures)

Scenario no.	P1	P2
Disruption	-	supplier no. 54
No.01 Tier 1 Tier 2 Tier 3 Tier 4 Tier 5	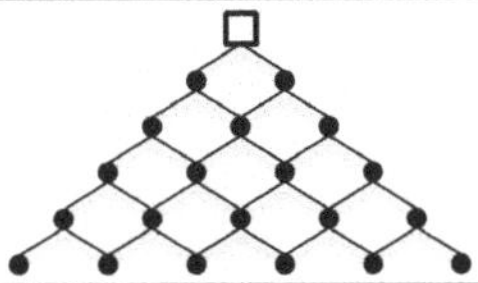	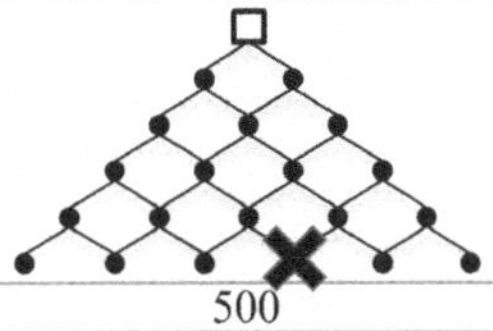
Demand_No.01	500	500
Order Ratio	50% each	50% each
Supply_Tier 5	16, 78, 156, 156, 78, 16	16, 78, 156, 0, 78, 16
Lead Time	all sections: 1 day	all sections: 1 day

Scenario	P3	P4
Disruption	no. 54 to no. 44 section	all sections
No.01 Tier 1 Tier 2 Tier 3 Tier 4 Tier 5	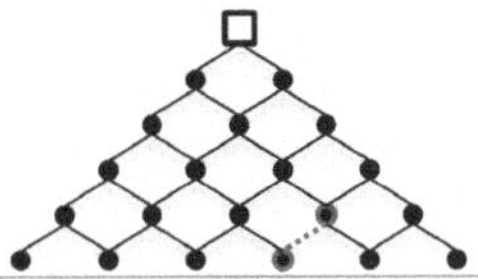	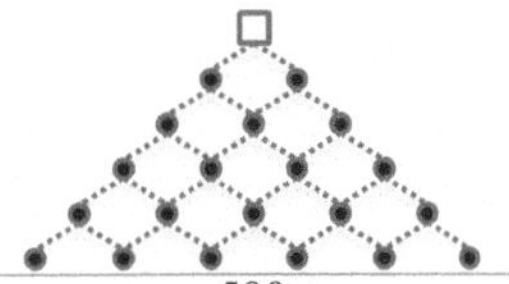
Demand_No.01	500	500
Order Ratio	50% each	50% each
Supply_Tier 5	16, 78, 156, 156, 78, 16	16, 78, 156, 156, 78, 16
Lead Time	no.54 to no.44 section: 2 days the other sections: 1 day	all sections: 2 days

3.2   The Simulation Results of the SCNs Disruption Model

The P1 and D1 Scenarios. The P1 and D1 scenarios assume that no disruptions occur. In Fig. 6, the horizontal axis indicates 'days' and ranges from five to 100. The vertical axis indicates 'units' and ranges from –50 to 50. When the vertical axis 'units' is positive, it indicates that supply exceeds demand, whereas when it is negative, it indicates that demand exceeds supply, thus implying a shortage of supply. In the P1 experiment results, minor effects due to rounding errors can be observed.

The P2 and D2 Scenarios. The P2 and D2 scenarios assume that supplier no. 54 stops production during emergencies. In Fig. 7, the vertical axis ranges from -25,000 to 50. D2 experiences a greater extent and range of damage due to disruptions compared to P2.

The P3 and D3 Scenarios. The P3 and D3 scenarios assume that transport delays on no.54 to no.44 section. In Fig. 8, the vertical axis ranges from –200 to 50. The damage from the section disruptions in P3 and D3 is much smaller than that from supplier disruptions in P2 and D2. In addition, since the delay in lead time only results in initial shortages, its impact on the entire supply chain remains minimal.

Table 2. Scenarios of the SCNs Disruption (Diamond Structures)

Scenario	D1	D2
Disruption	-	supplier no. 54
No.01 Tier 1 Tier 2 Tier 3 Tier 4 Tier 5		
Demand_No.01	500	500
Order Ratio	50% each	50% each
Supply_Tier 5	250, 250	250, 0
Lead Time	all sections: 1 day	all sections: 1 day

Scenario	D3	D4
Disruption	no. 54 to no. 44 section	all sections
No.01 Tier 1 Tier 2 Tier 3 Tier 4 Tier 5		
Demand_No.01	500	500
Order Ratio	50% each	50% each
Supply_Tier 5	250, 250	250, 250
Lead Time	no.54 to no.44 section: 2 days the other sections: 1 day	all sections: 2 days

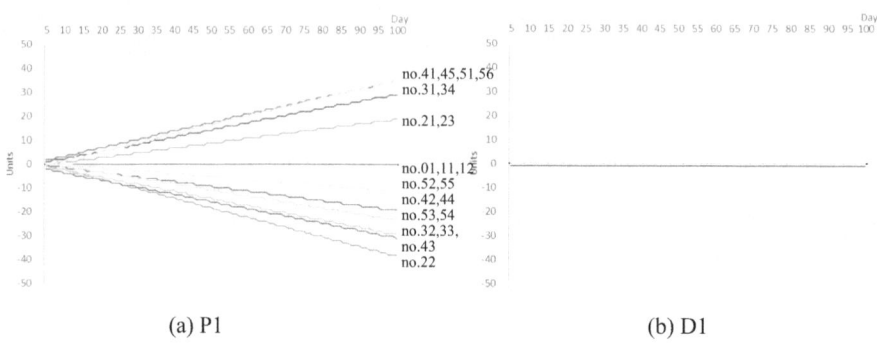

(a) P1   (b) D1

Fig. 6. The Experiment Results of No Disruption

The P4 and D4 Scenarios. The P4 and D4 scenarios assume that the transport delays on all sections. In Fig. 9, the vertical axis ranges from −2,500 to 50. There is a significant initial stockout, but it stabilizes to a certain level after ten days that corresponding to the number of days it takes the first supply from Tier 5 to reach the manufacturer in the top of structure. Therefore, it is possible to mitigate the risk of shortages by setting initial inventory levels relatively high.

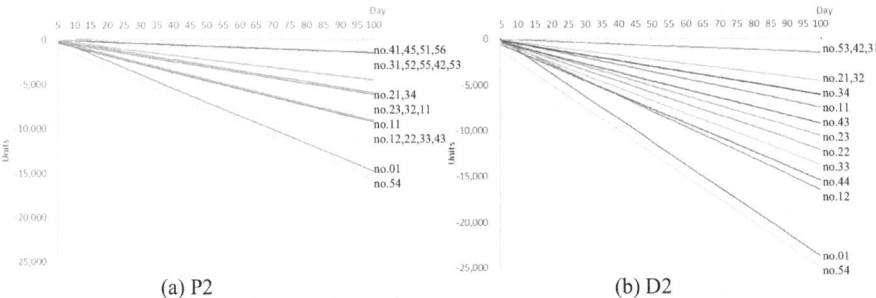

Fig. 7. The Experiment Results of Supplier No. 54 Disruption

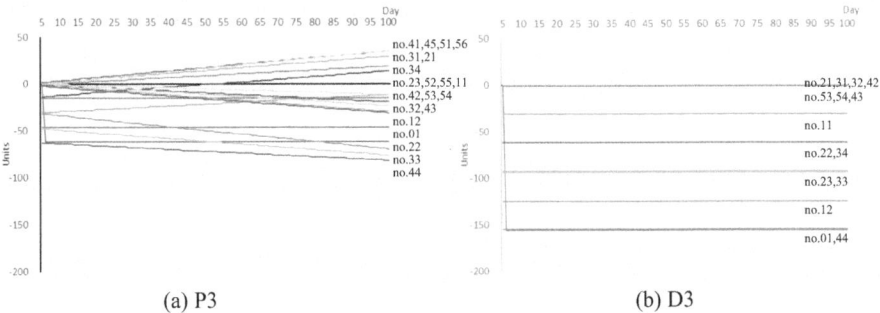

Fig. 8. The Experiment Results of the No.54 - No.44 Section Disruption

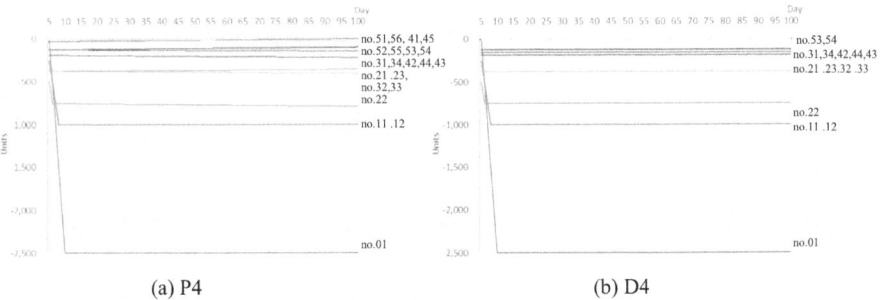

Fig. 9. The Experiment Results of the All-Sections Disruption

Average Units for Each Suppliers. Figure 10 shows the average daily units for each supplier over a period of 96 days, excluding the first four days, using a color scale. When comparing between both structures, the spread of impact is the same, but the degree of stockout is not the same and greater for the diamond structure. In the case of the no.54 disruption, the average units of the manufacturer are –7,409.27 in P2 and –11,875.00 in D2.

4   The SCNs Overlap Model

4.1   The Outline of the SCNs Overlap Model

The simulation experiments conducted using the conceptual model described earlier captured the spread and extent of the impact of disruptions that occur within SCNs, which consist of a single manufacturer and multiple suppliers. In this chapter, the examination of transaction structure alterations within SCNs formed by multiple manufacturers and multiple suppliers will be delved into.

In here, overlap is defined as a situation where a supplier supplies more than one manufacturer and indicates a situation where the supplier maintains existing relationships and builds relationships with new industries and sectors. Concentration also indicates the ratio of transactions within the same tier. The degree of overlap and concentration depends on the interval between manufacturers. This SCNs overlap model uses probabilities and indicators.

4.2   The Overlap and the Concentration of the Pyramidal Structures

In this model, three patterns with different spacing were established for each structure (see Tables 3 and 4). "Spacing" refers to the interval between manufacturers, and implies the nature of the relationship between them. If the relationship is competitive but close, the spacing is small.

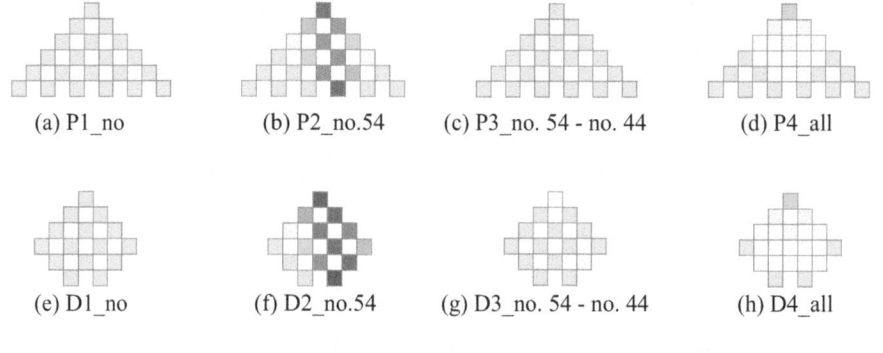

(a) P1_no   (b) P2_no.54   (c) P3_no. 54 - no. 44   (d) P4_all

(e) D1_no   (f) D2_no.54   (g) D3_no. 54 - no. 44   (h) D4_all

Average Units: ■ 15k ■ 12k ■ 9k ■ 6k ▨ 3k ☐ <1k☐ <150

Fig. 10.   Average Units.

Table 3 shows that the pattern of P_Dis. 6 is a configuration where multiple manufacturers form a pyramid structure, each with independent SCNs (Spacing of manufacturers is 6). P_Dis. 2 refers to the arrangement where the spacing of manufacturers is 2, while P_Dis. 1 denotes a spacing of one between manufacturers. In P_Dis. 6, there are no suppliers where two SCNs and three SCNs overlap. In the P_Dis. 2, two SCNs overlap occurs at 14 suppliers, and three SCNs overlap occurs at three suppliers. In P_Dis. 1, two SCNs overlap occurs at ten suppliers, and three SCNs overlap occurs at ten suppliers.

Figure 11 shows the concentration of each supplier of pyramidal structures. The higher the concentration, the darker the color; the lower the concentration, the lighter the color. In P_Dis. 6, since the SCNs are independent, transactions are limited to the same network, and the concentration follows a normal distribution within each tier of the SCN (see Fig. 11(a)). In P_Dis. 2, transactions beyond the conventional framework of SCNs occur, and at the center of the lower tier, there occur an overlap of three SCNs (i.e., the supplier group supplying to all manufacturers), but the concentration tends to be somewhat equalized in the center of each tier (see Fig. 11(b)). In P_Dis. 1, the range of overlap expands further, and the number of suppliers supplying to all manufacturers increases. As a result, the concentration of suppliers in the middle of each tier increases significantly (see Fig. 11(c)).

4.3 The Overlap and the Concentration of the Diamond Structures

Table 4 shows that the pattern of D_Dis. 4 is a configuration where multiple manufacturers form a diamond structure, each with independent SCNs (Spacing of manufacturers is 4). D_Dis. 2 refers to the arrangement where the spacing of manufacturers is 2, while D_Dis. 1 denotes a spacing of one between manufacturers. In D_Dis. 4, there are no suppliers where two SCNs and three SCNs overlap. In D_Dis. 2, two SCNs overlap occurs at 8 suppliers. In D_Dis. 1, two SCNs overlap occurs at 10 suppliers, and three SCNs overlap occurs at 4 suppliers.

Table 3. The SCNs Overlap (Pyramidal Structures)

Pattern	P_Dis. 6	P_Dis. 2	P_Dis. 1
Distance	6	2	1
2 Overlaps	0	14 Suppliers	10 Suppliers
3 Overlaps	0	3 Suppliers	10 Suppliers

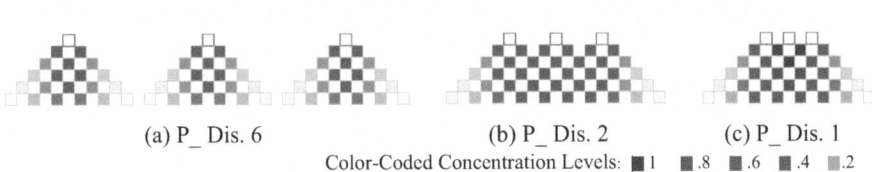

(a) P_Dis. 6   (b) P_Dis. 2   (c) P_Dis. 1

Color-Coded Concentration Levels: ■1 ■.8 ■.6 ■.4 ■.2

Fig. 11. The Concentrations of Pyramidal Structures

Figure 12 shows the concentration of each supplier of diamond structures. In D_Dis. 4, since the SCNs are independent, transactions are limited to the same network, and compared to P_Dis. 6, the concentration of lower-tier suppliers is relatively high (see Fig. 12(a)). In D_Dis. 2, transactions beyond the conventional framework of SCNs occur, but the concentration tends to be high even both ends of the lower tiers (see Fig.

Fig. 12(b)). In D_Dis. 1, at the center of the structure, there occur an overlap of three SCNs, and resulting in higher concentration of lower-tier suppliers compared to P_Dis. 1 (see Fig. 12(c)).

4.4 The Concentration in Each Tier

The term "HHI" means the Herfindahl–Hirschman Index, a commonly accepted measure of market concentration. The HHI takes into account the relative size distribution of the firms in a market. It approaches zero when a market is occupied by a large number of firms of relatively equal size and reaches its maximum of 1 (10,000 for %) when a market is controlled by a single firm. The HHI increases both as the number of firms in the market decreases and as the disparity in size between those firms increases [13].

In this case, the HHI is applied to assess the degree of concentration in the tier. It allows one to quantify the state of each tier (concentration of transactions). Table 5 shows HHI values for each tier in each pattern, the values are color-coded. In a pyramid structure, the HHI value decreases as the lower tiers. On the other hand, in the diamond structure, concentration occurs at the lower tiers even before the overlap occurs, and as the overlap increases, the HHI values increase significantly at all levels.

Table 4. The SCNs Overlap (Diamond Structures)

Pattern	D_Dis. 4	D_Dis. 2	D_Dis. 1
Distance	4	2	1
2 Overlaps	0	8 Suppliers	10 Suppliers
3 Overlaps	0	0 Suppliers	4 Suppliers

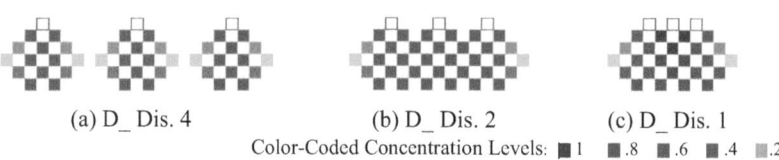

(a) D_ Dis. 4   (b) D_ Dis. 2   (c) D_ Dis. 1

Color-Coded Concentration Levels: ■ 1   ■ .8   ■ .6   ■ .4   ■ .2

Fig. 12. The Concentrations of the Diamond Structure

5 Discussion

In this study, by building the two types of conceptual models: the SCNs disruption model and the SCNs overlap model, it was possible to visualize how disruption and concentration affected the SCNs. These models are very simple and easy to understand the propagation of the supply chain disruptions and the concentration of the overlap. While it needs to be carefully considered for validation and verification when building models of real-world SCNs, which are very large, complex, and constantly changing.

Table 5. HHI Values

Pattern	P_Dis. 6	P_Dis. 2	P_Dis. 1	D_Dis. 4	D_Dis. 2	D_Dis. 1
Suppliers	60	40	30	42	34	24
2 Overlaps	0	14	10	0	8	10
3 Overlaps	0	3	10	0	0	4
No.01	1.00	1.00	1.00	1.00	1.00	1.00
(Suppliers)	(3)	(3)	(3)	(3)	(3)	(3)
Tier 1	0.50	0.50	0.83	0.50	0.50	0.83
(Suppliers)	(6)	(6)	(4)	(6)	(6)	(4)
Tier 2	0.38	0.46	0.75	0.38	0.46	0.75
(Suppliers)	(9)	(7)	(5)	(9)	(7)	(5)
Tier 3	0.31	0.44	0.69	0.31	0.44	0.69
(Suppliers)	(12)	(8)	(6)	(12)	(8)	(6)
Tier 4	0.27	0.42	0.64	0.34	0.44	0.66
(Suppliers)	(15)	(9)	(7)	(9)	(7)	(5)
Tier 5	0.25	0.41	0.60	0.50	0.50	0.75
(Suppliers)	(18)	(10)	(8)	(6)	(6)	(4)

Color-Coded HHI Values: 1 0.8 0.6 0.4 0.2 0

When applied to real-world transaction situations, more advanced modeling techniques and validation based on real data are required.

In the context of SCN disruptions, it is particularly important to connect early relief, early alternative production, and early alternative procurement in order to achieve rapid recovery and restoration. This requires preparedness not only during disasters but also during normal times, with a high level of awareness and recognition of the need to build inter-company networks, visualize SCNs, and formulate and revise the Business Continuity Plans (BCPs), and to translate this awareness into concrete actions. However, there is a large gap between raising awareness of SCN disruptions and translating that awareness into action. Furthermore, databases, information systems, and BCPs, once established, need to be constantly reviewed and updated, and there are real differences between companies (especially between large and small companies) in terms of awareness, actions, and continuity. The conceptual models developed in this study bridge the gap between "awareness and action" of supply chain risks, "establishment and continuity" of networks and BCPs, and disparities among companies.

6 Conclusion

In order to better understand the interactions and interrelationships of SCNs, this study used a conceptual model to clarify relationships and patterns that are difficult to capture using only numerical data. The conceptual model is expected to be useful not only in the transaction structure of SCNs, but also in various situations such as supply chain risk management, productivity improvement, and management strategy formulation.

The risk of supply chain disruption and concentration (competitive advantage) due to overlapping supply chain networks are closely related as shown in this study, and how to balance supply chain resilience and competitiveness is also an important issue.

References

1. Ahmadjian, C.L., Lincoln, J.R.: Keiretsu, Governance, and Learning: Case Studies in Change from the Japanese Automotive Industry. Institute of Industrial Relations University of California, Berkeley, Working Paper, No.76 (2000). https://www.jstor.org/stable/3086041
2. Shimokawa, K.: The Japanese Automobile Industry, Continuum International Publishing Group Ltd (2000)
3. Fujimoto, T.: Supply Chain Competitiveness and Robustness: A Lesson from the 2011 Tohoku Earthquake and Supply Chain 'Virtual Dualization'. Manufacturing Management Research Center Discussion Paper Series, No. 362 (2011). http://merc.e.u-tokyo.ac.jp/mmrc/dp/pdf/MMRC362_2011.pdf
4. Kraljic, P.: Purchasing Must Become Supply Management. Harvard Bus. Rev. **883509**, 109–117 (1983)
5. Christopher, M., Peck, H.: Building the resilient supply chain. Int. J. Logistics Manage. **15**(2), 1–13 (2004). https://doi.org/10.1108/09574090410700275
6. Yossi, S., Rice, J.B.: A Supply Chain View of the Resilient Enterprise. MIT Sloan Manage. Rev. **47**(1), 41–48 (2005). https://sloanreview.mit.edu/article/a-supply-chain-view-of-the-resilient-enterprise/
7. Tang, C.S.: Perspectives in supply chain risk management. Int. J. Prod. Econ. **103**(2), 451–488 (2006). https://doi.org/10.1016/j.ijpe.2005.12.006
8. Wagner, S.M., Bode, C.: An empirical investigation into supply chain vulnerability. J. Purch. Supply Manage. **12**(6), 301–312 (2006). https://doi.org/10.1016/j.pursup.2007.01.004
9. Klibi, W., Martel, A.: Scenario-based supply chain network risk modeling. Eur. J. Oper. Res. **223**(3), 644–665 (2012). https://doi.org/10.1016/j.ejor.2012.06.027
10. Kito, T.: Structural robustness of real-world supply chains: a complex network approach. J. Inf. Process. Trans. Math. Model. Appl. **6**(2), 174–181 (2013)
11. Kito, T.: Analysis of diversity of characteristics of auto-parts and product portfolios of suppliers by bipartite network projection. Trans. Jpn Soc. Artif. Intell. AI **30**(6), 721–728 (2015)
12. Okamoto, T.: Supplier overlap between Japanese automobile manufacturers: transition of affiliation network in three points in time. J. Product Dev. Manage. 12(2) (2015–2016)
13. U.S. Department of Justice. https://www.justice.gov/atr/herfindahl-hirschman-index

Modeling, Simulaiton, and Visualization of Digital Twin

A Digital Twin-Based Obstacle Avoidance System for Quadcopter UAV

Ziwen Zhan, Ke Fang[(✉)], and Ju Huo

Control and Simulation Center, Harbin Institute of Technology, Harbin, China
fangke@hit.edu.cn

Abstract. This paper introduces the development of a digital twin system designed for quadrotor UAVs (Unmanned Aerial Vehicle). It primarily utilizes AppDesigner and Unity to construct virtual models and a central server for the UAVs. Additionally, path planning for the UAVs is performed using the A-star algorithm, with improvements made to address potential collision issues in traditional algorithms. It brings forward experiments including virtual-real interaction and path flying which prove the effectiveness of the digital twin system. The system facilitates UAV control and obstacle avoidance research.

Keywords: Digital Twin · Simulation System · Quadrotor · A-star Algorithm · Obstacle Avoidance

1 Introduction

In recent years, the advancement of multi-rotor drones has been nothing short of remarkable, permeating various industries with unprecedented efficiency. Among these, quadrotors have emerged as a quintessential innovation, finding applications in agriculture [1], traffic surveillance [2], entertainment [3], and educational realms [4]. Recently, the research scope concerning quadrotors has spanned control, localization, simulation, and path planning [5]. Notably, path planning for quadrotors has garnered considerable attention among researchers. Particularly in an era characterized by extensive transportation networks, the imperative of devising safe, feasible, and energy-efficient flight paths for UAVs (Unmanned Aerial Vehicle) is self-evident. Commonly employed path planning methodologies for quadrotors encompass the A-star algorithm [6], the AFP algorithm [7], the minimum-snap algorithm [8], among others.The A-star algorithm is a search algorithm for finding the shortest path in a graph, usually applied to graph search and path planning problems. A-star algorithm has the characteristics of completeness and optimality, but it treats the UAV as a mass point, and the searched paths do not take into account the dimensions of the UAV itself [9], which may collide with the obstacles in the process of flight, causing a serious safety problem. In this paper, we propose a method to improve the A-star algorithm so that the planned paths are safe and collision free.

As we all know, an algorithm has a long way to go from being proposed to actually being applied to a real quadrotor UAV. It has to be validated by constant and repeated

S. Saito et al. (Eds.): AsiaSim 2024, CCIS 2170, pp. 301–312, 2024.
https://doi.org/10.1007/978-981-97-7225-4_23

experiments to test and improve the stability and reliability of the algorithm. This is a large and tedious task, but it is essential. To reduce the cost of validation, simulation has been continuously applied to quadrotor UAVs. However, simulation alone is not sufficient to validate the reliability of the method. Not only is the accuracy of the UAV modelling not up to the required level, but the fact that the UAV itself is only modelled, ignoring the environment in which it flies, does not really reflect the actual flight conditions. Furthermore, the simulation process does not really use the actual flight data, which also makes the simulation unconvincing.

Digital twin is a good solution to the above problems. Compared with simulation technology that lacks actual flight, digital twin makes good use of the actual flight data of UAVs; at the same time, digital twin can respond to the entire life cycle of UAVs, which is conducive to the maintenance and fault diagnosis of UAVs [10]. Since digital twin can realize two-way interaction between virtual space and physical space, which means that UAVs, after being trained and optimized in the virtual environment, can be directly applied to the actual UAV for seamless integration and migration. Since the concept of the digital twin was proposed by Prof. Michael Grieves of the University of Michigan [11], it has evolved significantly in recent years. It is generally recognised that a complete digital twin system consists of six parts [12]: virtual space, physical space, connectivity, data and algorithms, user interface and application scenarios. The virtual space is the virtual environment created by computer simulation; the physical space is the actual object, in this paper the application scenario refers to the quadrotor UAV; the connection part connects the virtual space with the physical space through network communication, data transmission and other means, so as to realise the interaction between the virtual space and the physical space, which is a major difference between digital twin and simulation; the data and algorithm part is responsible for collecting and processing the large number of digital twins generated by the digital twin system, including the historical flight data of the UAV and the simulation data generated by the simulation; the user interface provides a way for the user to interact with the digital twin system, which is usually a graphical interface or a command line window; for different application scenarios, it is necessary to customize different digital twins, and this paper proposes a digital twin system for a quadrotor UAV in the path planning application scenario.

The quadrotor UAV digital twin system constructed in this paper consists of actual quadrotor UAVs, as well as virtual quadrotor UAVs and virtual environments built using Unity. They communicate with the server center via the UDP protocol to achieve inter-action between the virtual and the real. To validate the effectiveness of this platform, we improved the traditional A-star algorithm to achieve collision-free path planning and integrated it into our digital twin platform.

2 Design of Digital Twin System

2.1 System Architecture

The digital twin architecture proposed in this paper is illustrated in Fig. 1. Physical UAVs transmit entity data to the server center, while virtual UAVs transmit simulation data to the server center. The server center receives and processes this data, performing data fusion. Additionally, through the server center, virtual UAVs are dynamically mapped onto physical UAVs in real time.

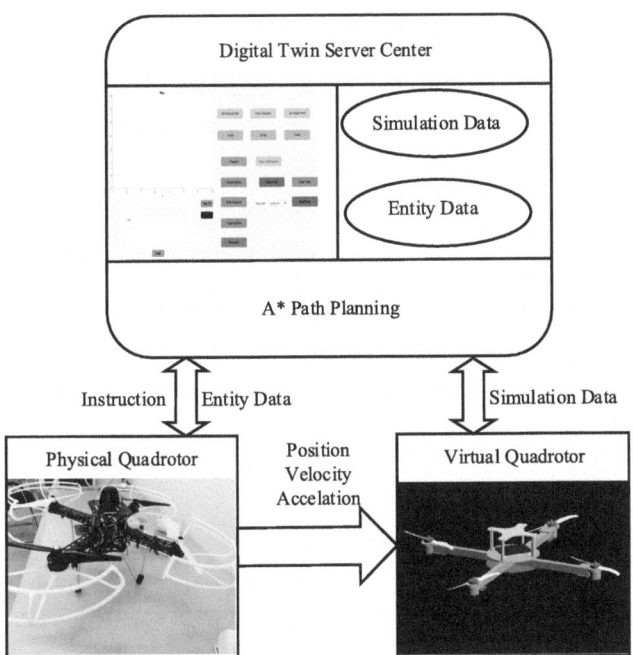

Fig. 1. Digital twin system architecture

The server center also deploys the improved collision-free A-star algorithm. Users can customize the starting point, destination, and obstacles in the user interface. After generating the path, the server center sends the path points to the flight controller at a fixed frequency.

2.2 Design of Physical Quadrotor

The quadrotor UAV used in this paper are as shown in Fig. 2. The flight controller utilizes an STM32F4 MCU (Microcomputer Unit) with a clock frequency of up to 168MHz, capable of meeting computational demands. The flight control algorithm employs cascaded PID control, as shown in Fig. 3.Cascade PID has two loops, the position loop and the attitude loop. The output of the position loop serves as the input for the attitude loop, while the output of the attitude loop serves as the control input for the system. This structure enables more precise response and stable control.

2.3 Design of Virtual Quadrotor

A virtual model includes geometric model and physical model.Geometric model is a precise description of the geometric characteristics of physical entities, which is used

Fig. 2. Physical quadrotor using in this system

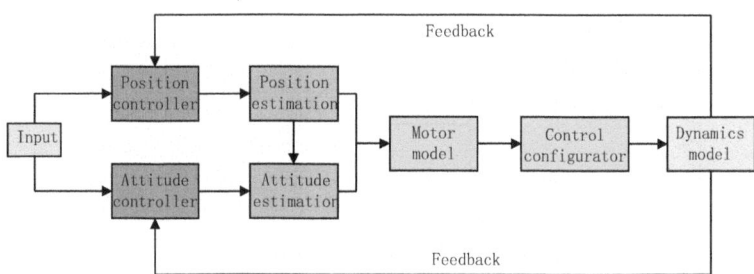

Fig. 3. Cascade PID Control

to support the visualisation of the virtual quadrotor. There are many ways to construct geometric models, and commonly used tools are Gazebo, Unity, etc. Unity is a cross-platform game development engine that is widely used in game development, virtual reality (VR), augmented reality (AR), 3D modelling and other fields. Due to the high-fidelity visualisation characteristics of Unity, this paper uses Unity to build the geometric model of the drone. The virtual environment of this paper is built based on Windows 10 operating system and unity, the geometric model of the quadrotor UAV is shown in Fig. 4.

The physical model of the quadrotor UAV comprises a flight dynamics model and a kinematics model. The kinematic equation is represented by Eq. 1. Illustrated in Fig. 5, the flight dynamics model elucidates the interaction between motor torque and paddle rotation, subsequently influencing the UAV's position and orientation through generated torque and lift. Parameters such as mass, rotational inertia, and pulling moment coefficient are typically acquired through measurement and parameter identification techniques. The resultant equations governing quadrotor UAV dynamics are presented in Eq. 2 and Eq. 3.

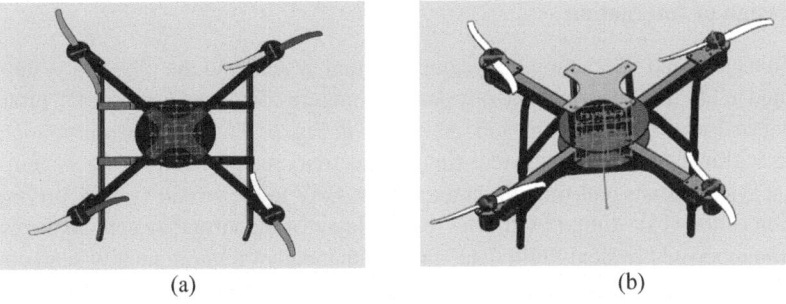

(a)   (b)

Fig. 4. Geometric model of the quadrotor.

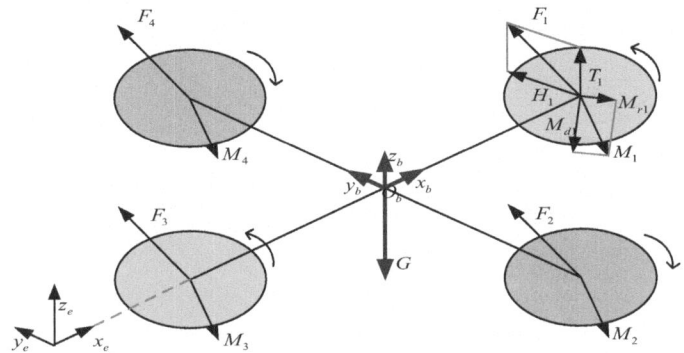

Fig. 5. Quadrotor Unmanned Aerial Vehicle Dynamics Model

$$\dot{\xi} = V \tag{1}$$

$$m\dot{V} + \omega \times mV = \sum_{i=1}^{4} (F_i + R + G) \tag{2}$$

$$J\dot{\omega} + \omega \times J\omega = \sum_{i=1}^{4} (M_i + F_i \times r_i) \tag{3}$$

The above model takes into account the gyroscopic moment, which has less effect on UAV flight, but results in greater computational effort, so the refined model of Eq. 4 and Eq. 5 is more commonly used in practice.

$$m\dot{V} = \sum_{i=1}^{4} (F_i + G) \tag{4}$$

$$J\dot{\omega} = \sum_{i=1}^{4} M_i \tag{5}$$

2.4 Design of Interaction

The connection and interaction between the virtual system and the physical system are illustrated in Fig. 7. The virtual and physical systems are connected via the UDP protocol, ensuring real-time data transmission. As shown in Fig. 6, the data transmission latency remains within 2 ms, meeting the real-time connectivity requirements of the system. The physical system sends real-time flight data of the UAV to the virtual system, driving the operation of the UAV digital twin. The service layer of the virtual system analyzes and filters the received physical flight data, and can manually or automatically send control commands to the physical system, promoting efficient system operation.

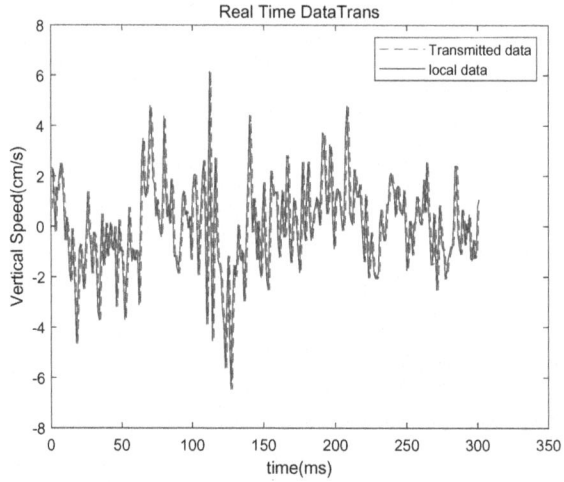

Fig. 6. Real-time data transmission

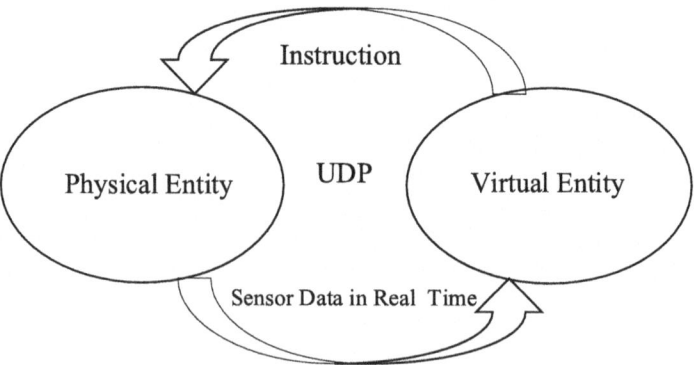

Fig. 7. Subsystem connection relationship

2.5 Design of User Interface

A graphical interface is designed as the user interface for the digital twin system. AppDesigner, a toolkit offered by Matlab, is employed for crafting interactive applications, leveraging its high compatibility and seamless deployment capabilities. Figure 8 delineates the user interface of the quadrotor UAV digital twin, meticulously crafted using AppDesigner. Through this interface, real-time control over UAV takeoff and landing is facilitated, enabling users to monitor the UAV's state and conduct simulation verification and path planning with ease.

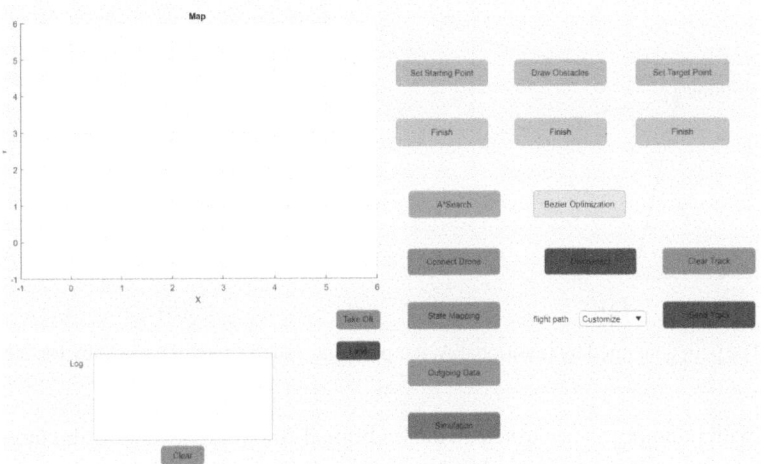

Fig. 8. User interface designed by AppDesigner

3 A-Star Algorithm and Improvement

3.1 A-Star Algorithm and Bezier Curve

The A-star algorithm, a global search algorithm, adeptly determines the optimal path by evaluating the cost function at each step.

$$f(n) = g(n) + h(n) \tag{6}$$

$$h(n) = \sqrt{(x_n - x_{goal})^2 + (y_n - y_{goal})^2} \tag{7}$$

The cost function, as expressed in Eq. 6, encompasses parameters such as the current node n, the cost function $f(n)$ associated with the node, the cost $g(n)$ from the starting point to the current node n, and the estimated cost $h(n)$ from the current node to the goal point n. In this equation, $h(n)$ represents the Euclidean distance, defined as illustrated in Eq. 7, where x_n and y_n denote the coordinates of the current grid center, while x_{goal}

and y_{goal} denote the coordinates of the target grid center. While the traditional A-star algorithm adeptly devises global paths, the resultant path often exhibit excessive folds, turns, and significant fold angles, as depicted in Fig. 9(a), rendering them unsuitable for UAV navigation. Bezier curves are generally used for the smoothing of polylines, as they are simple and effective to construct. In this paper, n-order Bezier curves are used to smooth the polyline paths generated by the A-star algorithm, making any point on the path n-order differentiable. The smoothed curve is shown in Fig. 9(b).

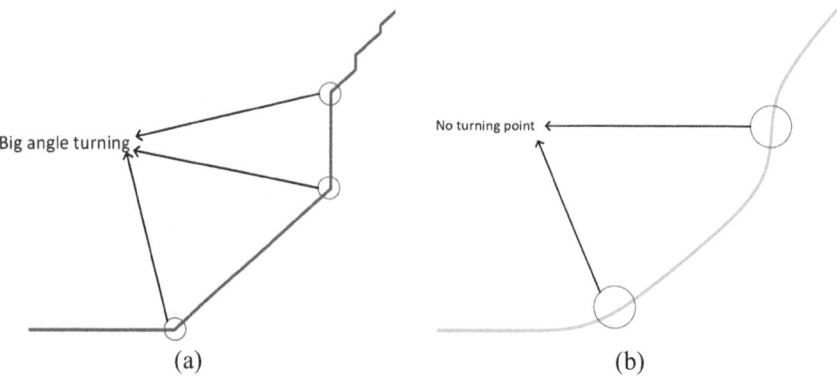

(a)   (b)

Fig. 9. The path generated by traditional A-star planning and the path optimized by Bezier curves

The path planning routes based on the traditional A-star algorithm exhibit large turning corners and multiple steering. As shown in Fig. 10, in comparison, the route improved by Bezier curves exhibits smoother and more optimized characteristics, significantly reduces the corners and turnbacks, and shows better path planning results.

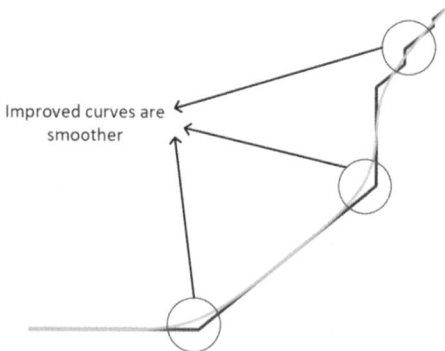

Fig. 10. Comparison between before and after improvement

3.2 Collision-Free A-Star Algorithm

The A-star algorithm employs an Open List to represent nodes awaiting exploration and a Closed List for nodes already explored or deemed unnecessary for exploration. As depicted in Fig. 11(a), by adding obstacle-representing nodes to the Closed List, A-star algorithm circumvents obstacles during map traversal, thus achieving obstacle avoidance. However, traditional A-star algorithms treat the UAV as a point mass, resulting in scenarios like Fig. 12(a), where the planned path narrowly skirts obstacle edges, risking collisions.

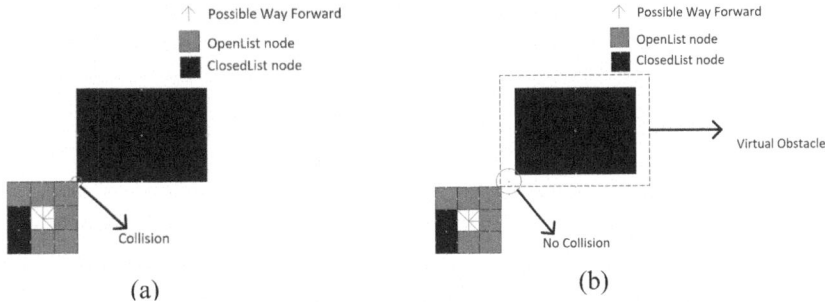

Fig. 11. Comparison between before and after improvement

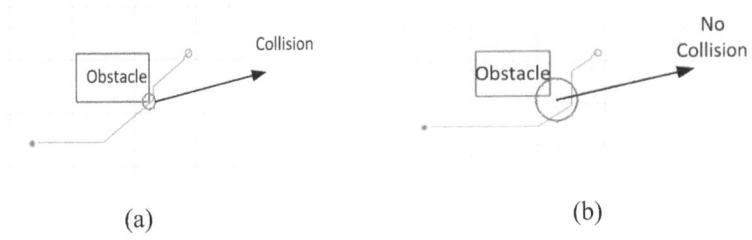

Fig. 12. Comparison between before and after improvement

Leveraging the fundamental obstacle avoidance principle of A-star, this paper proposes an improved A-star algorithm through virtual obstacle insertion. Illustrated in Fig. 11(b), this method entails expanding obstacle boundaries to create virtual obstacles. During optimal path traversal, A-star treats these virtual obstacles as real, adding their nodes to the Closed List, ensuring virtual obstacle avoidance. While contact with virtual obstacle edges may occur, since these are extensions of actual obstacles, planned paths avoid actual collisions. Thus, it constitutes a safe, collision-free planning algorithm as shown in Fig. 12(b).

4 Case Study

This section validates the performance and application effectiveness of the designed quadrotor UAV digital twin system through two experiments: virtual-real interaction and obstacles avoidance.

4.1   Virtual-Real Interaction Experiment

The server center establishes real-time communication with the UAV, receiving live data streams including key parameters such as position, velocity, and attitude. Subsequently, the server center transmits these data via UDP connection to the virtual space constructed by Unity, ensuring synchronization of the virtual UAV's state (position, velocity, attitude, etc.) with that of the physical UAV, as illustrated in Fig. 13. Additionally, the server center is equipped with the capability to send commands to the UAV, such as takeoff and landing. This facilitates real-time interaction and synchronization between the virtual space and the physical UAV.

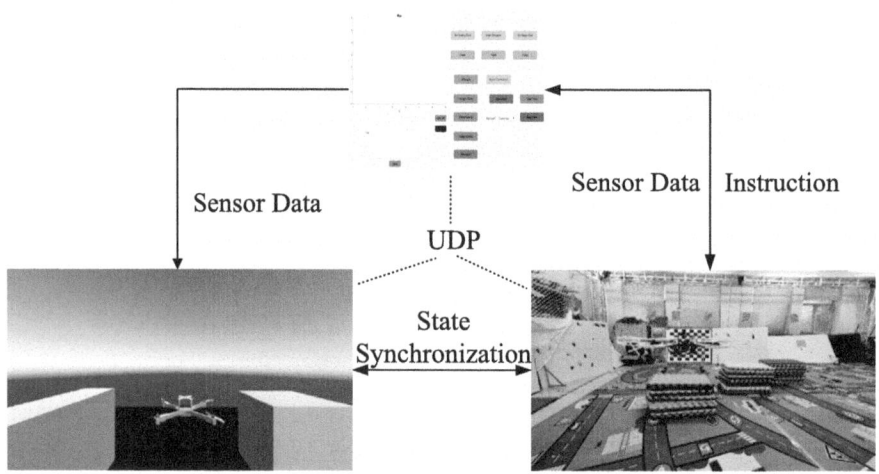

Fig. 13. Synchronization of physical and virtual UAV states

4.2   Path Following Experiment

The server center initially calibrates the starting and ending points of the UAV and utilizes the improved A-star algorithm to plan a collision-free path. Subsequently, the server center transmits the path points to the physical UAV, guiding it along the planned path from the starting point to the endpoint. Simultaneously, the virtual UAV within the virtual space follows the planned path from the starting point to the endpoint. By comparing the flight paths of the physical UAV and the virtual UAV in the virtual space, the effectiveness and reliability of the improved A-star algorithm in practical applications are analyzed and evaluated.

The planned paths and the actual flight paths of the UAV are depicted in Fig. 14, while the flight process is illustrated in Fig. 15. From the experimental results, it is evident that the improved A-star algorithm can effectively navigate around obstacles, thus enabling the digital twin system to efficiently guide the operation of the physical UAV.

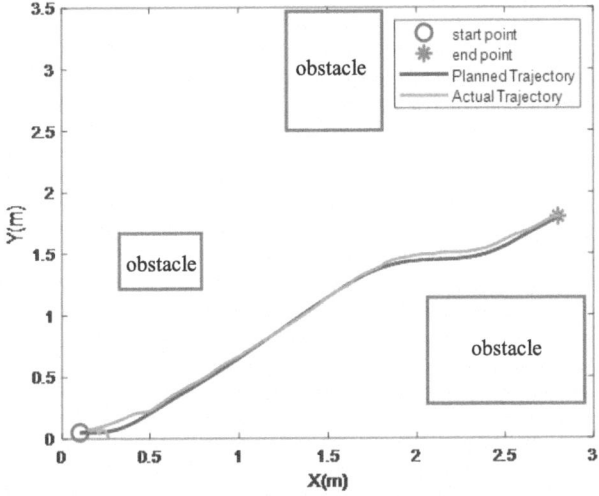

Fig. 14. Planned path and actual path

Fig. 15. Path following flight process

5 Conclusion and Future Work

In summary, this paper successfully establishes a digital twin system for quadrotor UAVs by utilizing AppDesigner and Unity to construct virtual UAVs and a central server. Path planning is conducted using the A-star algorithm, with improvements made to address collision issues inherent in traditional algorithms. Subsequent experiments involving virtual-real interaction and path flying validate the effectiveness of the developed quadrotor UAV system. This research provides substantial support for accelerating and guiding UAV research, laying a solid foundation for future exploration and development in this field. In the future, efforts will continue to strengthen research on virtual models of UAVs to reduce disparities between physical and virtual UAVs. Additionally, there will be a focus on researching more efficient variants of the A-star algorithm to shorten computation time. Furthermore, exploration will be conducted into the potential application of UAV digital twin in practical fields to enhance operational efficiency and safety.

References

1. Kim, J., Kim, S., Ju, C., et al.: Unmanned aerial vehicles in agriculture: a review of perspective of platform, control, and applications. IEEE Access **7**, 105100–105115 (2019)
2. Gupte, S., Mohandas, P.I.T., Conrad, J.M.: A survey of quadrotor unmanned aerial vehicles. Proc. IEEE Southeastcon **2012**, 1–6 (2012)
3. Kim, S.J., Jeong, Y., Park, S., et al.: A survey of drone use for entertainment and AVR (augmented and virtual reality). In: Augmented Reality and Virtual Reality: Empowering Human, Place and Business, pp. 339–352 (2018)
4. Giernacki, W., Skwierczyński, M., Witwicki, W., et al.: Crazyflie 2.0 quadrotor as a platform for research and education in robotics and control engineering. In: 2017 22nd International Conference on Methods and Models in Automation and Robotics (MMAR), pp. 37–42. IEEE (2017). http://www.springer.com/lncs. Accessed 25 Oct 2023
5. Mahony, R., Kumar, V., Corke, P.: Modeling, estimation, and control of quadrotor. IEEE Robot. Autom. Mag. (2012)
6. Seet, B.C., Liu, G., Lee, B.S., Foh, C.H., Wong, K.J., Lee, K.K.: A-STAR: a mobile ad hoc routing strategy for metropolis vehicular communications. In: Mitrou, N., Kontovasilis, K., Rouskas, G.N., Iliadis, I., Merakos, L. (eds.) Networking 2004. LNCS, vol. 3042, pp. 989–999. Springer, Heidelberg (2004). https://doi.org/10.1007/978-3-540-24693-0_81
7. Warren, C.W.: Global path planning using artificial potential fields. In: 1989 IEEE International Conference on Robotics and Automation, pp. 316–321. IEEE Computer Society (1989)
8. Mellinger, D., Kumar, V.: Minimum snap trajectory generation and control for quadrotors. In: 2011 IEEE International Conference on Robotics and Automation, pp. 2520–2525. IEEE (2011)
9. Ju, C., Luo, Q., Yan, X.: Path planning using an improved a-star algorithm. In: 2020 11th International Conference on Prognostics and System Health Management (PHM-2020 Jinan), pp. 23–26. IEEE (2020)
10. Tao, F., Xiao, B., Qi, Q., et al.: Digital twin modeling. J. Manuf. Syst. **64**, 372–389 (2022)
11. Liu, M., Fang, S., Dong, H., et al.: Review of digital twin about concepts, technologies, and industrial applications. J. Manuf. Syst. **58**, 346–361 (2021)
12. Tao, F., Zhang, H., Liu, A., et al.: Digital twin in industry: state-of-the-art. IEEE Trans. Ind. Inf. **15**(4), 2405–2415 (2018)

IR4AAS: An Identification Resolution-Enhanced AAS for Digital Twins Modeling

Mu Gu[1,2,3], Fangfang Gao[1,2(✉)], Lin Lin[3], Chunhui Su[1,2], Zhe Han[1,2], Hua Zhang[1,2], Yandong Li[1,2], Quanbo Lu[4], and Jiehan Zhou[5,6(✉)]

[1] CASICloud-Tech Co., Ltd., Beijing, China
gff0503@126.com
[2] Beijing Aerospace Intelligent Manufacturing Technology Development Co., Ltd., Beijing, China
[3] School of Mechatronics Engineering, Harbin Institute of Technology, Harbin, Heilongjiang, China
[4] School of Information Engineering, China University of Geosciences, Beijing, China
[5] School of Computer Science and Engineering, Shandong University of Science and Technology, Qingdao, China
jiehan.zhou@ieee.org
[6] Faculty of Information Technology and Electrical Engineering, University of Oulu, Oulu, Finland

Abstract. Existing Digital Twin (DT) models struggle to support interconnection and interoperability between DT applications. This paper proposes an identification resolution-enhanced Asset Administration Shell (AAS) for digital twins modeling (IR4AAS). IR4AAS features a unit-level DT model equipped with unique identification and a unified external service interface. IR4AAS builds a hierarchical DT model using business orchestration technique. Its effectiveness is validated through a case study in a precision machining workshop.

Keywords: Digital Twin Model · Business Orchestration · Interconnection · Asset Administration Shell · Identification Resolution

1 Introduction

As the manufacturing sector increasingly embraces digitalization, networking, and intelligence [1, 2], the global market for digital transformation in manufacturing is experiencing significant growth. This inevitably increases the demand for interconnected data exchange among Digital Twin (DT) models. However, there lacks uniformity in model construction and data exchange formats. While intra-enterprise interoperability is achievable, it does not support interconnectivity across enterprises, regions, and nations.

Addressing the challenge of information interconnectivity, extensive research has been undertaken on information models. Notably, the joint white paper from German Industry 4.0 and the Industrial Internet Consortium [3] underscores the convergence of DTs and Asset Administration Shells (AAS), promoting AAS as a practical approach. The German Industry 4.0 Platform has released a detailed description for AAS [4, 5]. There are works on developing AAS models using AutomationML or OPCUA, as detailed in [6, 7], and [8]. Lu et al. [9] studied a production line design platform based on AAS, while Tao et al. [10, 11] focused on the theoretical frameworks for DTs. Additionally, China Alliance of Industrial Internet released a white paper aiming at standardizing interoperable models [12]. However, there is still no relevant research on supporting multi-dimensional interconnection of DT models. This paper proposes an Identification Resolution-enhanced AAS model to improve interoperability between DT models.

The remainder of the paper is organized as follows: Section 2 introduces the Identification Resolution-enhanced AAS Model. Section 3 validates the effectiveness through a case study in a precision machining workshop. Section 4 draws a conclusion.

2 Identification Resolution-Enhanced AAS Model

2.1 Unit-Level DT Model Design Based on Identification Resolution

We employ a modeling approach based on the metamodel [13] to construct unit-level DT models tailored to describe the characteristics of production materials, equipment, production lines [14, 15]. The method utilizes the Industrial Internet Identification Resolution [12] to provide unique identities for physical objects. This enables indexing for the integration and invocation of DT models. The structure of the unit-level DT model consists of units, attributes, events, and behaviors. Each unit has its own attributes and behavior rules, and simulates the behavior of physical objects through the triggering of events and the execution of behaviors. The overall process is as follows: Firstly, unique identifiers are created for each unit and its attributes are determined based on the physical object. Then, possible events are defined and corresponding behavior rules are associated with each event. When an event occurs, the model triggers the corresponding behavior based on the identifier and current attributes of the unit. After the behavior is executed, the model can update the state and attributes of the unit, as well as the overall system's state and attributes. The entire process can be simulated using discrete time steps or continuous time simulation as needed. As shown in Fig. 1, the model is represented with the submodules of Identification, Properties, Events, Behaviors detailed as follows.

a) **Authenticatable unique identification code** serves as the basic index for the realization of the virtual-to-real mapping in a DT model, and it is also the key to the interconnection and interoperability between DT models. It consists of the identification code and the elements of DT models.

Identification Code of a DT Model: This is the unique identification code of a DT model, serving as the unique identification code for external models/systems to access and call this model.

Identification Code Describing the Elements of a DT Model: This refers to the identification code of an property parameter, and method in a DT model, serving as the unique identification code for external models or systems to refer the data of the DT model.

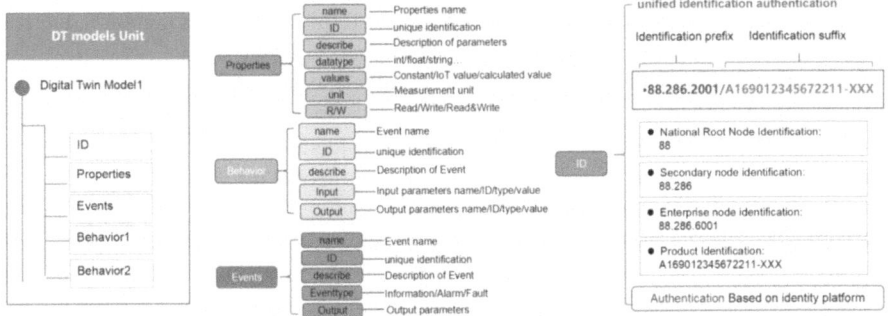

Fig. 1. Unit-level identifiable DT model.

Specifically, it consists of the identification prefix and suffix of a DT model. The identification prefix contains the national top node code, industry-level secondary node identification code, and the suffix contains the enterprise code, model/element code.

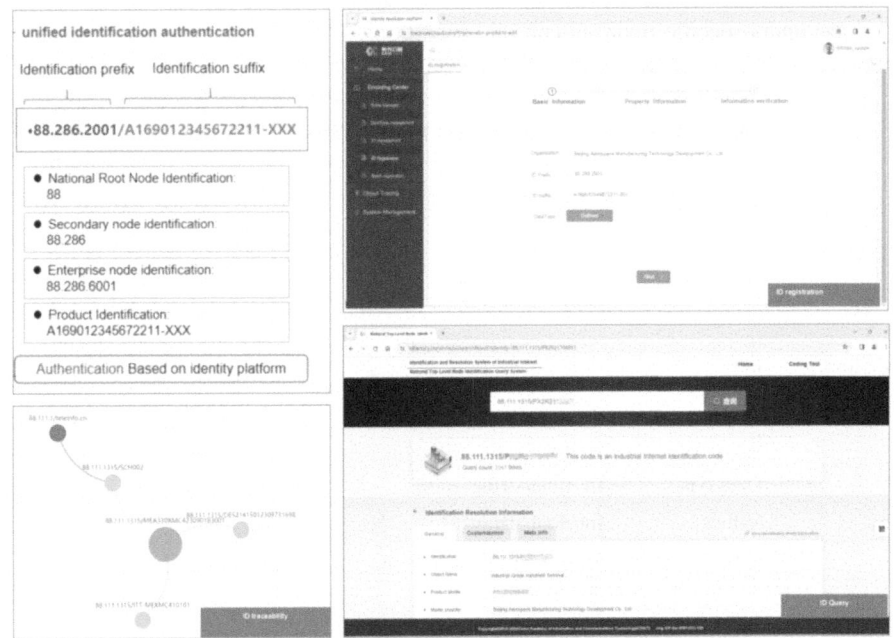

Fig. 2. Identification registration, querying, and authentication.

As shown in Fig. 2, the CASICloud has built a secondary node platform for identification resolution based on the INDICS platform [16, 17]. DT models are registered and generated as digital assets. After registration is completed and approved by the industry-level secondary node platform for identification resolution, a DT model will be assigned a unique ID. Users can perform unique authentication queries at the National Top-Level Node Identification Query System.

It should be noted that a DT model inevitably needs to integrate information models and simulation models provided by other model service providers. If the integrated model has already been registered on the identification resolution platform, this means that the integrated model is listed in the model directory of the platform, its existing identifier code will be used.

b) **Properties** refer to a series of parameters that describe the characteristics and functions of objects. The property elements are given by specifications such as IEC61360 [18], ecl@ss [19], etc. They map the operating data of physical objects in real time, including static properties like dimensions and manufacturers, and dynamic properties such as spindle speed and temperature. Table 1 lists the numerical and description types, including "Property Name", "Property representation code", "Data type", "Parameter value", and "Read/Write type".

c) **Event** describes the dynamic events of a target object, such as equipment failures, abnormal warning, equipment start-up, stop, etc. Additionally, events can be customized through association with behavioral models. Table 2 lists the property types of the "Event name", "Event ID", "Event type", "Output parameters", and others.

d) **Behavior** represents the functionalities and characteristics exhibited by the target object under specific environments and conditions, which can be represented through the description of data, geometry, knowledge, and mechanism. The resulted data of a behavior can be invoked by external applications. An object behavior is a function with inputs and outputs. Table 3 lists the property types of "Behavior name", "Behavior id", "Behavior type", "Input parameters (Name, ID, Parameter type, Parameter value)", "Output parameters (Name, ID, Parameter type, Parameter value)", required to describe an object behavior.

Table 1. Property description

Number	Item	Data types	Description
1	Name	String	Name of the property
2	ID	String	Unique identification code
3	Data types	Int/float/string…	Property value type
4	Values	Int/float/string…	Property value
5	R/W	Read/Write/Read&Write	Attribute readability

2.2 Hierarchical DT Model Design Based on Business Orchestration

A production system usually consists of multiple production units according to certain business logic. Therefore, a complex DT model is composed of a series of unit-level DT models, exhibiting a certain complexity in the interconnection and interoperability between sub-models. Figure 3 illustrates the integration of DT models based on business orchestration provided by the direct graph method [20], implementing a complex

Table 2. Event description

Number	Item	Data Types	Description
1	Name	String	Event of the property
2	ID	String	Unique identification code
3	Event types	Fault/Start/Stop…	Event type
4	Output parameter	Int/float/string…	Value parameter

Table 3. Behavior description

Number	Item	Data Types	Description
1	Name	String	Action of the property
2	ID	String	Unique identification code
3	Action description	String	Action description
4	Input parameter	String	Name of input parameter
5	Input parameter ID	String	ID of input parameter
6	Data type of input parameter	Int/float/string…	Data type of input parameter
7	Input parameter value	Int/float/string…	Input parameter value
8	Output parameter	String	Name of output parameter
9	Output parameter ID	String	ID of output parameter
10	Data type of output parameter	Int/float/string…	Data type of output parameter
11	Output parameter value	Int/float/string…	Output parameter value

hierarchical DT model. A graph element includes basic elements, interface extension classes, multidisciplinary simulation elements, and user-defined classes, which realize the business orchestration and interconnection among models, sub-models and model behaviors in the hierarchical DT model.

Interconnection Between Models: To facilitate the logical interconnection between various production units in a system, events triggered by the DT model of a processing equipment can initiate responses in the DT model of a delivery equipment. This ensures that production line equipment can effectively call on delivery equipment to deliver production materials in a timely manner during the manufacturing process. This can be achieved through visual drag-and-drop business orchestration using conditional control elements and object operation elements, as well as supporting the implementation of interconnection business orchestration through extension elements. Furthermore, the integration of DT models with third-party data/mechanism/simulation models can be realized through interdisciplinary simulation elements.

Interconnection Between Sub-models: By utilizing computational elements, conditional control elements, object operation elements, the interconnection business orchestration between sub-models can be established to ensure the collaboration of the entire production unit.

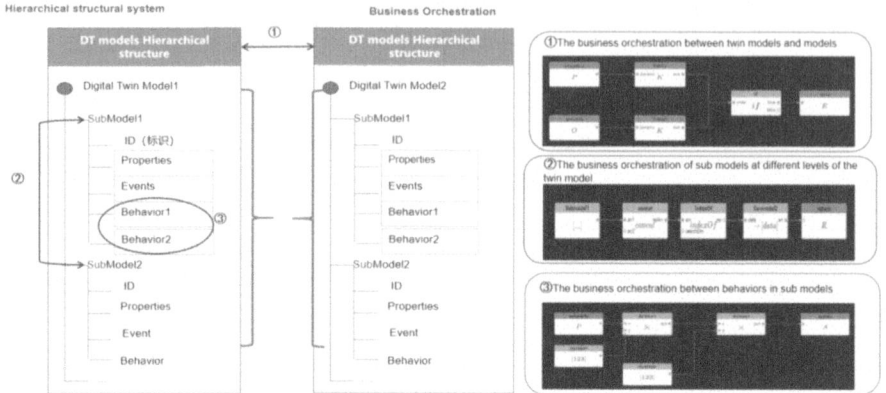

Fig. 3. A complex hierarchical DT model.

Trigger of Event Behaviors: The implementation of interconnection ultimately relies on triggering event behaviors between the sub-models. Utilizing elements such as attributes, behaviors, triggers, and return values in DT elements enables data calling and exchange between models to achieve business orchestration. During real-time operation, the predefined logic can be invoked and triggered.

2.3   Integration of DT Models

To facilitate iterative optimization between digital and physical systems, it is necessary to enable the connection among DT models and physical objects, and third-party models. We integrate AAS information exchange, encapsulating unit-level DT models into invocable components with standard API interfaces. This not only supports the hierarchical combination for constructing complex DT models but also enables model invocation through service calls. In this method, we employ RESTful APIs for data interaction and invocation of DT models, using unique IDs for objects/models/parameters as indexes.

3   Application Implementation

This application is developed based on the three.js engine. Firstly, we employ the IR-enhanced method to build DT models covering equipment-level, production line-level, and workshop-level models for a precision machining workshop. Their IDs are obtained from the identification resolution service provided by INDICS platform.

3.1 Implementation of the DT System for a Precision Machining Workshop

(1) Scene Modeling
Scene modeling involves in constructing a 3D digital environment mapping to the real scene. We use 3D modeling tools to build the digital scene, and then import it into the development engine three.js in formats such as fbx, obj, etc.
(2) DT Modeling, Data Binding, and Business Orchestration
   a. Property definition: Define static and dynamic properties of equipment, and their data interfaces based on the profile information and operating status of various workshop equipment.
   b. Event definition: Define events, such as faults, quality abnormalities, spindle collision warnings, event triggering conditions, which help monitor and display the operation events of equipment.
   c. Behavior definition: Define workshop operation and numerical control machine behavior, such as quality optimization, which are customized based on the behavior model library or based ondirected graph elements.
   d. Business orchestration: Users can visually drag and drop directed graph-based components into the editing area to configure and compose the business logic of DT models.
(3) Preview and Deployment of DT Applications
Figure 4 depicts the DT system for a precision machining workshop. Three.js is applied to develop the DT system. Three.js is a excellent tool with physical engine and real-time rendering engine. We develop scenes, UIs, and scripts with these softwares. The scene includes static environment models, dynamic 3D models, and animations. These 3D models and animations reconstruct the physical composition. We develop UIs to display supplementary 2D information for human-computer interaction, packed as a module and imported into the system. The MATLAB and MWorks provide the corresponding interface. Finally, we develop scripts to drive the DT scenes and update the UIs.

After the completion of scene modeling, DT modeling, and data binding, users can preview the 3D scene and deploy the DT application based on WebGL technology.

3.2 Interoperability and Optimization Analysis

The implemented application indicates that our method achieves virtual-to-real interconnection and mapping, supporting the transparent management of the production process to realize comprehensive perception and prediction of key production data on the workshop. Our method enables the coordination between the physical system and the digital system. Furthermore, as shown in Fig. 5, our method enables optimization analysis of manufacturing quality and prediction of equipment failures by integrating intelligent data analysis model, such as quality optimization model for dimensional errors based on feature engineering and time domain analysis model for spindle vibration in machining centers. This enhances four key aspects of workshop operations: production capacity, planning, delivery time, and product quality.

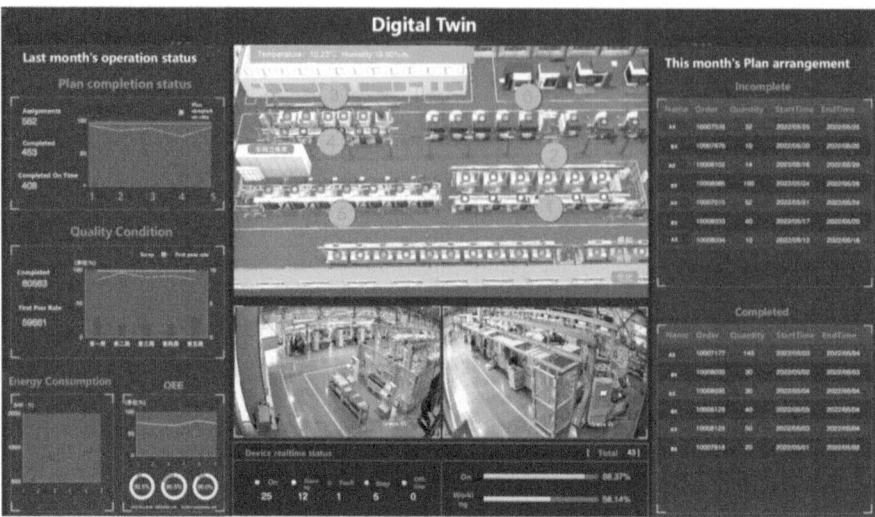

Fig. 4. The DT system for a precision processing workshop.

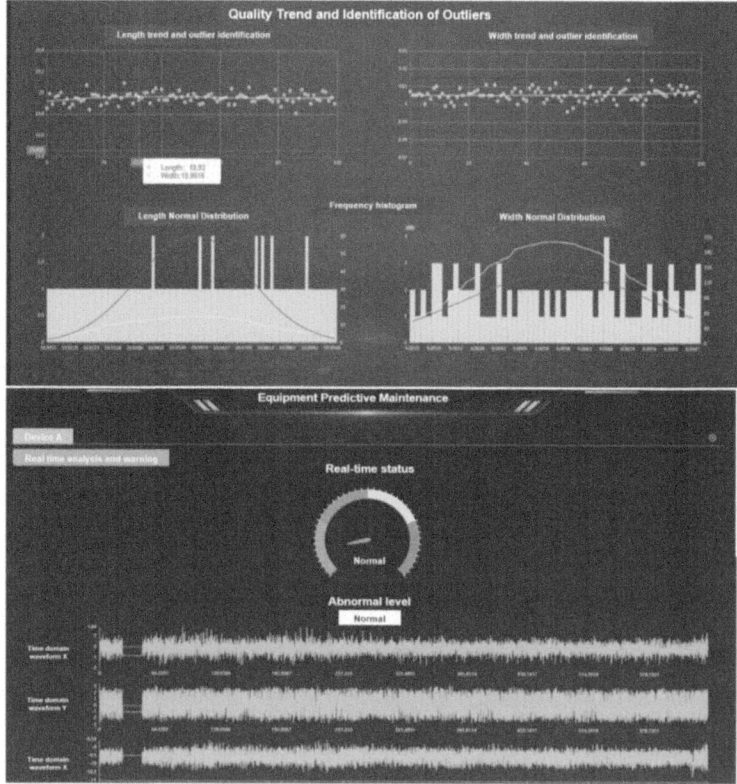

Fig. 5. Optimization analysis of manufacturing quality and prediction of equipment failures.

4 Conclusion

This paper introduces an Identification Resolution-enhanced Asset Administration Shell for Digital Twins modeling (IR4AAS). IR4AAS features a unit-level DT model equipped with unique identification and unified external service interfaces. We further develop a hierarchical DT model utilizing business orchestration techniques and validate its effectiveness through a case study in a precision processing workshop. Future research will focus on creating tools for constructing DT models integrated with industrial internet platforms.

Acknowledgments. This study was funded by "National Key Research and Development Program of China". (Grant number:2022YFE0197600).

Disclosure of Interests. The authors have no competing interests to declare that are relevant to the content of this article.

References

1. Shi, J.X., Zhu, J.C.: Characteristics, trends, and China's responses to the global industrial digital transformation. Econ. Vertical Horizontal **2022**(11), 55–63 (2022)
2. China Academy of Information and Communications Technology. White Paper on Global Digital Economy (2022)
3. Birgit, B., et al.: An industrial internet consortium and plattformindustrie 4.0 joint whitepaper. Digital Twin and Asset Administration Shell Concepts and Application in the Industrial Internet and Industrie 4.0
4. Details of the Asset Administration Shell Part1:The exchange of information between partners in the value chain of Industrie 4.0 (Version 3.0RC02)
5. Details of the Asset Administration ShellPart2:Interoperability at Runtime Exchanging Information via Application Programming Interfaces (Version 1.0RC02)
6. Zhao, M., Liu, F.W.: Returning to the source to see asset administrationshell—Industrial 4.0 components don digital attire. Knowl. Autom. (2017)
7. Yue, L., Liu, D., Fang, Y.F.: Asset administrationshell in industrial 4.0 components. Chin. J. Sci. Instrum. **2017**(12), 55–60 (2017)
8. Liang, Z.: Research and implementation of asset administrationshell information model based on OPC UA. Huazhong University of Science and Technology (2020)
9. Tao, F., et al.: Theory and application of digital twin model construction. Comput. Integr. Manuf. Syst. **27**(01), 1–15 (2021)
10. Tao, F., Zhang, H., Zhang, C.Y.: Advancements and challenges of digital twins in industry. Nat. Comput. Sci. **4**(3), 169–177 (2024)
11. Industrial Internet Industry Alliance: Industrial Internet Information Model White Paper (2020). https://www.aii-alliance.org/index/c316/n47.html
12. Liu, Y., Han, T.Y., Xie, B., Tian, J.: Data sharing mechanism based on industrial internet identification resolution system. Comput. Integr. Manuf. Syst. **25**(12), 3032–3042 (2019)
13. Yang, X.L., Liu, X.M., Zhang, H., Fu, L., Yu, Y.B.: Meta-model-based shop-floor digital twin architecture, modeling, and application. Robot. Comput.-Integr. Manuf. **84**, 102595 (2023)
14. Gao, F.F., Su, C.H., Wang, Y.G., Zheng, Y.: Modeling method and application of interconnected information model of production line virtual and physical systems based on cloud manufacturing. Manuf. Autom. **42**(03), 24–29+39 (2020)

15. Zhou, J.H., Zhang, S.H., Gu, M.: Revisiting digital twins: the origins, fundamentals and practice. Front. Eng. Manag. 1(1), 1–10 (2022)
16. Chai, X., Hou, B., Zou, P., Zeng, J., Zhou, J.H.: INDICS: an industrial internet platform. In: 2018 IEEE SmartWorld, Ubiquitous Intelligence & Computing, Advanced & Trusted Computing, Scalable Computing & Communications, Cloud & Big Data Computing, Internet of People and Smart City Innovation (SmartWorld/SCALCOM/UIC/ATC/CBDCom/IOP/SCI), Guangzhou China, pp. 1824–1828 (2018)
17. Chen, J., Zhou, J.H.: Revisiting industry 4.0 with a case study. In: 2018 IEEE International Conference on Internet of Things (iThings), Halifax, NS, Canada, pp. 1928–1932 (2018)
18. International Electrotechnical Commission. IEC 61360:2018 "Standard Data Elements and Generic Categories". International Standard (2018)
19. ecl@ss. The international classification for products and services [Version 10.0]. Berlin, Germany: ecl@ss Society (2019)
20. Xie, Y.X., Song, X., Tu, Y.C., Cui, Y., Zhou, J.H., Zhai, Y.J.: Research on modeling and simulation method of control system based on directed graph. In:China Simulation Society. Proceedings of the 34th China Simulation Conference and the 21st Asian Simulation Conference, Changsha China, pp. 80–95 (2022)

Research and Implementation of Digital Twin Modeling and Simulation Platform

Lingyun Yang[1,2], Junfan Zhang[3]([⊠]), Xiao Song[3], Qianchuan Zhao[1], Zhe Han[2],
Yuchun Tu[3], Yongxuan Xie[3], and Yuanyuan Hu[2]

[1] Department of Automation, Tsinghua University, Beijing, China
[2] Beijing Aerospace Smart Manufacturing Technology Development Company, Beijing, China
[3] School of Cyber Science and Technology, Beihang University, Beijing, China
918127890@qq.com

Abstract. Aiming at the development trend of scale and complexity of digital twin system, this paper proposes a general digital twin modeling and simulation platform framework from the perspective of system modeling. Based on the discrete event system specification (DEVS), we propose a definition for twins oriented to system simulation, and establish a unified modeling and simulation algorithm for hierarchical twins. This algorithm uses the graphical modeling method, supports the mixed simulation mechanism of discrete time and discrete event, and the twin formed has a clear logical structure and good reusability. Finally, using the proposed twin modeling algorithm, we establish the digital twin system of the converter shop. The establishment process of the system shows that the proposed method can easily model and simulate the complex digital twins.

Keywords: Virtual Entity · DEVS Theory · Simulation Model Definition · Converter Workshop

1 Introduction

With the rise of new technologies such as 5G, Internet of Things, artificial intelligence, blockchain, cloud computing, big data, and edge computing, some manufacturing powers such as the United States, Germany, and China have respectively proposed core concepts such as Industrial Internet, Industry 4.0, and Made in China 2025 at the national level. Intelligent manufacturing has become the development trend and goal pursued by the world's major manufacturing powers. Digital twin (DT), as a key enabling technology to solve the linkage between physical space and cyberspace of intelligent manufacturing, has received extensive attention and research in academia and industry. At present, major breakthroughs have been made in the fields of aerospace [1], intelligent manufacturing [2, 3], electric power [4], healthcare [5], shipping [6], smart city [7], and other industries.

In the era of Industry 4.0, digital twin technology has been widely used in the industrial field. At present, the research direction of digital twin is mainly divided into two categories: the top-level of system architecture and the underlying specific field in application. On the one hand, the overall architecture of digital twins is studied, such

S. Saito et al. (Eds.): AsiaSim 2024, CCIS 2170, pp. 323–336, 2024.
https://doi.org/10.1007/978-981-97-7225-4_25

as the components of 19 categories of digital twin technologies, like physical entity, virtual entity and physical environment, summarized by David [8]. Tao et al. [9] further abstracted the digital twin technology into five dimensions: physical entity, virtual entity, connection, data and service. On this basis, Li [10] and Fu [11] et al. further proposed a modeling and simulation platform design for digital twin applications. On the other hand, the specific industry application of digital twin technology is also discussed. For example, Cai [12] and Ge [13] et al. introduced the combination of digital twin and mining industry to monitor, simulate and optimize the motion state of the boring machine. Liu et al. [14] established a digital twin intelligent monitoring system for the Huaihe River, and realized heterogeneous parallel computing combining CPU and GPU, which could timely alarm water conservancy risks. Yin [15] and Miao [16] et al. used digital twin technology to monitor and simulate industrial robots, and integrated reinforcement learning with digital twin technology to realize the operation optimization of physical equipment.

At present, there are many theories related to digital twin, but most of them focus on the top-level logical framework or the specific domain implementation at the bottom, while the discussion and analysis of the system level are relatively few. Digital twin technology is showing a multi-domain and complicated development trend, how to describe the digital twin model from different fields in a unified and hierarchical way is still a big difficulty when applying digital twin technology. Therefore, this paper firstly analyzes the requirements of complex twins and gives the overall architecture of the platform. Based on the discrete event system specification (DEVS) [17] and the idea of hierarchical system modeling, the theoretical analysis of digital twin platform is given. Finally, a converter workshop case is established on the prototype platform, which proves the universality and convenience of the proposed method for the complex digital twin systems.

2 Overall Design of Digital Twin M&S Platform

2.1 Overall Architecture

The main goal of the digital twin is to establish the digital copy of the physical entity, simulate the digital model, and realize the monitoring and controling of the physical entity according to the results of simulation. Faced with the complex digital twin scenario of large-scale physical entities, the design and application architecture of the twin are further required. The current digital twin technology focuses on the application in a single field, but lacks the idea of system modeling, and the established twin entities have heterogeneous architecture, leading to poor reusability.

Therefore, this paper integrates the DEVS theory into the modeling and simulation technology of digital twins, divides the complex twin model into unified format atomic model and coupled model. With the function module in the way of system-of-systems [18], the complex twin can be described with basic model units. Thus, the unity and reusability of the twin model are realized.

In order to realize the goal of systematic modeling of digital twins, the platform is divided into three levels according to the process of modeling and simulation, including physical entity layer, twin mapping layer and simulation layer. The physical layer provides the basic interaction interface and connects physical devices to the digital twin platform. The twin mapping layer abstracts the physical entity according to the DEVS theory, and uses the DEVS modeling module to establish the twin composed of the atomic model and the coupled model. According to the established twin model, the simulation layer simulates its state according to the discrete time and discrete event mechanism, and controls or optimizes the twin according to the calculated results. Further divided by the specific functions, the digital twin modeling and simulation platform is composed of six core functional modules, which are entity connect module, twin mapping module, graphical modeling module, mechanism model library module, simulation module and display module. The relationship between these modules is shown in Fig. 1.

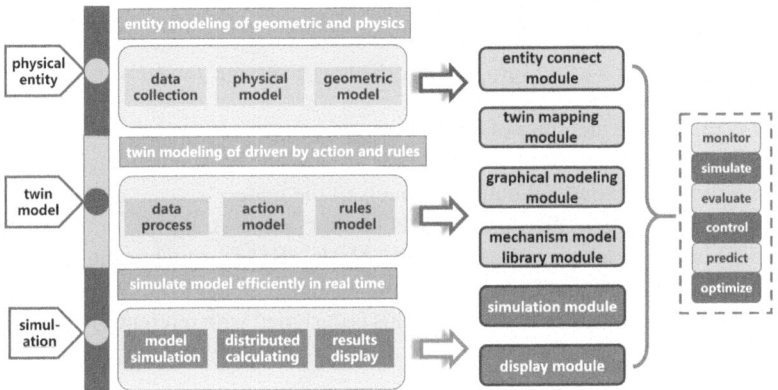

Fig. 1. Digital twin modeling and simulation platform architecture

2.2 Hierarchical System Modeling

It is difficult to model complex systems directly. In the modeling process, we can divide a complex system into subsystems and basic models, and then build the simulation system by encapsulating basic models as services and combining services as subsystems, as Fig. 2 shows.

In the modeling process, a complex system is represented as a directed graph. The directed graph $DG = (N, E)$, where N is a set of nodes, and E is a set of edges between the nodes, $Edge \subseteq \{(m, n)|m, n \in N\}$. Also, a graph can be described as an adjacency matrix, which includes n nodes using an $n \times n$ matrix, where the entry at (i, j) is 1 if there is an edge from node i to node j. Here, a node is either a basic model or coupled subsystem. An edge is a data connection between nodes.

The directed edge is defined as Eq. 1 shows.

$$Edge = <id, source, target> \qquad (1)$$

where id is the unique identifier, $source$ is the output port of a source node, and $target$ is an input port of a destination node.

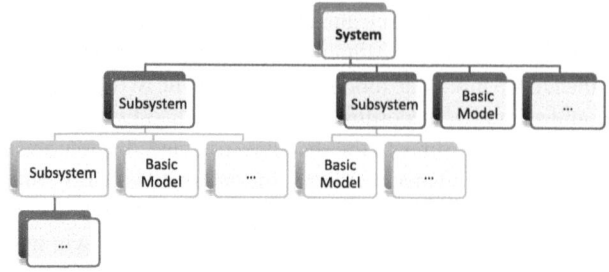

Fig. 2. Hierarchical System modeling, including subsystem and basic model

The node is defined as Eq. 2 shows.

$$Node = <id, graphical\_attr, ports, service\_model> \qquad (2)$$

where *id* is the unique identifier of the node, *graphical_attr* is the graphical attributes of the node, including the rendering information such as the node position and shape etc.

ports defines the input and output channels of the node.

service_model denote the basic models or subsystems in Fig. 2. In this paper, the basic model is implemented as Basic Service Model (BSM), which is the basic unit for building a complex system. The subsystem is designed as the Coupled Service Model (CSM), which is a composition of BSMs. This is designed with reference to the principle of DEVS.

2.3 Basic Service and Coupled Service Models

To extend the web service compatibility of DEVS models, BSM is proposed to be divided into two categories, i.e., the local service model (LSM) on the server and the remote service model (RSM). LSM is the traditional code model developed and uploaded by users. RSM is a user-defined API, which follows the resource-oriented principle with more compatibility, composability, flexibility and scalability.

The formal definition of BSM is shown in Eqs. 3, 4 and 5.

$$BSM = \{LSM, RSM\} \qquad (3)$$

$$LSM = <X, Y, S, BSM\_exector, \lambda, t_a> \qquad (4)$$

$$RSM = <X, X', Y, S, BSM\_exector, \lambda, t_a> \qquad (5)$$

where X is the set of external input events or data, Y is the set of model outputs, S is the set of internal states,

BSM_exector is an model executor including a state transition function. When there is an external or internal input that satisfies the specified conditions, *BSM_exector* will implement the state transition of the service model and then generate the model output y ($y \in Y$).

λ is the model output function, which is responsible for applying the model output y generated by *BSM_exector* to the models connected to it.

t_a is the time advance function, which is usually associated with the model's executor function *BSM_exector*.

For RSM, the *BSM_exector* of RSM, is essentially requesting a web API model on the remote network. X' is the set of internal input events or data of the RSM. Specifically, X' is consist of request address (URL), request method (Method), request header (Header), query parameters (Query), and request body (Body) of the API.

$$X' = <URL, Method, Header, Query, Body> \tag{6}$$

CSM is defined as Eq. 7 shows. A CSM is composed of one or more BSM and/or CSMs.

$$CSM = <X, Y, S, CSM\_Exector, \lambda, t_a> \tag{7}$$

where X, Y, λ and t_a are defined in the same way as in the BSM. S is the set of internal states of the model. Different from the BSM, the internal state of the CSM includes the management of all BSMs and CSMs that make up the CSM. *CSM_exector* is an executor function of the CSM and is also a simulator. When there is an external input that satisfies the specified conditions, the *CSM_exector* of the CSM simulates all its internal child elements, and then updates the state of all its internal BSMs, and finally outputs the user-defined outputs of the CSM.

Based on the BSM and CSM defined in this paper, a complex system can be composed of BSMs and CSMs, and the CSM can be split into some BSMs and CSMs. The structure of a complex system is shown in Fig. 3.

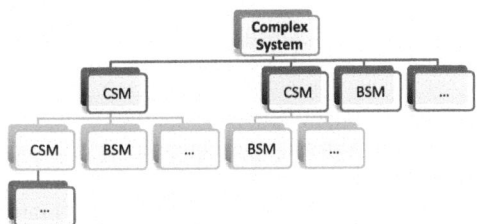

Fig. 3. The structure of a complex system

2.4 Simulation Engine of Directed Graph-Based System

Based on the proposed BSM and CSM, a complex system is modeled as a hierarchical directed graph (DG), in which the graph nodes are either BSMs or CSMs. In our implementation, these nodes are designed as graphic model block (GMB) in the canvas. GMB can be divided into two categories, basic model block and subsystem model block, corresponding to BSM and CSM respectively.

For instance, for an angle follow-up control system, it consists of 7 blocks, i.e., a given potentiometer, a feedback potentiometer, a torque motor, a tachometer motor, a payload and two amplifiers, these models are denoted by GMBs in Fig. 4.

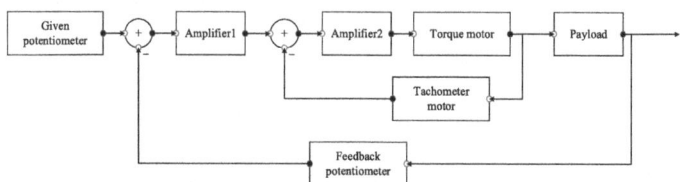

Fig. 4. A directed graph-based model of an angle follow-up control system

For implementation of this system model, there are often two simulation approaches, graph model compiler and graph model interpreter. The former is to compile the graph model into executable target code and then execute it [19]. The latter is to use a simulation engine to traverse the model blocks, schedule, interpret and execute them.

Compiling graph models into code is logically complex and difficult to implement. To design a simple yet efficient approach, we design a simulation engine, which implements a directed graph scheduling algorithm executing the service models and does not require compiling the graph into source code. The main procedure of the DG simulation engine is shown in Fig. 5 and explained as follows.

1. Initialization. Initialize all simulation parameters, such as total simulation time, simulation step size, etc. Initialize an empty executable service queue for the upcoming executable service models.
2. Loading service models. Traverse the model blocks in the graph, and instantiate the corresponding service model.
3. Establishing connections. Traverse the directed edges to bind the input-output connections of BSM and CSM. After the service model is executed, its output will be generated by the output function through these associations to other service models.
4. Updating executable service queue. If it is the first time to update this queue, add all self-executing service models, i.e., the service model with no external input ports, to the executable queue. Otherwise, add all executable service models, i.e., those with all external input ports activated, to the executable queue. An external input port being activated indicates that the port is available. In addition, self-executing service models are also executable service models.
5. Scheduling and executing. Pop a service model from the executable queue. If the service model is a BSM, call its executor function *BSM_executor* and then act the

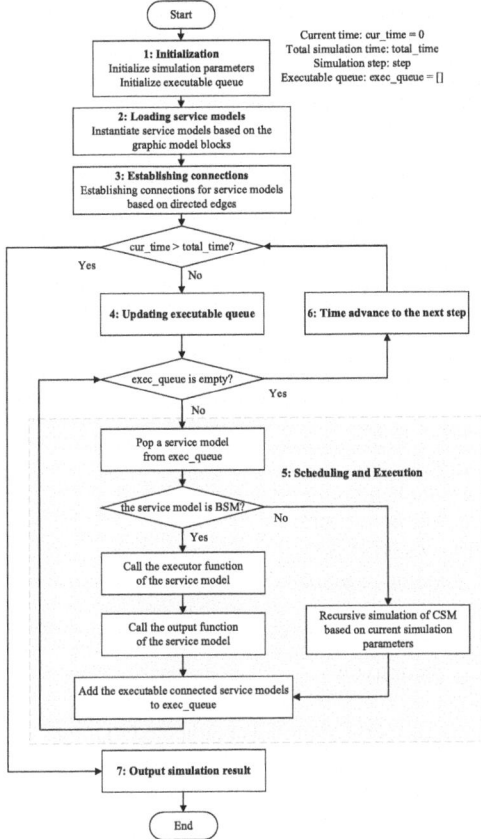

Fig. 5. Flowchart of the proposed interpreter

running results on the service models connected to it via its output function λ and activates the corresponding external input port. If the service model is a CSM, the executor function *CSM _executor* is called to recursively simulate the internal service models, and then act the running results on the service models connected to it via its output function λ and activates the corresponding external input port. After the service model executed, if the service model connected to it becomes executable, add it to the executable queue. Repeat step 5 until the executable queue is empty.

6. Time advance to the next time step. Call the time advance function of the executed service model one by one, and then advance the current time in the smallest step.

7. Output simulation result. If the current time is greater than the total simulation time, output the simulation result of the model, and then end the simulation.

The pseudo-code of the simulation strategy in the interpreter of the directed graph model is shown in Table 1.

Table 1. The pseudo-code of the proposed DG simulation engine

The pseudo-code of the proposed interpreter

Input: *nodes*: the model blocks in the directed graph
Input: *edges*: the edges in the directed graph
1: Initialize the simulation parameters. initialize current simulation time *cur_time* = 0, initialize total simulation time *total_time*;
2: Initialize an empty queue *exec_queue*;
3: for *node* in *nodes*:
4:   instantiate the corresponding service model based on *node*
5: end for
6: for *edge* in *edges*:
7:   establish input-output connections for the service models based on *edge*
8: end for
9: while *cur_time* < *total_time*:
10: update the *exec_queue*
11: while the *exec_queue* is not empty:
12:    *model* = *exec_queue*.dequeue()
13:    if isinstanceof(*model*, BSM):
14:      call the *BSM_exector* of *model*
15:      call the λ of the *model*
16:    else:
17:      simulate *model* recursively
18:    end if
19:    for *next_model* in the service models connected to *model*:
20:      if *next_model* is executable:
21:        *exec_queue*.enqueue(*next_model*)
22:      end if
23:    end for
24: end while
25: update current simulation time based on t_a, *cur_time* += Δt
26: end while
Output: output the simulation result

3   Application Cases of Digital Twin Converter System

3.1   Overview of Converter Workshop

In order to verify the effectiveness of the digital twin modeling and simulation method proposed in this paper, a converter workshop of an iron and steel group is taken as an example. The workshop contains three production lines, each with similar structure, which can be used for smelting steel, iron and other materials. The production line contains equipment or components such as converter, crane, fire door, scrap tank, hot metal ladle, oxygen gun, auxiliary gun, auxiliary silo, molten steel ladle, etc. And we model the digital twin of converter workshop based on the production line.

3.2 Twin Modeling in Converter Workshop

In the process of modeling converter workshop, the main problem is that the twin properties and actions are associated and complicated, so it is difficult to establish the digital twin model accurately and efficiently. In the converter workshop, the device data cannot be obtained directly from the data platform. It is necessary to request the melting number information of the production batch, then obtain the basic information such as the furnace number and product type according to the melting number, and finally obtain the oxygen consumption, the amount of auxiliary materials and other information by the furnace number and product type.

When requesting data from multiple related interfaces, a lot of logical judgments are involved, and complex rules need to be configured in the twins. If monitoring all the properties involved, the computing resource consumption is large and the efficiency is low. Based on DEVS theory, this paper develops a graphical modeling module to improve the model reusability. Meanwhile, the twin properties can be updated based on the discrete event input, greatly improving the computational efficiency of the simulation engine. The main details of establishing the converter shop twin are described in the following steps.

1) Establish twin structure tree

According to the production requirements of the converter workshop, the twin structure tree is created as Fig. 6. In the twin structure tree, three converters have similar structure, so we take the first furnace as an example to illustrate the modeling process. In the case of the first furnace twin, according to the different data sources, it can be divided into two sub-twins of real-time monitor and simulation prediction. The sub-twin of real-time monitor is responsible for managing the relevant quantity of smelting process. The sub-twins under the simulation prediction twin are responsible for invoking the mechanism model library for real-time prediction and analysis.

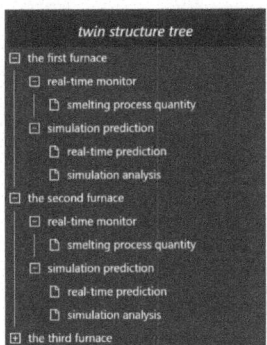

Fig. 6. Converter shop twin structure tree

In order to meet the monitoring and analysis requirements of various production tasks, a large number of properties need to be bound to the twins. However, in a single production task, not all properties are used, thus real-time updating of all properties results in a serious waste of computing resources. Therefore, the twin is dynamically updated based on the discrete event input, and only the changed properties are updated. Besides, multiple properties may come from the same data source, the reuse of properties can reduce redundant computation.

2) Establish twin discrete event input

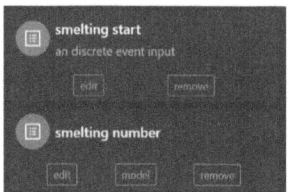

Fig. 7. Properties of the first furnace

As shown in Fig. 7, two properties of smelting start and smelting number are bound to the first furnace twin. The former is used to judge whether the smelting has started, and the latter is used to request the relevant information of the production task. The melting start property marks the start of smelting, so it can be defined as a discrete event input. The smelting number can only be obtained after the smelting starts, and the corresponding calculation logic is shown in Fig. 8.

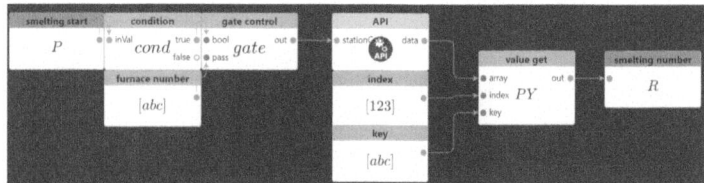

Fig. 8. Logic of smelting number property

For the smelting number property, graphical modeling module is used to model its logical rules. The logical entry is the conditions judgment of the smelting start property. If the smelting starts, the gate control model will pass the furnace number to the API model, invoking the API request to get the melting number and other relevant information. The return value of the API model is an array, which contains multiple data in JSON format. Therefore, the value get model is used to select the latest data from the array, and get the corresponding value to the key corresponding to the input key. The output of the value get model is the desired smelting number, which is used as the property value of the smelting number property. Through this logical structure, smelting start is defined as a discrete event input. When smelting stops or ends, the subsequent calculation will be automatically tailored to reduce the calculation amount.

3) Modeling the logic of bound properties

Oxygen amount, melted iron amount, scrap amount and other related properties are bound to real-time monitor twin. If each property separately request the data interface, it will consume a lot of network resources. Based on the graphical modeling module, a unified request interface is implemented as Fig. 9. A property is used as a unified entry to the interface data. When smelting starts, this property requests data API based on smelting number, and store the data as property value for other properties to use.

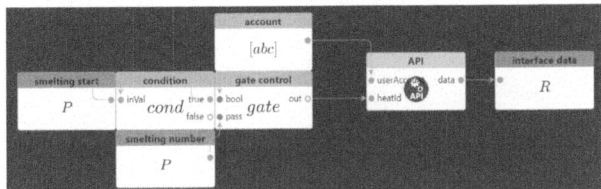

Fig. 9. Logic of unified data request property

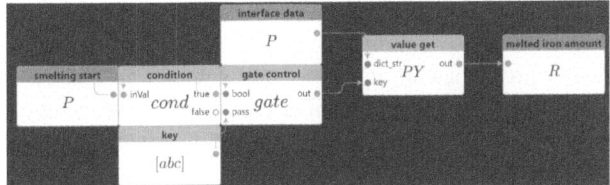

Fig. 10. Logic of melted iron consumption property

Figure 10 shows the logic of the melted iron amount property. When smelting starts, the gate control model passes the key to the value get model, and the value of melted iron consumption property is obtained from the unified data request property. Other properties such as oxygen amount and scrap amount are modeled similarly. By unifying the data request, network requests time usage is reduced and computational efficiency can be improved.

By the above steps, the twin model of the converter shop is created. Aiming at the characteristics of complex logic and many interface requests of the converter shop, we optimize the properties updating logic based on DEVS theory. The twin structure tree is established as a reasonable hierarchy structure, and special properties are defined as discrete events. Finally, different properties are bound to the corresponding twins. In simulation, only required properties are updated based on the discrete event input. Meanwhile, a property is established to manage the interface data in a unified manner, and its value is reused by other properties to avoid the repetitive network requests, so as to further improve the simulation efficiency.

3.3   Simulation and Effect Demonstration of Twins in Converter Workshop

While establishing the actions and rules model of the twin, the geometric model of the converter workshop is also established as Fig. 11. These 3D models are digital mappings of physical entities, reflecting the actual state of the converter shop from a geometric level. It can be bound with the twin properties data to drive the animation of the geometric model, so that the actual state of the physical entity can be displayed more vividly.

Fig. 11.   Geometric model of the converter workshop

After establishing the twin model and geometric model of the converter workshop, the twin will be simulated by the background simulation engine. Besides, smelting start property is used as a discrete event input. Properties related to smelting are updated by this event, and other properties are updated every simulation. Finally, the monitoring effect of the workshop digital twin is shown in Fig. 12. The average update delay of the twin data is about 0.4 s, and the charts and 3D animation reflect the state of the physical entity in real time. Based on the propsed digital twin modeling and simulation method, the monitoring, prediction, control and optimization of the converter shop are finally realized.

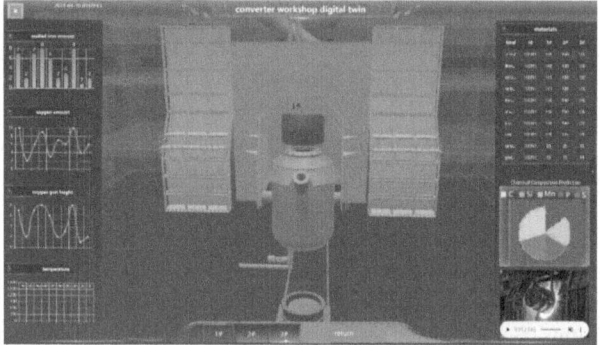

Fig. 12.   Converter workshop intelligent monitoring and real-time analysis

Applying the digital twin technology, the intelligent monitoring of the converter workshop is realized. Even if we are not in the workshop site, we can still fully understand the working situation of the converter. Based on the events and actions configured in the twin, the background simulation engine will simulate the converter workshop twin in real time. When the monitoring result of the properties or the prediction result of the mechanism model is abnormal, the digital twin platform will issue a warning to remind the user to take necessary measures. Besides, the properties configured on the twin are transmitted to the geometric scene in real time, which drives the geometric model to generate corresponding animations, and visualizes the data of properties in the form of charts. Finally, through the automatic monitoring and manual response, the converter can ensure the safety and long-term completion of the smelting work.

4   Conclusion

Based on DEVS theory, this paper proposes an universal digital twin modeling method for complex systems with multiple fields, disciplines and functions. It can easily describe the corresponding twins of various complex systems with a unified graphical language. Applying this method, a digital twin modeling simulation prototype platform is established. Through the simulation of the converter workshop, results show that this digital twin architecture can organize the model and data well, and has good scalability.

By the converter workshop digital twin case, this paper introduces the modeling and simulation idea of complex digital twin model. The platform realizes the hybrid simulation mechanism of discrete time and discrete event, dynamically updating the properties to improve the simulation efficiency. Meanwhile, based on the DEVS theory, the twin model has high reusability. For example, repeated network requests are modeled as a unified data request property to be reused for other properties, so as to avoid the repeated network request and improve the simulation speed.

From the perspective of system modeling, this paper studies the key points and difficulties faced by complex systems in digital twin modeling and simulation. In view of these difficulties, a digital twin modeling simulation method based on discrete event system specification is proposed, which can accurately and conveniently model the complex system digital twin.

Acknowledgments. This work was supported by the National Key Research and Development Program of China under Grant 2023YFB3308200 and Beijing Natural Science Foundation (L233005).

References

1. Li, L.: Digital twin in aerospace industry: a gentle introduction. IEEE Access **10**, 9543–9562 (2021)
2. Zhou, G.: Knowledge-driven digital twin manufacturing cell towards intelligent manufacturing. Int. J. Prod. Res. **58**(4), 1034–1051 (2020)
3. He, B.: Digital twin-based sustainable intelligent manufacturing: a review. Adv. Manuf. **9**(1), 1–21 (2021)
4. Sifat, M.M.H.: Towards electric digital twin grid: technology and framework review. Energy and AI **11**, 100213 (2023)

5. Aydın, Ö.: OpenAI ChatGPT generated literature review: digital twin in healthcare. Emerg. Comput. Technol. **2** (2022)
6. Lee, J.H.: Real-time digital twin for ship operation in waves. Ocean Eng. **266**, 112867 (2022)
7. Deng, T.: A systematic review of a digital twin city: a new pattern of urban governance toward smart cities. J. Manag. Sci. Eng. **6**(2), 125–134 (2021)
8. Jones, D.: Characterising the digital twin: a systematic literature review. CIRP J. Manuf. Sci. Technol. **29**, 36–52 (2020)
9. Tao, F.: Five-dimension digital twin model and its ten applications. Comput. Integr. Manuf. Syst. **25**(1), 1–18 (2019)
10. Li, L.: Multidisciplinary collaborative design modeling technologies for complex mechanical products based on digital twin. Comput. Integr. Manuf. Syst. **25**(06), 1307–1319 (2019)
11. Fu, Z.: Flexible discrete-event system modelling and simulation platform design oriented to digital twin. Comput. Integr. Manuf. Syst. 1–25 (2024)
12. Cai, X.: Research on digital twin technology of coal mine tunneling machine system. J. Syst. Simul. 1–13 (2024)
13. Ge, S.: Digital twin: meeting the technical challenges of intelligent fully mechanized working face. J. Mine Autom. **48**(07), 1–12 (2022)
14. Liu, C.: Research and application of digital twin intelligent flood prevention system in Huaihe River Basin. China Flood Drought Manag. **32**(1), 47–53 (2022)
15. Yin, G.: Monitoring system for shield machine tool changing robot based on digital twin. Comput. Integr. Manuf. Syst. 1–20 (2024)
16. Miao, T.: Research on digital twin simulation method of industrial robot integrated withreinforcement learning. J. Mine Autom. 1–12 (2024)
17. Zeigler, B.P., Praehofer, H., Kim, T.G.: Theory of Modeling and Simulation. Academic Press (2000)
18. Ackoff, R.L.: Towards a system of systems concepts. Manag. Sci. **17**(11), 661–671 (1971)
19. Ehsan, A.: An environment for developing simulatable AADL-DEVS models. Simul. Model. Pract. Theory **123**(102690), 1–19 (2023)

Noise Transparentization Approach for High-Precision Visualization of 3D Scanned Point Clouds

Qingyu Mao[1(✉)], Tomomasa Uchida[2], Kyoko Hasegawa[3], Satoshi Takatori[4], Liang Li[5], and Satoshi Tanaka[5]

[1] Graduate School of Information Science and Engineering, Ritsumeikan University, Kyoto, Japan
is0597rs@ed.ritsumei.ac.jp
[2] FUJIFILM Business Innovation Corp., Minato City, Japan
[3] School of Information and Telecommunication Engineering, Tokai University, Shibuya City, Japan
[4] Research Organization of Open Innovation and Collaboration, Ritsumeikan University, Kyoto, Japan
[5] College of Information Science and Engineering, Ritsumeikan University, Kyoto, Japan

Abstract. A point-based deep peeling method applicable to processing three-dimensional (3D) scanned point clouds is proposed in this paper, which has been originally used in the field of polygon processing. With this method, the task of photorealistic, i.e., opaque visualization of a 3D scanned point cloud, is achieved by extracting visible points and making noise points transparent, i.e., invisible. Additionally, this method does not result in loss of clarity and can finely visualize the detailed structure of the object. It allows the method to achieve high-precision visualization of 3D scanned point clouds. By taking advantage of the low color stability of the noise, our method ensures that the visualized object itself is made opaque and visible while the noise points are made invisible. The effectiveness of transparentizing noise is demonstrated using actual 3D scanned point cloud data for a high-value cultural heritage object, and the proposed method may be a ideal solution of opaque visualization of 3D scanned point clouds.

Keywords: 3D scanned point cloud · depth peeling · measurement noise

1 Introduction

With advancements in scanning methods [1,2] (laser scanning and photogrammetry [3]), 3D scanned point clouds are increasingly used in cultural heritage preservation, virtual reality (VR) [4,5], and many other fields. It is important to preserve cultural characteristics, such as the features of museum artifacts and historical buildings that can easily deteriorate over time, especially for those objects and structures situated outdoors. In other areas of computer vision research, there have been many studies [6–8] related to cultural heritage preservation. Digital archives [9,10] have become a solution for preserving cultural heritage as digital data for computer processing, allowing visualization without degradation of the object visualization. High-precision 3D point clouds have become more feasible to generate with advancements in measurement technologies and the use of drones [11]. However, 3D scanned point clouds often contain measurement noise [12] caused by environmental and weather conditions during data collection, which reduces visualization quality. The visualization of 3D scanned point clouds therefore needs to address the challenge of noise.

There are already some methods [13] for addressing the problem of noise, but they may reduce accuracy. For efficient high-precision visualization, we apply deep peeling, which has thus far been used for polygon data, to 3D point clouds and successfully extract parts of point clouds close to viewpoints. We also take advantage of the color instability of the noise to make the noise transparent, i.e., invisible, without affecting the rendering of the target object to be visualized.

The method proposed in this paper makes the noise statistically transparent during visualization, enabling high-precision and high-speed visualization of 3D scanned point clouds. We neither extract the noise points explicitly nor remove them by hand.

2   Noise in 3D Scanned Point Clouds

Typically, there are two types of measurement noise found in 3D scanned point clouds: outlier noise and Gaussian noise.

2.1   Outlier Noise

Outlier noise often appears at a substantial distance from the scanned target object. For example, outlier noise often occurs when tree leaves cause the light of a laser scanning device to scatter. It is also likely to occur in areas where dust is in the air, such as inside an iron factory. The outlier noise is detected as low-density points, with these parts of the image containing only a few points per pixel.

Nonetheless, this noise significantly reduces the visibility of the image generated from a 3D scanned point cloud (Fig. 1).

2.2   Gaussian Noise

Gaussian noise is distributed near the surface of a scanned object. This noise also significantly affects the visibility of visualized images. For example, when the wind shakes a tree, noise appears on the leaves and branches of the tree. Unlike outlier noise, Gaussian noise severely blurs the edges and surfaces of visualized objects, significantly reducing visibility (Fig. 2).

Fig. 1. The point cloud for a Gion Festival float (Kyoto, Japan) in which outlier noise points have been artificially added as 1% of the total points.

2.3 Characterization of Noise Points

Here, we briefly summarize the characterization of the noise points.

The most apparent characteristic of noise is the low point density and the low point count in comparison with the non-noise parts of the scanned object. However, the effect of noise is not small. For a large-scale 3D point cloud, the number of points comprising the image of an object can reach tens of millions or even billions. Given this high number of points, a mere 1% outlier noise, for example, in a 3D scanned point cloud significantly impacts our observations, as shown in Fig. 1. Notably, the noise points are distributed not only on the surface and inside the object but also everywhere in space. This sometimes leads to a noise point density that is not much lower than that of the object itself.

The noise points near the scanned object should be similar in color to the nearby portion of the scanned object. However, the noise points have low color stability because they are regarded as being randomly moved from the right places. This feature is more evident for outlier noise. Gaussian noise also shows low color stability, although it usually has a point density closer to that of the non-noise parts.

3 Transparent Visualization Method

Here, we briefly review stochastic point-based rendering (SPBR) [14], which is the inspiration for our noise transparency method. SPBR is a method for achieving high-quality transparent visualization of a 3D scanned point cloud.

3.1 Stochastic Point-Based Rendering (SPBR)

Large-scale point cloud data for large-scale cultural sites, for example, contain complex but valuable information about the external and internal structure of

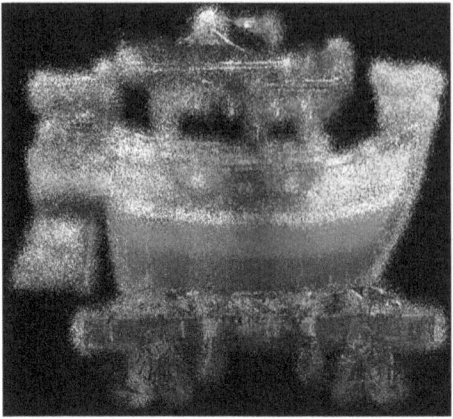

Fig. 2. The point cloud of a Gion Festival float (Kyoto, Japan), for which Gaussian noise points are artificially added as 1% of the total points.

the object or structure being measured. Therefore, we often use transparent visualization so that both the external and internal structures can be observed simultaneously. We have proposed a transparent visualization method called SPBR, which enables fast and precise transparent visualization without the need for depth sorting of point clouds, in contrast to previous methods [15].

The steps for executing the SPBR are as follows (Fig. 3):

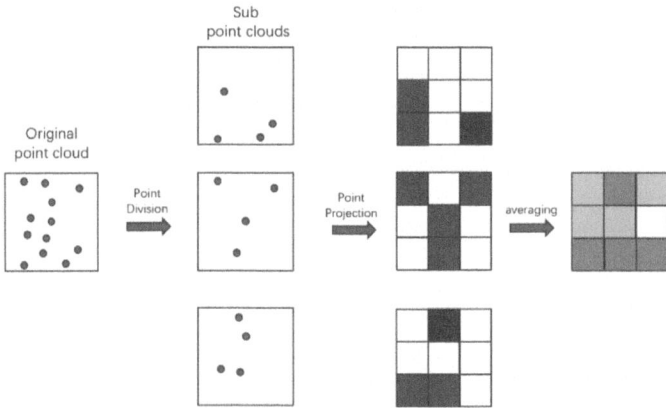

Fig. 3. The flow chart of the SPBR method

Point Division. First, we randomly divide the original point cloud into multiple point subsets. Each subset has the same number of points and is statistically independent of other subsets.

Point Projection. Second, the point subsets obtained above are independently projected onto the image plane to produce multiple intermediate images. When projecting the points onto the image plane, pixel-by-pixel hidden point processing is performed to reflect the opacity of the points.

Ensemble Averaging. Finally, we calculate an average color at each corresponding pixel of the intermediate images to create an average image. This average image becomes the image resulting from transparent visualization.

3.2 Transparentization Method

Because the standard point density of the scanned object is much greater than that of the noise points, a large number of points are projected to each pixel, even for a large number of point subsets of approximately several hundred. Each point subset contains sufficient data that describe the full information of the original point cloud. Therefore, the ensemble average of the point subsets successfully smooths the color at each pixel. In ensemble averaging, the effect of a noise-point image, whose probability of appearance is small, is negligible. Therefore, the resultant pixel color becomes the noise-eliminated object color (see Fig. 4). Note that the background color also affects the ensemble average in transparent visualization.

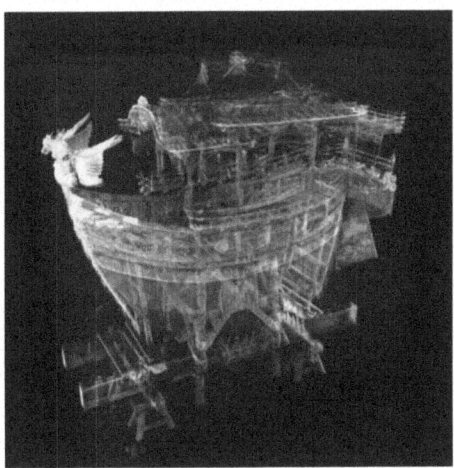

Fig. 4. Transparent image created by SPBR for the point cloud from Fig. 1 (the number of point subsets is 50).

As shown in Fig. 4, an SPBR can make noise transparent or invisible in transparent visualization [16]. The purpose of the current work is to achieve similar noise transparency in photorealistic images, i.e., opaque visualization.

The difference between transparent and opaque visualization is in whether or not the internal structure is incorporated. Therefore, our task is to remove the effect of the points describing the internal structure from the given point cloud data. To achieve this goal, we developed "point-based depth peeling" inspired by conventional polygon-based depth peeling (see the next section).

4   Polygon-Based Depth Peeling

Depth peeling [17,18] is a transparency processing technique initially developed for polygon rendering. Specifically, it involves peeling off layers of polygons that overlap from nearest to farthest, one by one.

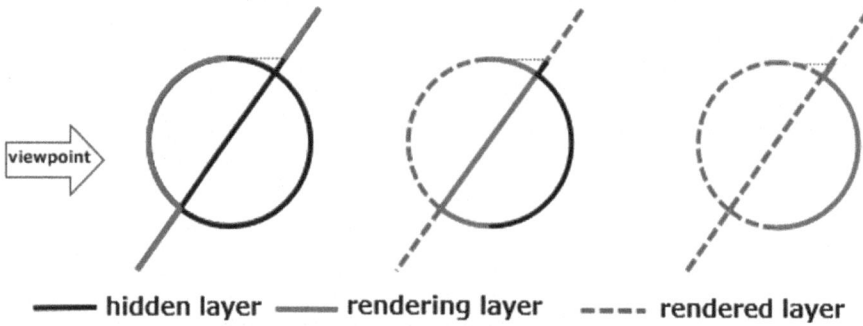

Fig. 5. The polygon-based depth peeling process

Figure 5 shows the polygon-based depth peeling process. It is assumed that the viewpoint is from the left. The removal of the hidden surface is conducted using the depth buffer per pixel, peeling the surface portion closest to the viewpoint. The recursive repetition of this peeling creates multiple layers, peeling each layer one by one. Applying alpha blending to all the layers results in a transparent image. On the other hand, visualizing only the first layer generates an opaque image, neglecting the hidden parts of the polygon mesh. We aim to develop a technique whose results are similar to those of the latter case for point cloud data.

5   Proposed Method: Point-Based Depth Peeling

5.1   Idea of Point-Based Depth Peeling

A large-scale 3D scanned point cloud has a very high point density. This is because point acquisition is executed repeatedly for the same small local portion during scanning. Therefore, the variation in the point depths becomes very small for each small local portion. In addition, in each small local portion, where 3D points are projected onto the same pixel, points should be assigned almost the same color, apart from color fluctuation, i.e., color noise. Thus, since 3D scanned points have strong local color stability, we can eliminate color fluctuations by averaging, i.e., smoothing local point colors per projected pixel. This strategy for obtaining the true point color can be carried out by executing point-based depth peeling per pixel for points only near the surface closest to the viewpoint. Note that our point-based depth peeling method aims to achieve opaque, i.e., photorealistic, rendering, which differs from conventional polygon-based depth peeling for transparent rendering.

The strategy of our point-based depth peeling method is as follows. The 3D points are projected onto the 2D image plane. Corresponding to each pixel in the image plane, we appropriately select 3D points close to the viewpoint and average their colors to determine the true color of the pixel. Namely, we select and use points with the best color stability so that the effect of noise can be avoided mitigated as much as possible.

5.2   Steps of the Point-Based Depth Peeling

Figure 6 illustrates how to select points with the best color stability and use them for ensemble averaging, i.e., pixel-oriented color averaging. In this example, there are $L_{max} = 12$ layers for a focused pixel. Since the majority of points are blue and exhibit better stability, the true color is to be blue. The following steps describe the process of assigning the true color to the focused pixel.

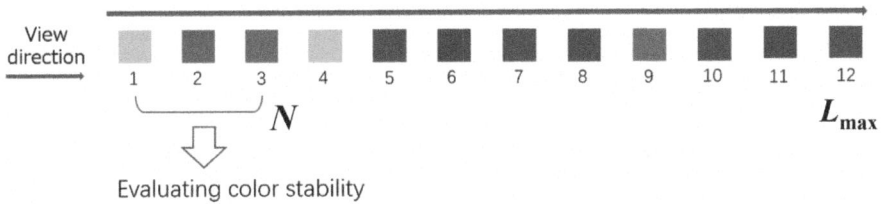

Fig. 6. Point-based depth peeling is used to extract the true color, eliminating color noise.

Step 1. Set the maximum layer count L_{max}, which defines the possible maximum depth of the object surface.

Step 2. Set the layer-range number N, which defines the number of layers over which the color stability is evaluated.

Step 3. Evaluate the color variance over N layers, starting from the layers closest to the viewpoint and terminating at the L_{max}-th layer. To evaluate the color variance, we use the sum of the variances of the red, green, and blue colors.

Step 4. Determine the pixel color as the median value of the N layers that have the lowest variance in Step 3.

5.3   Parameters

There are two main important parameters in this method. The layer count L_{max} and the resolution R of the output image.

The Maximum Layer Count L_{max}. The number of layers parameter refers to the maximum number of layers to be extracted from each pixel. Small adjustments to this parameter will not have a large impact on the visualization results, as we will only use the part of each pixel that has the highest color stability. However, it is not a parameter that can be set arbitrarily. Ideally, if the density of the point cloud is high enough, the larger the number of layers is, the better it is to exclude the effect of noise, but in practice, we have to consider the low point density parts of the object, such as the edges. If the number of layers is too large, it can lead to some of the points of the scanned object also being transparent or obtaining points inside the object. On the other hand, if the number of layers is too few, for some points, cloud data with considerable noise may be insufficient to transparentize the outlier noise in the background or may result in only the part where the noise is mixed with the points of the scanned object being obtained, resulting in some noise remaining.

In short, this parameter can be understood simply as the intensity of the transparency. The more layers there are, the better the transparency. To achieve the best results, we need a good balance between the noise transparency effect and the degradation of the object visualization itself.

The Resolution R of the Output Image. This parameter affects the resolution of our output image and impacts the noise transparency.

Figure 7 shows the projection of a 3D point cloud onto a 2D plane. We need to first determine the viewpoint and the direction, and then project the 3D point cloud onto the visual plane (which can be understood as taking a photo of the point cloud from the viewpoint in the direction). We can think of the points as

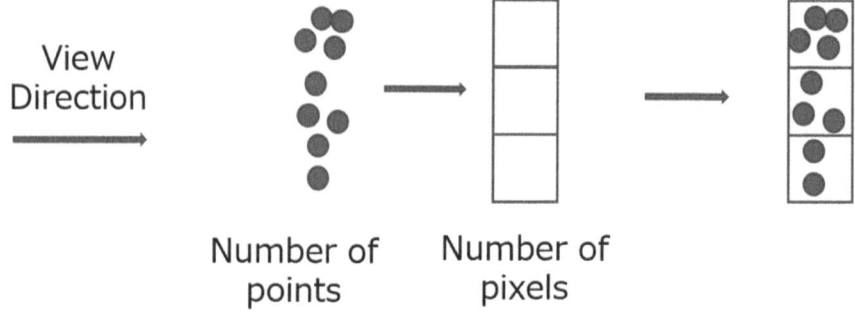

Fig. 7. The projection of a 3D point cloud onto a 2D plane

small balls, and the pixels can be thought of as boxes. All we have to do is to put the ball into the box at the corresponding position.

The point cloud itself determines the number of balls, while the resolution determines the number of boxes. Therefore, if the resolution is too large, there will be too few points in each pixel, making it difficult to transparentize the noise. Conversely, if the resolution is too small, the final output image will be too blurry, making it difficult to use.

In practice, we find that as the resolution decreases, the effect of noise transparency tends to improve, especially for noisy data.

The reason for this is that as the number of points in each pixel increases, our method enables the stable part of the color to be more easily identified. In addition, a larger number of points allows the parameter of the number of layers to be adjusted over a wider range, making it easier to identify the most appropriate value.

6   Visualization Experiments

6.1   Results of Noise Transparention

We apply the proposed method to the 3D scanning points for the festival float of the Gion Festival (Kyoto, Japan) and demonstrate the effectiveness of the method. Noise points that are classified as outlier noise tend to be widely scattered across the point cloud, and the point count is low. On the other hand, noise points classified as Gaussian noise cluster closer to the object, and the point count and point density are greater than those of the outlier noise, as described in Sect. 2.

To evaluate the effectiveness of our proposed method, we conducted tests using a point cloud contaminated with 1% outlier noise points (Fig. 1) and another point cloud with 10% Gaussian noise points (Fig. 2).

All of the following experimental results were generated with $L_{max} = 30$ and a resolution of R=1024×1024 using point cloud data containing 25,427,464 points. With this parameter, it is possible to keep the results minimally damaging to the object, while effectively making the noise transparent.

Outlier Noise Point Cloud Data. Figure 8 shows the results for a point cloud containing outlier noise. Almost all of the noise is well transparentized, and the object itself remains opaque.

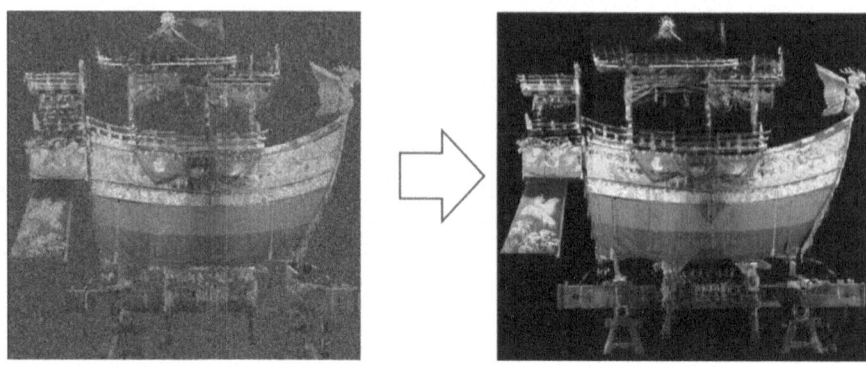

Fig. 8. Outlier noise processing results

Gaussian Noise Point Cloud Data. Figure 9 shows the results for a point cloud containing Gaussian noise. Similar to Fig. 8, almost all of the noise is made transparent, and the object itself remains opaque. To date, Gaussian noise has been difficult to transparentize because it has a higher density and larger number of points than outlier noise. This makes it difficult to distinguish noise from real points when describing a scanned object.

6.2   Quantitative Evaluation of Noise Transparentization

Here, we conduct a quantitative evaluation of the images created by our proposed method.

We prepare a reference point cloud with minimal inclusion of noise points, i.e., a point cloud in which noise points are removed as much as possible. Then, we prepare an experimental point cloud in which noise points are added. By applying the proposed method to both images, we create reference image A

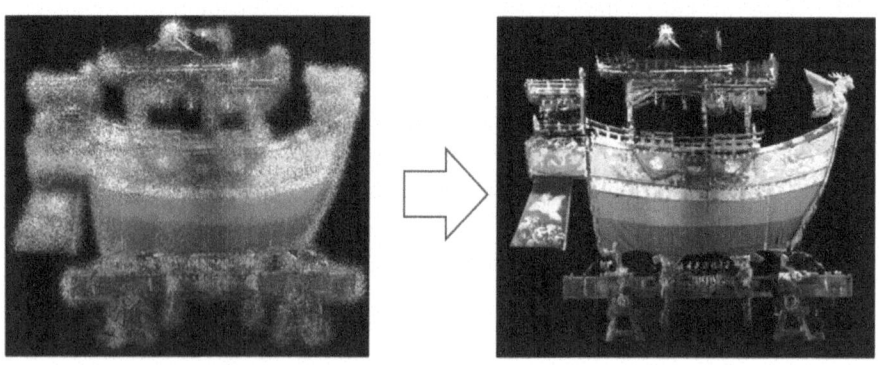

Fig. 9. Gaussian noise processing results

and evaluation image B. For the evaluation of image B, we use the following color-space distance at each corresponding pixel:

$$d = \sqrt{(R_A - R_B)^2 + (G_A - G_B)^2 + (B_A - B_B)^2} \qquad (1)$$

where (R_A, G_A, B_A) and (R_B, G_B, B_B) are the RGB colors of images A and B at the corresponding pixel, respectively. The effectiveness of the noise transparentization is evaluated based on how few pixels there are with a distance d smaller than a given threshold d0. We define the evaluation metric p as follows:

$$p = \frac{n_{d<d_0}}{n_{pixel}} \qquad (2)$$

where n_{pixel} is the total number of pixels and $n_{(}d < d_0)$ is the number of pixels with $d < d_0$.

We use the image from Fig. 10 as reference image A. By applying the right-hand images from Figs. 8 and 9 as evaluation images, p = 0.85 in both cases, which means that 85% of the noise is made invisible. This approach satisfies what is needed and is useful for practical visual analysis.

Fig. 10. The comparison image

7   Conclusions

In this paper, we proposed a method called point-based depth peeling, which utilizes the concept of depth peeling commonly employed in polygon processing. This method enables high-quality and photorealistic (opaque) visualization of 3D scanned point clouds. The key idea is to extract the surface portion of the point cloud nearest to the viewpoint, identify color-stable layers, and make the noise transparent, i.e., invisible, through pixel-oriented image averaging. We demonstrated the effectiveness of the proposed method by applying it to real 3D scanned point cloud data for a festival float with high cultural value. We showed that both outlier and Gaussian noise can successfully be made invisible.

Acknowledgments. We would like to thank the Funehoko Preservation Society for their generous cooperation. We would also like to thank Prof. Kozaburo Hachimura, Prof. Hiromi T. Tanaka, Prof. Takanobu Nishiura, Prof. Keiji Yano, Mr. Atsushi Okamoto, and Dr. Hiroshi Yamaguchi for their cooperation and valuable advice.

References

1. Vosselman, G., et al.: Recognising structure in laser scanner point clouds. Int. Arch. Photogramm. Remote Sens. Spat. Inf. Sci. **46**(8), 33–38 (2004)
2. Linsen, L., Prautzsch, H.: Local versus global triangulations. Eurographics (Short Presentations) (2001)
3. Baltsavias, E.P.: A comparison between photogrammetry and laser scanning. ISPRS J. Photogramm. Remote. Sens. **54**(2–3), 83–94 (1999)
4. Discher, S., Masopust, L., Schulz, S.: A point-based and image-based multi-pass rendering technique for visualizing massive 3D point clouds in VR environments (2018)
5. Thiel, F., et al.: Interaction and locomotion techniques for the exploration of massive 3D point clouds in VR environments. Int. Arch. Photogramm. Remote Sens. Spat. Inf. Sci. **42**, 623–630 (2018)
6. Aicardi, I., et al.: Recent trends in cultural heritage 3D survey: the photogrammetric computer vision approach. J. Cult. Heritage **32**, 257–266 (2018)
7. Guidi, G., et al.: Virtualizing ancient Rome: 3D acquisition and modeling of a large plaster-of-Paris model of imperial Rome. In: Videometrics VIII, vol. 5665. SPIE (2005)
8. Dylla, K., et al.: Rome reborn 2.0: a case study of virtual city reconstruction using procedural modeling techniques (2010)
9. Koller, D., Frischer, B., Humphreys, G.: Research challenges for digital archives of 3D cultural heritage models. J. Comput. Cult. Heritage (JOCCH) **2**(3), 1–17 (2010)
10. Gomes, L., Bellon, O.R.P., Silva, L.: 3D reconstruction methods for digital preservation of cultural heritage: a survey. Pattern Recogn. Lett. **50**, 3–14 (2014)
11. Heritage, G., Hetherington, D.: Towards a protocol for laser scanning in fluvial geomorphology. Earth Surface Processes Landforms J. Br. Geomorphol. Res. Group **32**(1), 66–74 (2007)
12. Masuda, H.: Challenges and technological trends in large-scale point cloud processing based on 3D measurement. J. Jpn. Soc. Precision Eng. **79**(5), 384–387 (2013). (in Japanese)
13. Blas, M.R., et al.: Fault-tolerant 3D mapping with application to an orchard robot. IFAC Proc. Volumes **42**(8), 893–898 (2009)
14. Tanaka, S., et al.: See-through imaging of laser-scanned 3D cultural heritage objects based on stochastic rendering of large-scale point clouds. ISPRS Ann. Photogramm. Remote Sens. Spat. Inf. Sci. **3**(5), 73–80 (2016)
15. Zhang, Y., Pajarola, R.: Deferred blending: image composition for single-pass point rendering. Comput. Graph. **31**(2), 175–189 (2007)

16. Uchida, T., et al.: Noise-robust transparent visualization of large-scale point clouds acquired by laser scanning. ISPRS J. Photogramm. Remote Sens. **161**, 124–134 (2020)
17. Bavoil, L., Myers, K.: Order independent transparency with dual depth peeling. In: NVIDIA OpenGL SDK, vol. 1, p. 12 (2018)
18. Liu, B., Wei, L.-Y., Xu, Y.-Q., Wu, E.: Multi-layer depth peeling via fragment sort. In: 2009 11th IEEE International Conference on Computer-Aided Design and Computer Graphics, pp. 452–456. IEEE (2009)

Research on Remote Operation and Maintenance of Heat Treatment Factory Integrated with 5G and Digital Twin

Ying Cui[1], Lin Qin[2], Xiao Song[1(✉)], and Junfan Zhang[1]

[1] School of Cyber Science and Technology, Beihang University, Beijing 100191, China
songxiao@buaa.edu.cn
[2] Beijing Starter Technology Co., Ltd., Beijing 100191, China

Abstract. In recent years, the mature application of 5G technology and the introduction of policies related to the "Industry 4.0" era in intelligent manufacturing have greatly promoted the exploration of new modes of intelligent manufacturing, thereby facilitating the upgrading and transformation of the manufacturing industry. Through research on 5G technology and digital twin technology, the automation and monitoring of inspection operations information in heat treatment factories have been achieved. Addressing the current issues of low transparency, single-mode, poor real-time performance, and lack of models in the production process monitoring of heat treatment furnaces, this paper proposes a research scheme for remote operation and maintenance of heat treatment factories integrated with 5G and digital twin technology. Based on a web-designed graphical model-oriented digital twin application system, it realizes unified monitoring of multi-dimensional production states with full-domain, 3D visualization, and ultra-low-latency data collection and transmission based on 5G. Application results in a certain enterprise's heat treatment workshop show that this scheme can effectively meet the enterprise's production monitoring needs, improve factory operation and maintenance efficiency, and product qualification rate, and achieve paperless, automated, digitalized, and intelligent operation and maintenance of heat treatment factories.

Keywords: 5G · Digital Twin · Heat Treatment · Remote Operation and Maintenance

1 Introduction

With the continuous advancement of industrialization and the rapid development of information technology, heat treatment factories play a critical role as one of the key manufacturing sectors [1]. However, traditional heat treatment factories face many challenges in their production process, such as the production environment being invisible [2], insufficient monitoring points [3], and untimely inspections [4]. These issues not only affect product quality and production efficiency but may also lead to safety hazards and increased production costs [5].

The production status of heat treatment furnaces directly relates to the quality of products [6]. Currently, there is a widespread issue of too few monitoring points, which results in an uneven distribution of temperature and carbon potential, affecting the qualification rate and scrap cost of products [7]. Heat treatment furnaces are high-temperature, sealed, and energy-consuming equipment [8]. During internal operation, the temperature can reach up to 1200 °C, and the vacuum can reach 0.64 kPa when evacuated. These factors prevent factories from observing the distribution of furnace temperature, gas, carbon potential, etc. in real-time and accurately predicting product quality, making process quality control particularly challenging [9]. Moreover, the factory's established inspection plan of once every two hours often fails due to the high temperature of the on-site working environment, limited staff, and heavy inspection tasks [10]. As a result, employees frequently fail to conduct inspections on time, which prevents mastering the product's production status promptly [11]. The timeliness and efficiency of inspections are hard to guarantee, posing safety hazards.

To solve these problems, remote monitoring and inspection have become a matter of urgency [12]. Digital twin technology, as a bridge connecting virtual space and physical space, has gradually become a research hotspot for scholars both domestically and internationally in recent years [13]. By facilitating full interaction between virtual models of equipment and physical objects, digital twin technology enables intelligent perception of physical equipment and achieves high-fidelity mapping in the information space. It facilitates data accumulation, mining, and predictive analysis throughout the lifecycle of physical equipment, providing a novel solution for production monitoring and operation and maintenance (O&M) in heat treatment factories [14]. Therefore, the introduction of digital twin technology addresses the insufficient interaction between virtual and physical aspects of heat treatment equipment, enabling product status monitoring. This has significant implications for enhancing product heat treatment efficiency and ensuring the efficient and stable operation of production lines. However, traditional wired or 4G network connections are limited by bandwidth and stability, making it hard to meet the requirements for high-definition presentation and immersive experiences at industrial sites, therefore a new type of network connection that can adapt to industrial environment requirements is needed [15]. In recent years, new infrastructure such as 5G and big data has entered a fast lane of development [16, 17]. Through the services provided by new infrastructure construction, such as digital transformation, intelligent upgrading, and integrated innovation, the manufacturing industry has been propelled from digitization and networking towards intelligence [18, 19]. The country vigorously promotes the deployment of "new infrastructure", with 5G being mentioned twice, indicating its pivotal role in the seven major areas of "new infrastructure". Compared to traditional communication technologies, 5G boasts better scalability, energy efficiency, and reliability. Additionally, with a peak transmission rate of 10 Gbit/s, an increase in device connectivity density by 10 to 100 times, and advantages such as ultra-low air interface latency of just 1 ms, it meets industrial-grade transmission requirements. This makes it possible to achieve full connectivity of factory equipment and real-time perception of production status, thus facilitating the realization of smart manufacturing [20].

This paper utilizes a combination of 5G and digital twin integration technology to achieve real-time transmission of data from heat treatment factories and dynamically monitor the production status of products. To address the two types of equipment critical to product quality in the factory, namely the heat treatment equipment and gear grinding equipment, we aim to set up data collection points inside the furnaces. Through these points, we will conduct process quality modeling to visualize the in-furnace production process, enabling remote inspection, monitoring, and analysis of equipment operation and product quality. The goal is to improve equipment maintenance efficiency, enhance product qualification rates, and reduce safety hazards. The specific solutions are as follows:

- Setting up data collection monitoring points inside the furnace. A total of 510 data collection points is installed inside the furnace. The collected data is transmitted to the server via 5G network modules. When production process parameters exceed the limits, timely alarm notifications are issued to enhance the controllability of temperature and carbon potential distribution in the furnace, thereby improving the product qualification rate.
- Visualizing remote inspection inside the furnace. Leveraging digital twin technology to visualize the temperature field distribution inside the furnace and the carburization process of products, enabling remote inspection and timely adjustment of heat treatment parameters. This facilitates early prediction of product quality under current production conditions, thereby reducing the occurrence of safety accidents.

2 Related Work

2.1 5G Features

The 5G network, characterized by its high speed, massive connectivity, and low latency, is a crucial foundation for advancing China's 'Broadband China' strategy and supporting the development of the digital economy [21]. 5G networks continue to evolve in terms of capabilities and flexibility. Network design has surpassed consumer-focused mobile broadband services by introducing tailored functionalities for the Internet of Things (IoT), Industrial IoT (IIoT), and Cyber-Physical Systems (CPS) [22, 23]. The Industrial IoT entails a communication network for industrial objects (machines, equipment, and processes) to reliably exchange monitoring/control data, while CPS employs Industrial IoT to provide a consistent (synchronous and interactive) digital description of objects. In fact, 5G introduces three main connectivity services: Enhanced Mobile Broadband (eMBB) [24], Massive Machine Type Communication (mMTC) [25], and Ultra-Reliable and Low Latency Communication (uRLLC) [26]. By combining these services, 5G can support extreme variations in IIoT applications [27]. In this direction, there is a need to explore dedicated (non-public) 5G network modes and related spectrum licensing options to achieve specialized, customizable, and cost-effective services. In heat treatment plants, these features meet the demands for real-time data transmission and remote control [28]. Traditional factory networks often fall short in supporting large-scale data transmission and high-speed communication. Introducing 5G technology enables more efficient data exchange and remote operations. For example, in heat treatment plants, the real-time data collected by various sensors on temperature, pressure, humidity, etc., need to be transmitted to control centers for analysis and processing [29]. The high-speed transmission capability of 5G ensures the timeliness and accuracy of data, thus enhancing production efficiency and quality.

2.2 Digital Twin

Digital Twin (DT) technology is currently a research hotspot in the integration of information technology and industry [30]. Researchers have studied the definition and application of digital twins from aspects such as product design [31], manufacturing [32], and operation and maintenance [33]. A typical application of digital twin technology is remote operation and maintenance of products. By embedding sensors in products, key operational parameters of the products are collected and transmitted back to the product operation and maintenance center. Through data processing and analysis, real-time monitoring and prediction of product status are achieved, and maintenance is organized promptly based on the product status to reduce the risk of product failures and downtime [34].

Currently, scholars both domestically and internationally have conducted extensive research on visualization monitoring systems from various perspectives. In 2017, Tao Fei and others first proposed the concept of a digital twin workshop, elucidating the system composition, operating mechanisms, characteristics, and key technologies of digital twin workshops [35]. Building on this, they explored the theory and implementation methods of interaction and integration between the physical world of the workshop and the information world based on twin data. At the same time, they considered data as the core driving force of DT technology, viewing it as one dimension. They expanded the three-dimensional conceptual model to a five-dimensional model, including physical entities, virtual models, services, twin data, and the connections among the aforementioned four parts. In 2020, the Industrial Internet Consortium (IIC) proposed DT from the perspective of the industrial internet, asserting that DT consists of data, models, and service interfaces. Data is the essential foundation, models are the key to implementation, and services are the means of presentation and expression. It also emphasized the interoperability between twins, the multiple components within a twin, and the combination of DTs to construct complex systems. The technological framework has more guiding significance for the establishment of DT, but the connections between the various parts, implementation techniques, and standards are not explicitly provided. The academic community attempts to summarize all application scenarios of the new generation of information technology with DT. The engineering community has soberly adopted the concept of DT, which is more an evolution of thought.

3 Overall Architecture of 5G and Digital Twin

During the heat treatment process, carburizing processes, phase transformations of metal materials, and uneven heating and cooling can all lead to gear deformation or insufficient hardness, thereby shortening the service life of the gears. Therefore, during heat treatment, it is necessary to strictly control the heat treatment history of the gear and related process parameters such as carburization. Traditional monitoring of temperature curves and carbon potential curves cannot accurately reflect the uniformity of temperature and carbon potential distribution inside the furnace, which poses significant challenges to precise control of the heat treatment process.

Leveraging the high bandwidth and low latency characteristics of 5G networks, the 5G and digital twin technology can be used to simulate and view monitoring points inside the furnace, as well as the real-time temperature field and carbon potential distribution. At the same time, with the aid of industrial big data mining technology, and by combining expert knowledge models, equipment operation models, process mechanism models, etc., a web-based digital twin application system oriented towards graphical models is designed. This system allows for the real-time monitoring of product production status, as shown in Fig. 1.

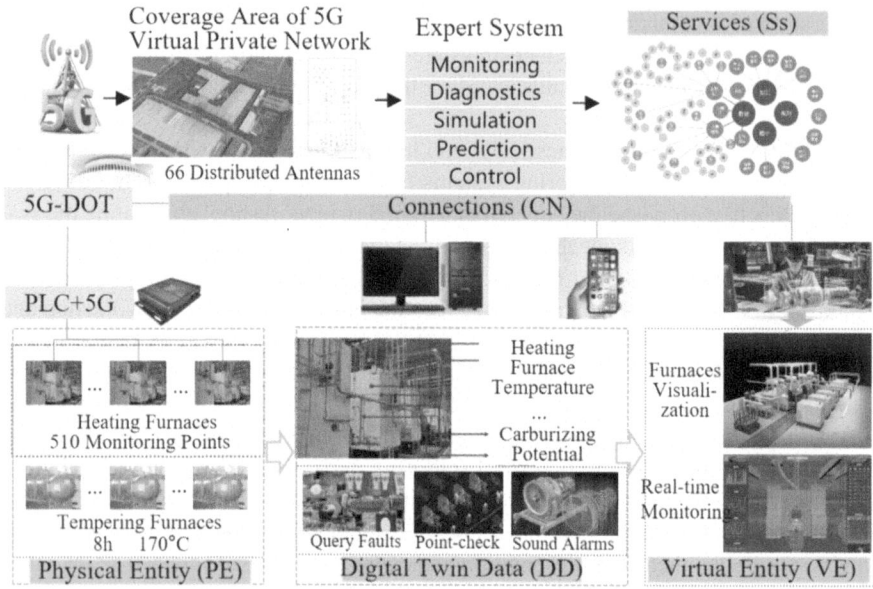

Fig. 1. Digital twin application system topology diagram

The topology diagram of the heat treatment system shows the architecture M_{DT} of the five-dimensional model of the digital twin, as shown in Eq. (1):

$$M_{DT} = (PE, VE, Ss, DD, CN) \tag{1}$$

where *PE* represents a physical entity, *VE* represents a virtual entity, *Ss* represents services, *DD* represents digital twin data, and *CN* represents the connections between components.

As shown in Eq. (2), VE includes a geometric model (G_v), a physical model (P_v), a behavior model (B_v) and a regular model (R_v), which can describe and characterize *PE* from multiple time and spatial scales.

$$VE = (G_v, P_v, B_v, R_v) \tag{2}$$

where G_v is a 3D model describing the geometric parameters and relationships of *PE*, which has good spatiotemporal consistency with *PE*. P_v adds information such as the physical properties, constraints, and characteristics of PE on the basis of G_v. BV describes evolutionary behavior over time, dynamic functional behavior, performance degradation behavior, etc. R_v includes regular rules based on historical correlation data.

DD integration integrates the information data and physical data of the furnace to meet the consistency and synchronization requirements of the information space and the physical space. As shown in Eq. (3), DD mainly includes PE data (D_p), VE data (D_v), Ss data (D_s), knowledge data (D_k), and fusion derived data (D_f).

$$DD = (D_p, D_v, D_s, D_k, D_f) \tag{3}$$

CN realizes 5G interconnection between physical entities, virtual entities, services, and data, thereby supporting real-time interconnection and integration of virtual and real entities.

3.1 5G Virtual Private Network Connection

Constructing a 5G virtual private network that covers the entire factory, relying on the low latency, high reliability, and ubiquitous connectivity of the 5G network to ensure the timeliness, security, and reliability of the data transmission from the heat treatment furnace. It can also serve as the operational network for the factory. The total coverage area of the 5G virtual private network is approximately 11,900 square meters, with the factory building having a height of 20 m.

A 5G indoor distributed pico site system is adopted, with a dipole antenna node (DOT) in Distributed Antenna System (DAS) installed every 15 m at the top of each of the three workshops, totaling 66 DOTs evenly distributed. Every eight DOT antennas are controlled by one Indoor Radio Unit (IRU), and each DOT antenna is connected to the IRU via Category 6 network cables, with nine IRUs installed in total. Every three IRUs are controlled by one Baseband Unit (BBU), with fiber optic cables laid between the IRUs and the BBU for connectivity, resulting in a total of three BBUs installed. Each BBU constitutes one 5G site, and three 5G sites are used to cover the three workshops.

By equipping sensors, Programmable Logic Controllers (PLCs), Personal Computers (PCs), mobile devices, and Virtual Reality (VR) glasses present at the production site with 5G communication modules to connect to the 5G network, data can be transmitted wirelessly, at high speeds, and in real-time, eliminating the need to rewire the production site. After the data is processed through simulation calculations by the expert system, the calculated results and the rendered high-definition model are transmitted back to the digital twin application system for display via the 5G network. This system construction plan supports the connection of equipment to the 5G network without modification, saving on equipment procurement costs, while also ensuring the secure and controllable transmission of on-site information.

3.2 Digital Twin Application System

The system is equipped with a complete set of heat treatment equipment and has built-in monitoring points inside the furnace to monitor parameters such as furnace temperature and carbon potential in real time. A web-based digital twin mapping model is constructed to enable real-time monitoring, early warning, and control of the production status of the actual heat treatment physical equipment.

Heat Treatment Equipment. The complete heat treatment equipment includes 5 heating furnaces and 6 tempering furnaces, with a tempering duration of 8 h at a temperature of 170 °C, utilizing gas radiant tube heating. The PLC is centrally located in the control room, and a data acquisition and display system has been developed to show monitoring parameters of the furnace, including the temperature of the heating furnace and carbon potential. Each furnace is equipped with 5 thermocouples for real-time temperature measurement, and 510 monitoring points for temperature and carbon potential are set up, evenly distributed along the inner wall of the furnace chamber to better simulate the temperature and carbon potential distribution inside the furnace. In addition, the system can query faults, provide point inspection reminders, and is equipped with flashing lights and sound alarms. It models the technological processes affecting product phase transformation, surface morphology, and performance, such as temperature field, carbon concentration distribution, and stress within the heat treatment furnace, to assist in process optimization. This is to enable real-time monitoring of the production environment inside the furnace and predict the quality performance of the products, with predictive analysis and early warning in case of anomalies.

Digital Twin Mapping Model Construction. The Web-based Digital Twin Mapping application primarily focuses on the data-level twin mapping of actual physical heat treatment line equipment, without considering the display of three-dimensional visualization. The Web-based Digital Twin Mapping application is the result interface after the Web-based Digital Twin Mapping model is published and operational. It serves as a large-screen application interface for real-time monitoring, early warning, and even control of the real physical heat treatment equipment. The Web-based Digital Twin Mapping model is defined as a combination of identifiers, attributes, events, and behaviors, along with the inclusion of geometric, mechanistic, data, and knowledge models, as shown in Fig. 2.

Identifier. ID is unique and used to distinguish the Web-based Digital Twin Mapping model.

Attributes. It characterizes the status of the real physical equipment. To represent the various states, characteristics, and indicators of the equipment, this paper defines three types of attributes: IoT (Internet of Things) attributes, constant attributes, and calculated attributes.

Events. It refers to activities passively triggered by the Digital Twin Mapping model when attribute data changes. For the events of a Web-based Digital Twin Mapping model, it is usually necessary to consider which attributes they are related to, that is, which attribute data changes may trigger events. This corresponds to the dependencies of the event.

Behaviors. It which are active operations on physical equipment. Typically, when an emergency event is triggered, certain rescue measures need to be taken, or when users want to change attribute data after observing the trend of attribute data changes, the activities they perform correspond to the behaviors of the Web-based Digital Twin Mapping model.

Geometry model. Provides accurate 3D geometric representations of physical entities, ensuring that virtual models match the shape and size of the entity.

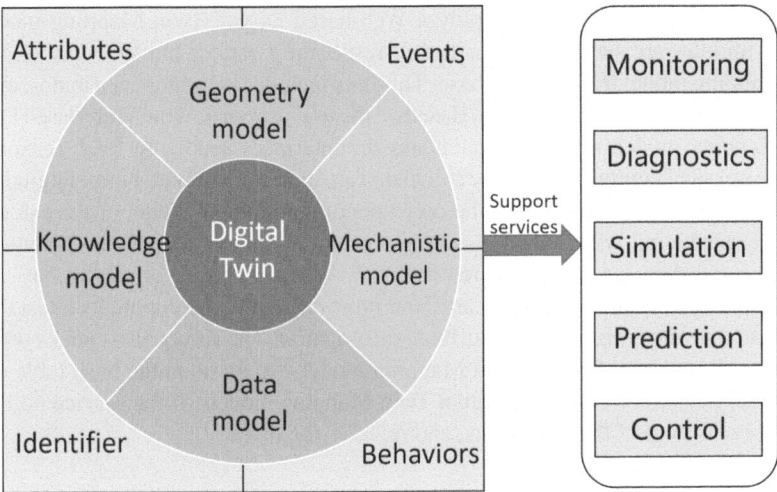

Fig. 2. The digital twin supports the service mapping diagram

Mechanistic model. Based on the laws of physics and engineering principles, it describes the inner workings of the system.

Data model. Utilize sensor data, operational records, and more to simulate the real-time state of an entity.

Knowledge model. It is used to enhance the decision-making ability of the system and optimize operations.

For the construction of the Web-based Digital Twin Mapping model, it essentially involves breaking down the representational logic of the real physical equipment into a hierarchical mapping model. Then, through an integrated system modeling and simulation environment, the attributes, events, and behaviors between the twin mapping models are orchestrated logically, which can provide services for monitoring, diagnostics, simulation, prediction, and control.

Digital Twin Mapping Model Operation. Based on the definition of the Web-based Digital Twin Mapping model in Previous section, to ensure that the model can monitor and issue warnings about the real physical equipment in real-time, the operation of the Web-based Digital Twin Mapping model must satisfy certain conditions. These include the ability for changes in IoT attributes to notify the update of associated calculated attributes, for changes in attributes to notify the update of associated events, and for updates in attributes and events to be promptly notified to the digital twin application system.

The operation of the Web-based Digital Twin Mapping model is primarily implemented through a middleware server. The middleware server maintains a hash table for the running services of Web-based Digital Twin Mapping models. When a user deploys a Web-based Digital Twin Mapping model to run, the middleware server retrieves the corresponding model from the database and then instantiates it as a continuously running service.

When inspection personnel deploy a Web-based Digital Twin Mapping model to run, the middleware server retrieves the corresponding data of the Web-based Digital Twin Mapping model from the database. This data includes identifiers, attributes, events, behaviors, and potential sub-models. Based on the IoT attributes of the Web-based Digital Twin Mapping model, the server calculates the interfaces needed to fetch sensor data from the physical equipment. For the calculated attributes of the Web-based Digital Twin Mapping model, the server parses the corresponding graphical model of the calculated attributes, analyzing which attributes they depend on. These dependent attributes are the monitored data for the calculated attribute. When a change occurs in any of the dependent attributes, the calculated attribute must reactively recompute its latest output value to update the corresponding attribute information. The server also checks whether the current Web-based Digital Twin Mapping model still exists in the hash table of the running services for Web-based Digital Twin Mapping models. If the service no longer exists, it breaks out of the loop, exits, and releases the thread.

4 Results Analysis

Through the graphical modeling environment of the digital twin application system, a Web-based Digital Twin Mapping model for a heat treatment line of a furnace is constructed. In the construction interface of the Web-based Digital Twin Mapping model, there are three main sections: the Web-based Digital Twin Mapping Model Structure Tree, the Web-based Digital Twin Mapping Model Editing Area, and the Graphical Modeling Environment, as shown in Fig. 3. In the operational interface of the Web-based Digital Twin Mapping model, there are four main sections: the Web-based Digital Twin Mapping Model Structure Tree, the Web-based Digital Twin Mapping Model Basic Information Area, the Web-based Digital Twin Mapping Model Attribute Monitoring Area, and the Web-based Digital Twin Mapping Model Behavior Control Area, as depicted in Fig. 4.

Under normal circumstances, the heat treatment production line for manufacturing gears in a blast furnace needs to be kept in a safe state at all times to ensure product quality. The staff only need to monitor the cloud-based data in real time to understand the status of the equipment. To ensure the quality of gear products produced by the blast furnace heat treatment line, the heating furnace must meet specific working conditions.

In the experiment, the temperature of the heating furnace is controlled between 2000 to 3000 °C, and the pressure working range is from 10 to 26 bar. Moreover, a certain reductive atmosphere needs to be maintained inside the heating furnace. Among them, the concentration of hydrogen is kept between 40% to 80%, the concentration of carbon monoxide is between 4% to 6%, and the concentration of methane is maintained between 20% to 30%. At the same time, the concentration of inert atmospheres within the heating furnace, such as the concentration of carbon dioxide, is controlled between 1.5% to 3.5%. In addition to the working conditions of the heating furnace, the gears produced on this production line must meet a series of standards. The surface hardness must be controlled between 40 to 60 HRC, the core hardness needs to be within the range of 32 to 42 HRC, the case hardening depth must be between 2 to 3 mm, the maximum residual austenite content should be controlled between 9 to 11%, the martensite content

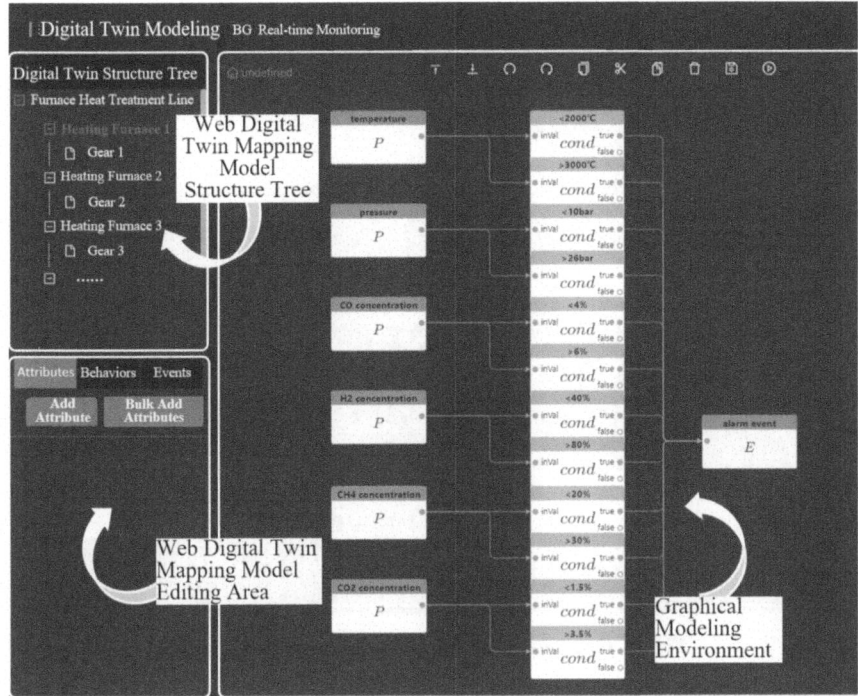

Fig. 3. Construction of the digital twin mapping model

needs to be between 1 to 5%, and the ferrite level must be controlled between levels 1 to 5. Additionally, the carbide content also needs to be controlled within 1 to 5%. The system interface displays curves for the heating furnace and the tempering furnace, as well as relevant information about the gear products, as shown in Fig. 5.

Based on the production environment parameters within the furnace such as CO, CO_2, CH_4, H_2, O_2, and H_2O, chemical reaction equations are constructed, and an iterative residual is set to analyze convergence. A big data model for product quality analysis and prediction is used to numerically simulate the combustion process of gases in the heating furnace. Experimental results show that after 375 steps of iterative calculation, the residual reaches the convergence limit, as shown in Fig. 6.

From the temperature cloud map inside the heating furnace, it can be seen that as the reaction proceeds, the temperature gradually increases from the nozzle to the interior of the burner, and the temperature in the middle area is the highest, reaching 2500K, as shown in Fig. 7.

Based on the production environment parameters within the furnace, the temperature distribution and carbon potential distribution on the product's surface are simulated to timely predict the product's quality indicators, such as case depth, microstructure, martensite, etc. If there are any anomalies in product quality, the system can provide early warnings and assist maintenance personnel in adjusting parameters, as shown in Fig. 8.

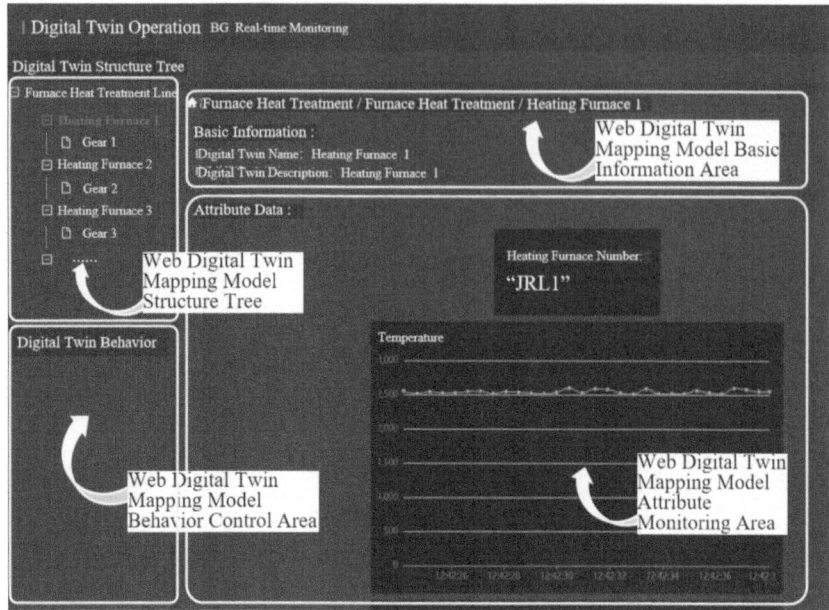

Fig. 4. Operation of the digital twin mapping model

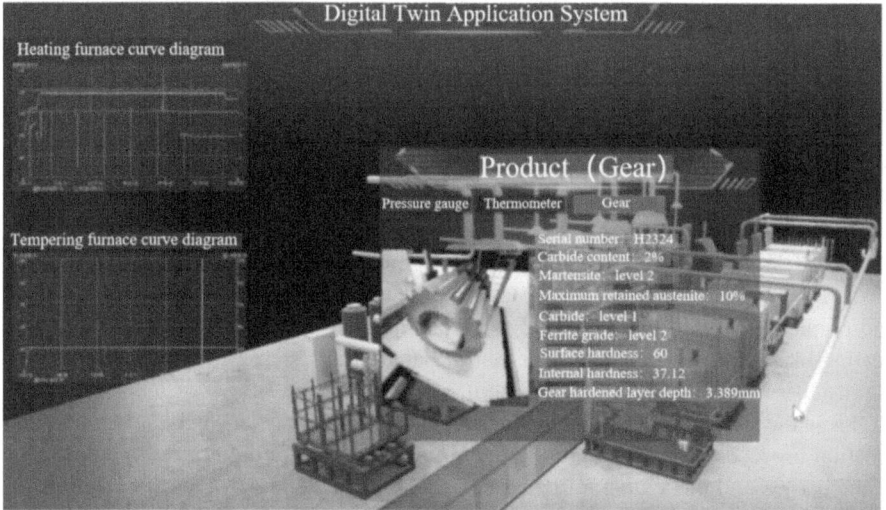

Fig. 5. Digital twin application system information visualization

When attributes within the heating furnace twin mapping model fall outside of the valid range, it triggers a safety warning event, notifying staff that the furnace may be malfunctioning. Similarly, when attributes within the gear twin mapping model are out of the valid range, it triggers a quality warning event, informing staff that the product with

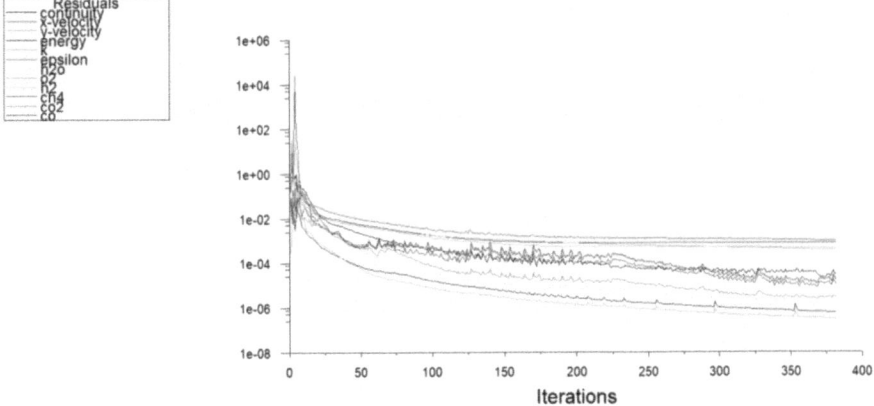

Fig. 6. Heating furnace internal parameters iterative residual curve

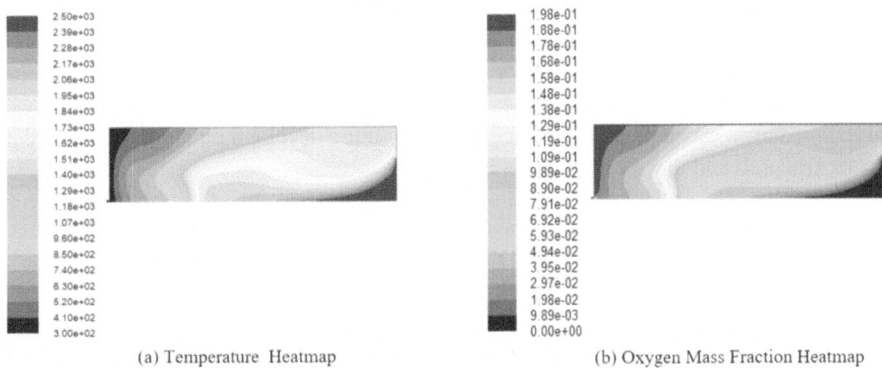

(a) Temperature Heatmap   (b) Oxygen Mass Fraction Heatmap

Fig. 7. Heating furnace parameter cloud map

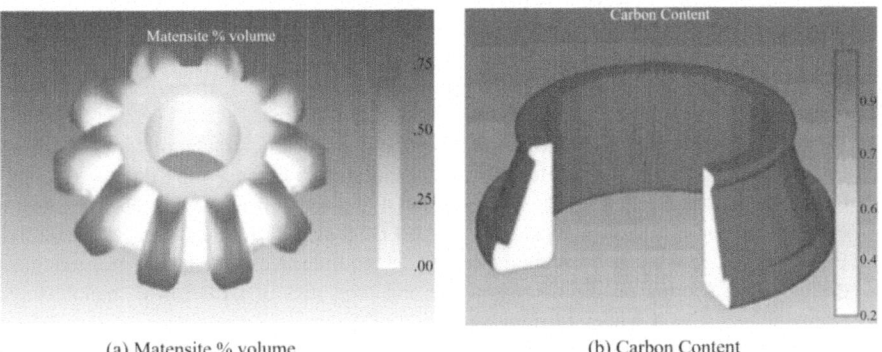

(a) Matensite % volume   (b) Carbon Content

Fig. 8. Product quality prediction model analysis schematic

the corresponding number may be defective, and prompting an investigation into possible production line issues. By applying 5G and digital twin technology to the operation and maintenance optimization of new general-purpose heat treatment furnaces, operators can perform remote equipment inspections and maintenance, reducing the time staff spend on tasks and improving equipment operational efficiency, significantly reducing safety risks. Through the visualization of process data and real-time analysis and prediction alarms for product quality, staff can precisely control the heat treatment process, improving the product pass rate.

5 Conclusion

This paper introduces a method to improve heat treatment processes. By adding more data collection points inside furnaces and using 5G and digital twin, it makes furnace production visible and allows for remote inspections. This boosts efficiency, improves product quality, and reduces safety risks. The approach solves common issues in gear manufacturing and can be applied across the industry to raise overall technological standards. Although the application results in a certain enterprise's heat treatment workshop are favorable, the paper does not mention the feasibility and effectiveness of its application in other types of enterprises or enterprises of different scales, lacking broad validation across various application scenarios. In future developments, we will consider further research and discussion on this matter.

Acknowledgments. This work was supported by the National Key Research and Development Program of China under Grant 2023YFB3308200 and Beijing Natural Science Foundation (L233005).

References

1. Laleh, M., Sadeghi, E., Revilla, R.I., et al.: Heat treatment for metal additive manufacturing. Progress Mater. Sci. (2023)
2. Xiang-Hai, D., Xin-Li, W., Xin-Yuan, Z., et al.: A study on impact of urban production environment on flexible employment of manufacturing factories. J. Hangzhou Dianzi Univ. (Soc. Sci.) (2018)
3. Taniguchi, K., Umegaki, S., Ueno, K.: Application of quenching simulation for CVT pulley. In: International Federation for Heat Treatment and Surface Engineering Congress (2014)
4. Omar, S.M.T., Plucknett, K.P.: The influence of DED process parameters and heat-treatment cycle on the microstructure and hardness of AISI D2 tool steel. J. Manuf. Process. (2022)
5. Hentschel, O., Krakhmalev, P., Fredriksson, G., et al.: Influence of the in-situ heat treatment during manufacturing on the microstructure and properties of DED-LB/M manufactured maraging tool steel. J. Mater. Process. Technol. (2023)
6. Albaplant, B.G.: Industrial furnaces and heat treatment technologies. Alluminio e leghe: A&L (2022)
7. Shivakumar, M., Hamritha, S., Shilpa, M., et al.: Optimization of heat treatment parameters to improve hardness of high carbon steel using Taguchi's orthogonal array approach. Key Eng. Mater. **933**, 129–136 (2022)

8. Samoilov, V.M., Nakhodnova, A.V., Osmova, M.A., et al.: Use of Raman spectroscopy for determination effective heat treatment of carbon materials in high-temperature furnaces (2021)
9. GMP&A Group: The energy crisis in an energy-hungry glass industry. Glass Mach. Plants Accessories (2023)
10. Wei, H., Fengtian, Y., Jingsheng, W., et al.: Factor analysis and cooling technology study on the thermal environment of the high temperature mining's working face. IEEE (2024)
11. Fouaidy, M., Chatelet, F., Drean, D.L., et al.: Recent results of high temperature vacuum heat treatment program of SRF resonators at IJCLab. IEEE Trans. Appl. Superconductivity **31**(5) (2021)
12. Ranfft, A.: Experience & innovation in connecting man to machine. Wire Cable Technol. Int. Serving Manuf. Specifiers Users Wire Cable (2023)
13. Goodwin, T., Xu, J., Celik, N., et al.: Real-time digital twin-based optimization with predictive simulation learning. J. Simul. **18**(1), 47–64 (2024)
14. Javaid, M., Haleem, A., Suman, R.: Digital twin applications toward industry 4.0: a review. Cogn. Robot. **3**, 71–92 (2023)
15. Huang, J.S., Jan, Y.H., Yu, D., et al.: Manufacturing excellence and future challenges of wireless laser components for 4G/5G optical mobile fronthaul networks. 1–2 (2018)
16. Bounegab, A.: Performances d'un système FBMC pour la 5G PRESENTATION (2019)
17. Rehman, S., Shahriar, F.: 5G Multi Input Multi Output (MIMO) Presentation (2019)
18. Gibbins, J.: BIZ: investment/manufacture: Bosch introducing 5G technology across all manufacturing plants. Truck Bus Builder: Int. Newslett. Commercial Veh. Manuf. Dev. (12), 42 (2020)
19. Kulkarni, S.S., Bavarva, A.A.: A survey on various handover technologies in 5G network using the modular handover modules. Int. J. Pervasive Comput. Commun. (2023)
20. Salahdine, F., Han, T., Zhang, N.: 5G, 6G, and beyond: recent advances and future challenges. Ann. Telecommun. **78**(9), 525–549 (2023)
21. Xiao-Jian, Z.: Introduction of key technologies in 5G network and exploration of industry application integration. Value Eng. (2024)
22. Sisinni, E., et al.: Industrial internet of things: challenges, opportunities, and directions. IEEE Trans. Ind. Inform. **14**(11), 4724–4734 (2018)
23. Mahmood, A., et al.: Industrial IoT in 5G-and-beyond networks: vision, architecture, and design trends. IEEE Trans. Ind. Inform 1 (2021). https://doi.org/10.1109/TII.2021.3115697
24. Othman, A., Nayan, N.A.: Public safety mobile broadband system: from shared network to logically dedicated approach leveraging 5G network slicing. IEEE Syst. J. **15**(2) (2021)
25. Gomes, R., Vieira, D., Ghamri-Doudane, Y., et al.:Network slicing for massive machine type communication in IoT-5G scenario. In: IEEE Vehicular Technology Conference. IEEE (2021)
26. Brighente, A., Mohammadi, J., Baracca, P., et al.:Interference prediction for low-complexity link adaptation in beyond 5G ultra-reliable low-latency communications. IEEE Trans. Wirel. Commun. (2022)
27. Gundall, M., et al.: Introduction of a 5G-enabled architecture for the realization of industry 4.0 use cases. IEEE Access **9**, 25 508–25 521 (2021)
28. Karow, V.: Innovative plants for recycling and heat treatment from Tenova LOI Thermprocess. Aluminium (2022)
29. de Jesus Benevides, C.M., Costa, C.C.M., Cardoso, Y.P., et al.:Heat treatment effect study on bioactive compounds of unconventional food plants. Res. Soc. Dev. **2020**(11) (2020)
30. Kolisch, G., Hobus, I., Hansen, J., et al.: Development and testing of a multi-criteria expert system for the real-time energetic optimization of wastewater treatment plants – EOS (2024)
31. Tao, F., Sui, F., Liu, A., et al.: Digital twin-driven product design framework. Int. J. Prod. Res. **57**(12), 3935–3953 (2019)

32. Kritzinger, W., Karner, M., Traar, G., et al.: Digital twin in manufacturing: a categorical literature review and classification. IFAC-Papers Online **51**(11), 1016–1022 (2018)
33. Lu, Q., Xie, X., Parlikad, A.K., et al.: Digital twin-enabled anomaly detection for built asset monitoring in operation and maintenance. Autom. Constr. **118**, 103277 (2020)
34. Hassan, M., Svadling, M., Björsell, N.: Experience from implementing digital twins for maintenance in industrial processes. J. Intell. Manuf. **35**(2), 875–884 (2024)
35. Tao, F., Zhang, M.: Digital twin shop-floor: a new shop-floor paradigm towards smart manufacturing. IEEE Access **5**, 20418–20427 (2017)

Research on Worker Action Recognition and Evaluation in Intelligent Manufacturing Training Based on Industrial Metaverse

Gang Wu[1], Tan Li[1(✉)], Yuqi Zhou[1], Jin Guo[1], Jingyu Zhu[1], Nanjiang Chen[1], Weining Song[2], Yalan Xing[3], Xianghui Meng[1], Yanwen Lin[4], Qi Wang[4], and Runqiang Li[4]

[1] NanChang University, Nanchang 330031, China
litan@ncu.edu.cn
[2] East China University of Technology, Nanchang, Jiangxi, China
[3] Beijing University of Aeronautics and Astronautics, Beijing, China
[4] QuikTech Co., Ltd., Beijing, China

Abstract. Industrial Metaverse plays an increasingly important role in Intelligent Manufacturing Training, improving the user immersion as well as reducing the cost and risks in training. Action recognition technology can effectively detect human actions, help construct the industrial metaverse space of Human-in-the-loop, and empower the low-intrusion, personalized intelligent manufacturing training system. Intelligent Manufacturing Training-Worker Action Recognition and Evaluation (IMTware) is researched to improve the human-manchine interaction and intelligent guidance in industrial training. In terms of action recognition, IMTware extracts the skeletal point sequences of human practice training actions in the video via OpenPose network and inputs them into ST-GCN network for action recognition. In terms of action evaluation, a method combining Dynamic Time Warping (DTW) keyframe matching with cosine similarity solving is proposed to compute the action similarity between the skeletal point sequences and the standard skeletal point sequences, and evaluate them according to the similarity to give personalized correction suggestions. Subsequently, the performance of the system is validated using overhead crane command actions training in converter steelmaking as an example, achieving satisfactory results.

Keywords: Intelligent Manufacturing Training · Worker Action · Action Evaluation · Industrial Metaverse

1 Introduction

Intelligent manufacturing training is a practical training and education method based on intelligent manufacturing technology, aiming at simulating the real industrial production environment, training trainees to master knowledge and skills related to intelligent manufacturing [1], which is of great significance for vocational skills training to control the quality of talents, improve the training efficiency, and update the way of talent training.

S. Saito et al. (Eds.): AsiaSim 2024, CCIS 2170, pp. 365–380, 2024.
https://doi.org/10.1007/978-981-97-7225-4_28

The industrial metaverse, which integrates hot technologies such as virtual reality, digital twins, and deep learning, is attracting widespread attention for its highly interactive and immersive features, and its core concept of "Human-in-the-loop" [2] is becoming a new trend in the field of empowering intelligent manufacturing training.

Workers are the main carriers of industrial activities, and in some industrial activities, specific actions have clear standards and norms and represent special meanings. Therefore, standardized and regulated instruction of trainees' movements is crucial for practical teaching. Most of the current practical teaching adopts the one-to-many classroom teaching mode, which can't provide detailed guidance to the trainees' movement specification. The industrial metaverse of "virtual-reality mapping and virtual-reality interaction" can provide trainees with low-invasive, intelligent and personalized practical training. The two teaching modes are compared, as shown in Table 1.

Table 1. Table of comparison of the two teaching styles.

	Traditional teaching	Intelligent Evaluation System
Action Standards	Different teachers, Different experience Hard to standardize	Recorded by professionals Scientific standardized actions
Training Assessment	Teacher Ratings Highly subjective	Action Comparison Objective quantification
Action Guidance	One-to-many teaching	One-to-one teaching
Training Demands	Time-space limit	Anytime and Anywhere
Interactivity	Weak interaction	Personalized instruction

The core of the intelligent evaluation system is to accurately identify and analyse the trainee's operating actions and determine whether the trainee's movements meet the standardized requirements. Action recognition technology can be used to detect movements through computer vision, deep learning and other technologies. However, there are still some difficulties in applying the current action recognition technology and evaluation technology to the intelligent manufacturing training scene. Intelligent manufacturing training is often carried out in factory workshops, where the light interference is serious and the background is complex, which causes great trouble to recognize workers' actions. For this kind of situation, skeleton-based action recognition technology, which has the advantages of less interference, smaller data volume, and more closely matching the actual movement of the human body, has become a popular direction for scholars to study. Yan et al. proposed a novel model of dynamic skeletons called Spatial-Temporal Graph Convolutional Networks (ST-GCN) [3]. This network constructed multiple layers of spatial temporal graph convolution by using human body joints as vertices of the graph and connections between joints as edges of the graph. This model allows information to be integrated automatically along both the graph and the temporal dimension, and achieves superior performance. Li et al. propose the actional-structural graph convolution network (AS-GCN), which stacks actional-structural graph convolution and temporal convolution as a basic building block. This network model utilizes sub-networks to capture the relationships between each joint point, obtaining richer and more complete joint relationships, and extends the ST-GCN to represent higher-order dependencies [4].

As a downstream task of action recognition, action evaluation is applied to several scenarios. Chen developed a system for evaluating action of golf swing, which can provide the action evaluation of trainer and the standard, and the difference between the postures of trainer and the standard in the key frames are given. In addition, this system uses Dynamic Time Warping (DTW) to align two videos with inconsistent length, which solves the problems of inconsistency in the speed and duration of movements in the two videos [5]. Qian et al. designed a rehabilitation training system based on Kinect motion sensor camera. The system can collect human real-time skeleton data with Kinect, calculate the angle between feature nodes, form the angle sequence, and compare the similarity between the measured angle sequence and the standard action sequence of the action library [6]. However, there is still a shortage of research in practical training scenes, and problems such as individual differences in trainees and differences in the speed of movement execution bring large errors to the evaluation of actions, so it is necessary to design a more reasonable method to circumvent these problems.

Intelligent Manufacturing Training-Worker Action Recognition and Evaluation (IMTware) is researched to improve the human-manchine interaction and intelligent guidance in industrial training. In terms of action recognition, IMTware extracts the skeletal point sequences of human practice training actions in the video via OpenPose network and inputs them into ST-GCN network for action recognition. In terms of action evaluation, a method combining Dynamic Time Warping (DTW) keyframe matching with cosine similarity solving is proposed to compute the action similarity between the skeletal point sequences and the standard skeletal point sequences, and evaluate them according to the similarity to give personalized correction suggestions. Subsequently, the performance of the system is validated using overhead crane command actions training in converter steelmaking as an example, achieving satisfactory results.

2   Research on IMTware

Intelligent Manufacturing Training-Worker Action Recognition and Evaluation (IMTware) consists of two basic modules: action recognition and action analysis. After the trainee uploads a practical training video, the action recognition module identifies the category of the trainee's actions in the video and forwards the result to the action analysis module. The action analysis module then compares the trainee's actions with the corresponding actions in the standard action library, ultimately outputting the trainee's action score and providing corresponding guidance suggestions (Fig. 1).

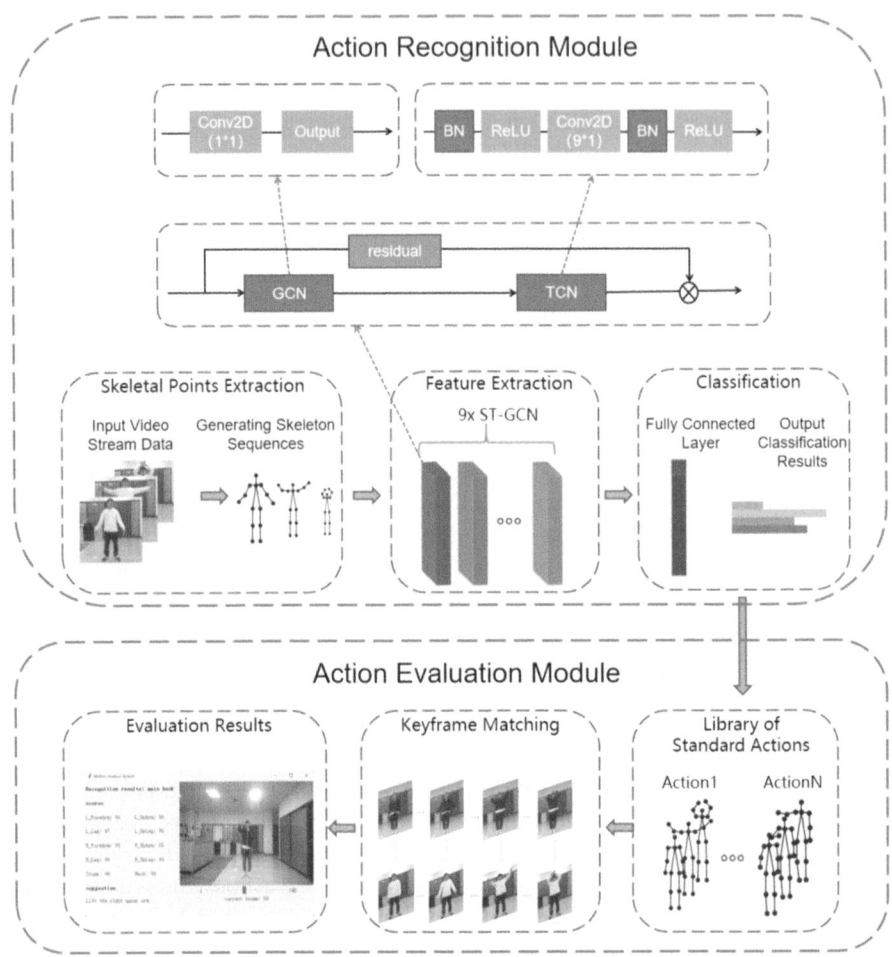

Fig. 1. Overall framework of IMTware

2.1 ST-GCN Based Action Recognition

Intelligent manufacturing training sites are often located in factory workshops, where the background is complex and lighting interference is severe, posing significant challenges for action recognition. In such circumstances, skeleton-based action recognition offers significant advantages. Therefore, the first step is to convert RGB video data into skeleton point data, which is then input into the action recognition network. After research and comparison, this paper selects the Spatial-Temporal Graph Convolutional Network (ST-GCN) as the base network for action recognition.

Skeleton Points Extraction. In this paper, the OpenPose model is used for human skeleton point extraction. Based on the human skeleton point model provided by the COCO dataset, the human body is divided into 18 key points. These key points are illustrated below (Fig. 2).

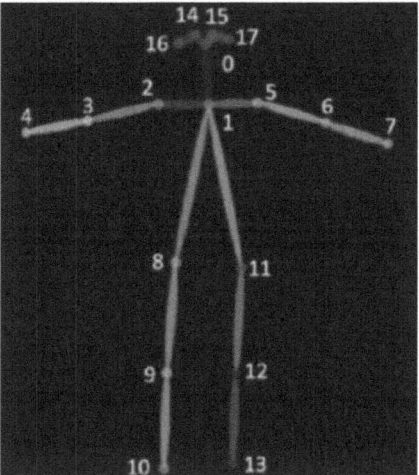

Fig. 2. Human skeletal point models.

Using the OpenPose model, it is possible to extract the coordinates of key points of the human body in each frame of a video and record them. In this way, the changes in movement of the human body in the video can be converted into a sequence of coordinate points for 18 skeletal points.

ST-GCN. Utilizing spatiotemporal convolutional networks, it extracts both temporal and spatial features of the skeletal point data during movement, culminating in the prediction of scores for each type of action. The specific implementation process is outlined below:

Using the pre-trained OpenPose model, RGB video data is converted into skeletal point sequence data. Concurrently, a spatiotemporal graph is created. This graph represents the connectivity of skeletal points across different frames, consisting of two main components: points and edges. The points encompass the coordinate information of all skeletal points in each frame, while one part of the edges represents the natural articulations of the human body, the other part serves to establish connections between identical nodes across different frames. Once the skeletal graph is constructed, the adjacency matrix A of the topological graph can be built, with a size of (18, 18). To better express motion, nodes are categorized as the node itself, the central node, and the eccentric node, resulting in the final shape of A being (3, 18, 18).

The feature extraction module mainly includes spatial feature extraction and temporal feature extraction. Spatial feature extraction is mainly achieved using graph convolutional networks (GCN). GCN can aggregate information from self-skeletal points and adjacent skeletal points, thereby updating skeletal information. The process is as follows in the following formula:

$$H^{l+1} = \sigma\left(D^{-\frac{1}{2}}AD^{-\frac{1}{2}}H^lW^l\right) \tag{1}$$

In the task of action recognition, where A represents the adjacency matrix constructed in the previous context, D denotes the degree matrix of A, and W represents the weight parameters, we utilize a 2D convolution for weight allocation in this paper. The network takes as input the output information from OpenPose and the adjacency matrix A. The output matrix format of OpenPose is (N, C, T, V, M), where N is the number of videos, C represents the information of skeletal points (in this case, 2D pose estimation is used, so C includes the x and y coordinates of skeletal points and confidence score), V denotes the number of skeletal points (which is 18 in this paper), and M indicates the number of people in each frame. Before inputting into the network, the skeletal point information is transformed into the format (N × M, C, T, V). The following diagram illustrates the process of information aggregation using GCN with an input of size (C, T, V) (Fig. 3).

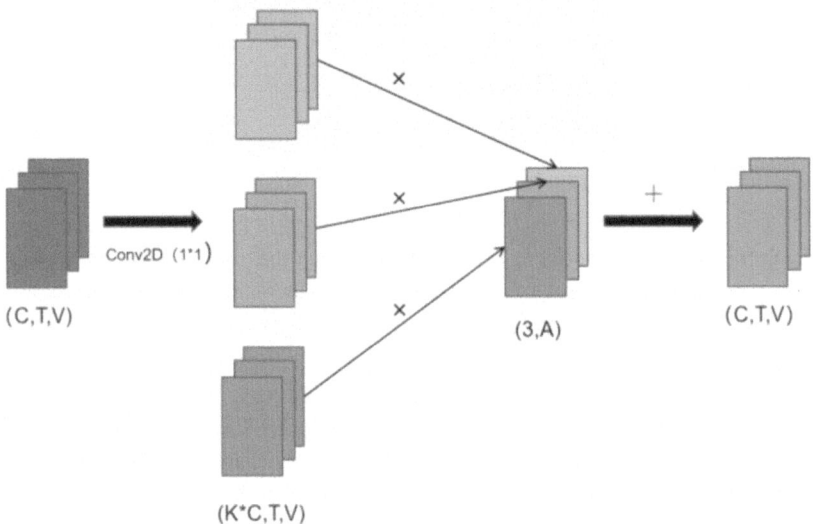

Fig. 3. Information processing of GCN.

After inputting a set of (C, T, V) skeletal point data, a one-dimensional convolution with a 1x1 kernel is applied, dividing the data into three groups. Each group is then multiplied by one of three adjacency matrices. Finally, the results are summed up to obtain the aggregated data after information fusion.

The extraction of temporal features (TCN) is performed using a two-dimensional CNN, where the convolutional kernel size is 9 × 1. Each time, 9 consecutive frames of a node are used for feature extraction. The convolutional kernel moves one frame at a time until all frame information of a node has been extracted, and then the next frame's information is extracted (Fig. 4).

Finally, scores for each action are obtained through a fully connected layer. The category with the highest score will serve as the output of the action recognition module.

2.2 DTW and Cosine Similarity Based Action Evaluation

Action evaluation is used to quantify the quality of action execution. By comparing the extracted skeletal point sequences with standard skeletal point sequences, action

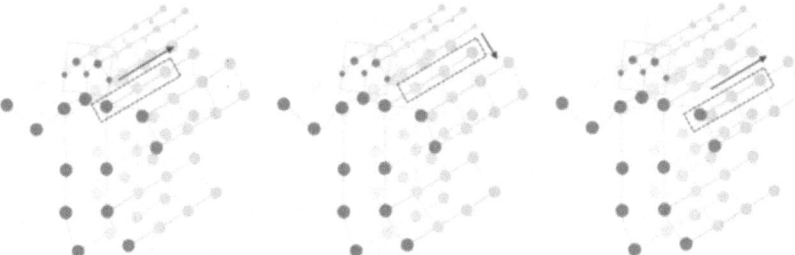

Fig. 4. The implementation process of TCN.

similarity is calculated. Based on this similarity, actions are evaluated, and personalized correction suggestions are provided.

The Method for Measuring Similarity. Typically, similarity is quantified using different forms of distance measures. The Euclidean distance is the most intuitive and commonly used distance calculation method. For skeletal point data in coordinate point format, it can directly calculate the direct line distance between the trainee's skeletal points and the standard skeletal points. The formula is as follows:

$$d = \sqrt{(x_1 - x_2)^2 + (y_1 - y_2)^2} \tag{2}$$

This calculation method is very simple and intuitive, but it has certain limitations. The distance between coordinate points can be affected by the distance from which the video is shot, and differences in body proportions can also influence the distances, for instance, some individuals having longer arms. Consequently, even if two people perform the same action, the scores may not be perfect.

This paper proposes the use of cosine similarity to calculate the similarity of actions for various body parts. Human joint coordinates can be transformed into structural vectors of the human body. For example, the coordinates of the left elbow and left wrist can form the vector of the left forearm. In this way, each pose can be transformed into a set of vectors.

Assume that the forearm vector of the action to be evaluated is $\mathbf{a}_1 = (x_1, y_1)$ and the forearm vector of the standard action is $\mathbf{a}_2 = (x_2, y_2)$. Then, the similarity between them can be calculated using the cosine similarity formula as follows:

$$Similarity = \frac{\mathbf{a}_1 \cdot \mathbf{a}_2}{|\mathbf{a}_1||\mathbf{a}_2|} = \frac{x_1 x_2 + y_1 y_2}{\sqrt{x_1^2 + y_1^2} \times \sqrt{x_2^2 + y_2^2}} \tag{3}$$

Cosine similarity values range from −1 to 1, with values closer to 1 indicating greater similarity. Multiplying the similarity by 100 yields the corresponding score, allowing for the quantification of differences between the action to be evaluated and the standard action. To reduce computational load, only ten key body structure vectors will be retained: left forearm, left upper arm, right forearm, right upper arm, left lower leg, left upper leg, right lower leg, right upper leg, neck, and trunk.

Keyframe Matching Alignment. Due to variations in the speed at which students perform actions, differences in bone point data offset and length may occur, leading to significant errors in the direct calculation of similarity. In the figure below, the blue and black lines depict the variation over time of the y-coordinate of the skeletal point at the right wrist for two individuals performing the same action (Fig. 5).

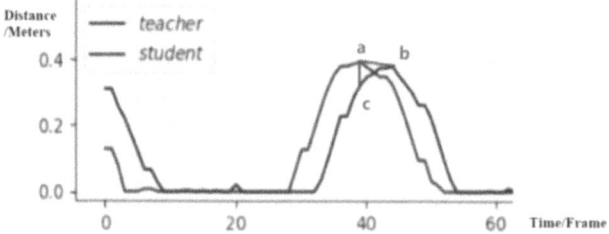

Fig. 5. Comparison of right wrist y-coordinates during the same action

We can observe that the actions performed in these two sequences are nearly identical. However, when compared frame by frame, there are significant discrepancies. For instance, in the figure, the corresponding point to point 'a' in the blue line is point 'b' on the black line, rather than point 'c' which is at the same frame rate as point 'a'. This discrepancy leads to inaccurate results.

Dynamic Time Warping (DTW) can solve the aforementioned problem by finding the optimal matching path that minimizes the distance between two sequences. The matching process is illustrated in the following figure: the blue curve represents action sequence X ($x_1, x_2,..., x_6$), and the red curve represents action sequence Y ($y_1, y_2,...,$ y_7). The dashed lines between the two sequences represent similar actions. The shortest distance is calculated using the following formula:

$$D(m, n) = dist(x_m, y_n) + \min(D(m - 1, n), D(m, n - 1), D(m - 1, n - 1)) \quad (4)$$

In this formula, D (m, n) represents the cumulative distance between the first m actions of sequence X and the first n actions of sequence Y, while dist (x_m, y_n) represents the distance between the m-th action in X and the n-th action in Y (Fig. 6).

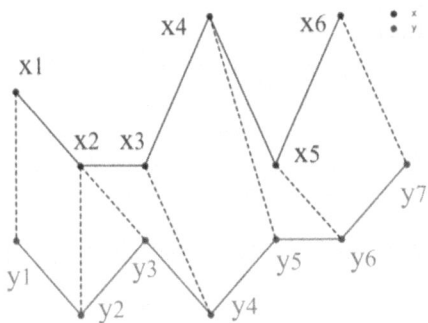

Fig. 6. DTW Matching Path.

Action Correction. Based on the comparative analysis of actions, it is possible to identify body parts with significant errors. For instance, when evaluating forearm movement at a specific frame, as shown in the diagram below (Fig. 7):

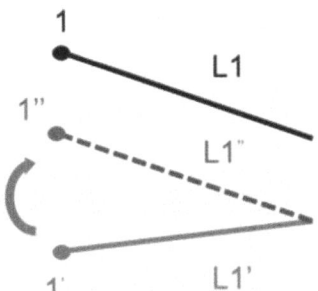

Fig. 7. Action correction process.

Where L1 and L1′ represent the forearm vectors of the standard action and the trainee's action, respectively. Adjusting the position of skeletal point 1 can reduce the gap. By using the cosine formula, the correct position of skeletal point 1 can be calculated. This allows us to provide specific guidance to the trainee, such as raising the left forearm to align the corresponding skeletal point with its correct position.

3   Demonstration of IMTware in Overhead Crane Command Actions

According to the manufacturing training project, a standard action library is formed based on action analysis. When the trainee uploads his/her practice video, the action recognition system identifies the action categories and matches them with the actions in the standard action library. If a matching action is found in the action library, the action evaluation system will compare the standard action with the trainee's action, provides an evaluation score and output personalized corrective suggestions (Fig. 8).

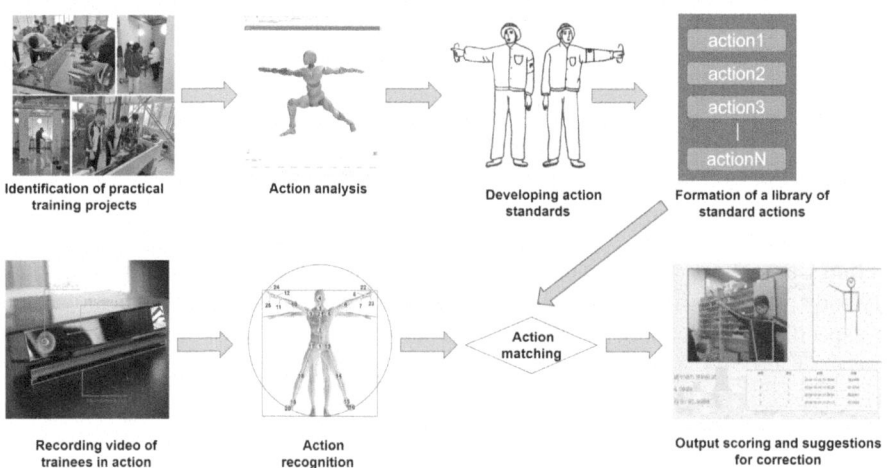

Fig. 8. The implementation process of the training system.

In this paper, the converter steelmaking digital twin system project is selected as the testing background. The project aims to utilize deep learning, digital twin and industrial metaverse technologies to achieve process optimization and operation guidance for the steel smelting process. During the project research, it was found that the steel manufacturing process is characterized by high environmental complexity, strong noise interference, large material volume and high risk factors, which make the overhead crane an indispensable lifting equipment. The crane command operation has a set of standard actions and gestures. Accurate and clear crane command actions can not only overcome noise interference and deliver timely and accurate operation information to improve production efficiency, but also avoid workers from being exposed to the dangerous production environment to ensure the smooth progress of production and workers' safety. Therefore, the IMTware can be effectively used to assist the training of relevant personnel to quickly master standard, professional crane command maneuvers (Fig. 9).

In order to test the evaluation performance of IMTware, this paper selects five actions in the overhead crane command as examples, and the action details are shown in Table 2:

Fig. 9. Operation of overhead cranes in the converter steelmaking process.

Table 2. Details of actions.

Action category	Action details	Examples of actions
1.Stop	Horizontal placement of the forearm in front of the chest, fingers spread apart, palm facing downwards, horizontally swinging to one side	
2.Asking for the main hook	Make a natural fist with one hand and place it over your head, lightly touching the top of your head	
3.Asking for the secondary hook	One hand clenched into a fist, forearm raised upward without moving, the other hand extended, palm lightly touching the elbow joint of the front hand	
4.Moving the hook horizontally	The forearm is extended straight upward to the side, fingers together, palm facing outward, directed towards the direction of the load to be moved, swinging downward to a position level with the shoulder	
5.Ending the work	Both hands with fingers spread apart, crossed in front of the forehead	

3.1 Production of Data Sets and Pre-processing

The self-constructed dataset is captured by a video camera and 50 videos are recorded for each action as a training set and 20 as a test set. In order to ensure the generalization ability of the recognition system, the dataset is produced by choosing different angles and distances for filming. The videos are processed to meet the input requirements of the network, first, the videos are trimmed and compressed to ensure each segment lasts approximately 4 s with a resolution of 340 × 256. Subsequently, the OpenPose network was utilized to convert the human body actions in the videos into skeletal point sequence data, which is then stored in JSON format files. At the same time, a tag file is made, containing the naming of each video and the corresponding action category information. Recording a standard action video for each action and saving it into the standard action library.

3.2 Training the Network

Due to the limited number of videos in the dataset, direct training may not achieve satisfactory results. Therefore, this study adopts a transfer learning approach to train the network. Initially, the network is trained on the public dataset Kinetics400, which contains approximately 300,000 video clips covering up to 400 human activities. The pre-trained model is then saved, and further training is conducted on our own data based on this pre-trained model.

The overhead crane command actions are relatively simple, and the model training has yielded excellent results, achieving a recognition accuracy of 95% for actions. The trained model is then saved for later use in behavior recognition tasks. The recognition results are shown in the following figure (Fig. 10):

3.3 Visualization of Results

As shown in the figure below, once the trainee uploads an exercise video, the system identifies the actions within the video and visually presents the categories on the user interface. Subsequently, the corresponding standard actions will be called up and the scores will be displayed by comparing the vector sequence of each part of the standard actions with the trainee's actions. Simultaneously, the video actions are analyzed frame by frame. Scores below 90 are considered inadequate. In each frame of the video, body parts performing excellently are highlighted in green, while those with significant errors are marked in red. Additionally, corrective suggestions are provided based on the deviation from the standard actions (Figs. 11, 12 and 13).

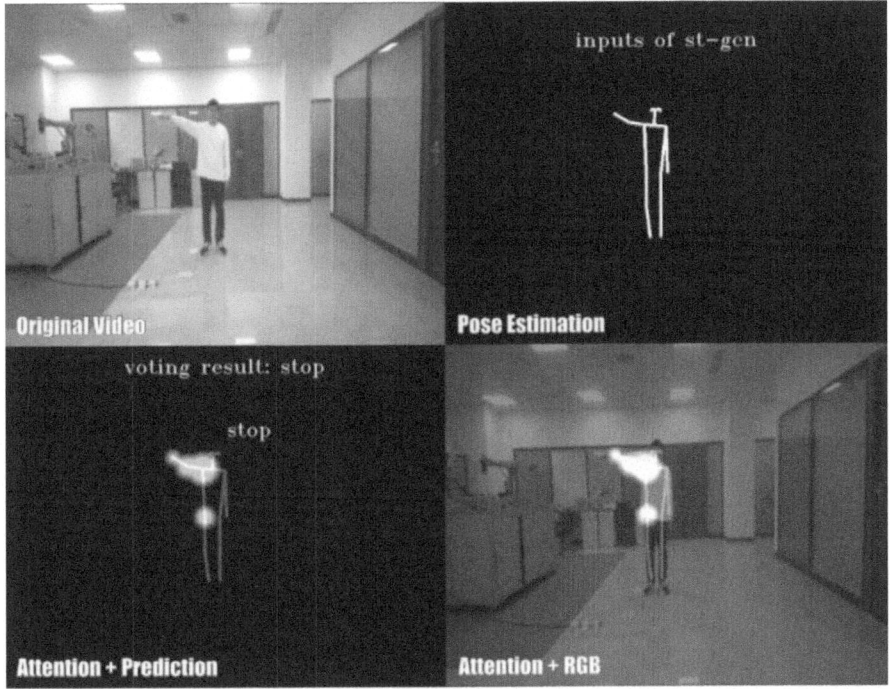

Fig. 10. Action Recognition Visualization Interface

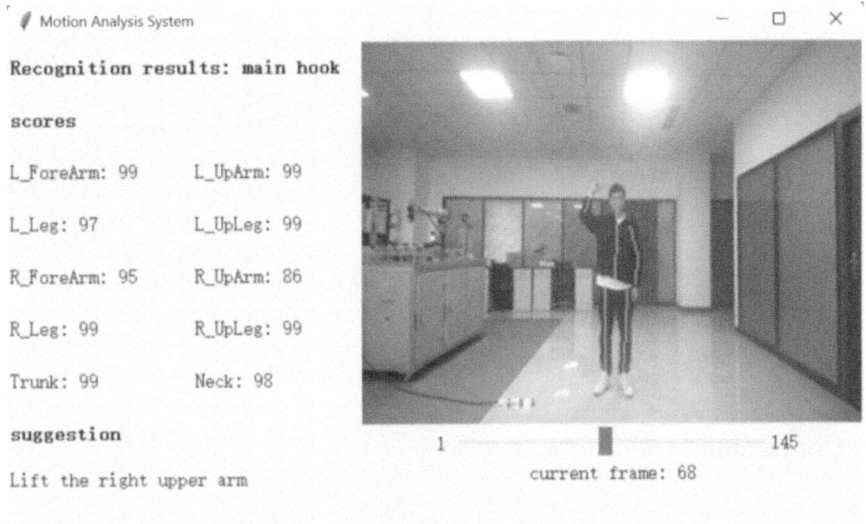

Fig. 11. The evaluation results for action 1

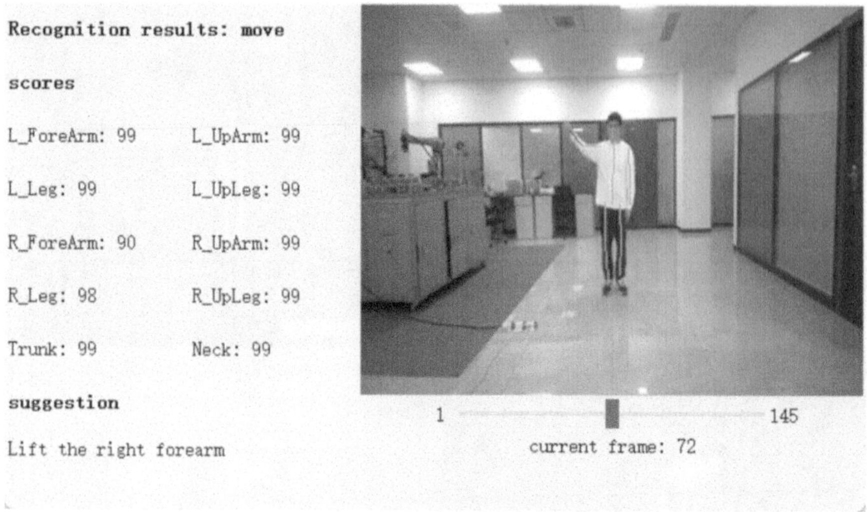

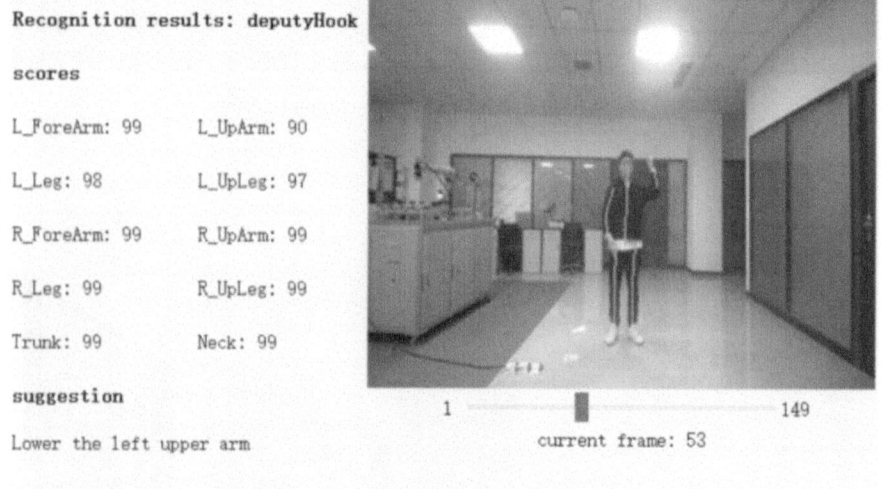

Fig. 12. The evaluation results for action 2

Fig. 13. The evaluation results for action 3

4   Conclusion

Standardization of worker actions is crucial in certain industrial operations, imposing high demands on worker training. Traditional practical training methods exhibit significant limitations. This paper proposes intelligent Manufacturing Training-Worker Action Recognition and Evaluation (IMTware), which can accurately identify trainee actions, quantify action quality, and provide personalized guidance.

Initially, this paper conducted a comparative analysis of methods for action recognition and evaluation. For action recognition, the OpenPose network was employed as the skeleton point extraction network, with the ST-GCN network serving as the foundational model. In the realm of action evaluation, various similarity measurement methods were explored, ultimately selecting cosine similarity as the final assessment criterion. To address differences in individual action speed and start times, a dynamic time warping algorithm was employed for keyframe matching. Appropriate methods were devised to identify areas with significant errors and provide corrective recommendations.

To validate the system's effectiveness, this paper selected overhead crane command action training in converter steelmaking as a case study. The overhead crane command action training system was designed, and the corresponding dataset was prepared and preprocessed. The system demonstrated excellent performance on the self-made dataset, boasting high action recognition rates and the ability to differentiate action quality. Lastly, a user interface was designed to present scores for each body part in every frame, visualize areas with significant errors.

References

1. Zhou, Y., et al.: Research on intelligent manufacturing training system based on industrial metaverse. In: Hassan, F., Sunar, N., Mohd Basri, M.A., Mahmud, M.S.A., Ishak, M.H.I., Mohamed Ali, M.S. (eds.) AsiaSim 2023. CCIS, vol. 1911, pp. 28–43. Springer, Singapore (2024). https://doi.org/10.1007/978-981-99-7240-1_3
2. Zheng, Z., et al.: Industrial metaverse: connotation, features, technologies, applications and challenges. In: Fan, W., Zhang, L., Li, N., Song, X. (eds.) AsiaSim 2022. CCIS, vol. 1712, pp. 239–263. Springer, Singapore (2022). https://doi.org/10.1007/978-981-19-9198-1_19
3. Yan, S., Xiong, Y., Lin, D.: Spatial temporal graph convolutional networks for skeleton-based action recognition. In: Proceedings of the AAAI Conference on Artificial Intelligence, vol. 32, no. 1 (2018)
4. Li, M., Chen, S., Chen, X., et al.: Actional-structural graph convolutional networks for skeleton-based action recognition. In: Proceedings of the IEEE/CVF Conference on Computer Vision and Pattern Recognition, pp. 3595–3603 (2019)
5. Chen, X.: Action evaluation system based on human three-dimensional posture. Zhejiang University (2018). (in Chinese)
6. Qian, C., Zhang, X., Tao, J., et al.: Design and research of Kinect-based rehabilitation training system. J. Jilin Univ. (Inf. Sci. Ed.) **38**(01), 92–98 (2020). https://doi.org/10.19292/j.cnki. jdxxp.2020.01.013. (in Chinese)
7. Shi, L., Zhang, Y., Cheng, J., et al.: Two-stream adaptive graph convolutional networks for skeleton-based action recognition. In: Proceedings of the IEEE/CVF Conference on Computer Vision and Pattern Recognition, pp. 12026–12035 (2019)
8. Shi, L., Zhang, Y., Cheng, J., et al.: Skeleton-based action recognition with directed graph neural networks. In: Proceedings of the IEEE/CVF Conference on Computer Vision and Pattern Recognition, pp. 7912–7921 (2019)
9. Parmar, P., Tran Morris, B.: Learning to score Olympic events. In: Proceedings of the IEEE Conference on Computer Vision and Pattern Recognition Workshops, pp. 20–28 (2017)
10. Bruce, X.B., Liu, Y., Chan, K.C.C., et al.: Skeleton-based human action evaluation using graph convolutional network for monitoring Alzheimer's progression. Pattern Recogn. **119**, 108095 (2021)

11. Wang, J., Du, Z., Li, A., Wang, Y.: Assessing action quality via attentive spatio-temporal convolutional networks. In: Peng, Y., et al. (eds.) PRCV 2020, pp. 3–16. Springer, Cham (2020). https://doi.org/10.1007/978-3-030-60639-8_1
12. Pan, J.H., Gao, J., Zheng, W.S.: Action assessment by joint relation graphs. In: Proceedings of the IEEE/CVF International Conference on Computer Vision, pp. 6331–6340 (2019)

Author Index

GPSR Compliance

The European Union's (EU) General Product Safety Regulation (GPSR) is a set of rules that requires consumer products to be safe and our obligations to ensure this.

If you have any concerns about our products, you can contact us on ProductSafety@springernature.com

In case Publisher is established outside the EU, the EU authorized representative is:

Springer Nature Customer Service Center GmbH
Europaplatz 3
69115 Heidelberg, Germany